Recent Trends in Human and Animal Mycology

Karuna Singh • Neelabh Srivastava
Editors

Recent Trends in Human and Animal Mycology

Springer

Editors
Karuna Singh
Department of Zoology, Mahila Mahavidyalaya
Banaras Hindu University
Varanasi, Uttar Pradesh, India

Neelabh Srivastava
Department of Zoology, Mahila Mahavidyalaya
Banaras Hindu University
Varanasi, Uttar Pradesh, India

ISBN 978-981-13-9434-8 ISBN 978-981-13-9435-5 (eBook)
https://doi.org/10.1007/978-981-13-9435-5

This Springer imprint is published by the registered company Springer Nature Singapore Pte Ltd.
The registered company address is: 152 Beach Road, #21-01/04 Gateway East, Singapore 189721, Singapore

Foreword

It is a pleasure for me to write a foreword for the book *Recent Trends in Human and Animal Mycology* edited capably by Dr. Karuna Singh of the Department of Zoology (MMV), Banaras Hindu University, India. In a time when the concept of **One Health** is well accepted, this book takes things to the next level by considering fungal disease entities within the conceptual framework of **Planetary Health**, taking into account the environment, animals, humans and plants, plus their fungal pathogens, and taking a holistic view of how they all interplay. Not only are the authors focused on human and animal disease, but they are interested in crop diseases, environmental health and food security and what strategies are affordable in the developing as well as the developed world.

Marauru, Chermette and Guillot from France cover the superficial fungal diseases of companion animals, with special emphasis on dermatophytes and *Malassezia*, common and important pathogens of pedigree dogs and cats. Signs, diagnostic approaches and therapy are covered in an overarching up-to-date review. Singh and Gumsta from India review the phaeohyphomycosis, uncommon but challenging causes of infections which are characterized by the elaboration of melanin within lesions, this pigment being a critical virulence factor. Clinical disease, diagnostic testing and future areas for research are all covered. Ovchinnikov and Vasyliev from Russia cover the *Chrysosporium*-related fungi, the major cause of mycotic disease of reptiles worldwide. Interestingly, it is the popularity of exotic reptiles as pets that has facilitated worldwide spread of these infections in recent years, a failure of international biosecurity akin to the situation for chytrid disease of frogs.

de Moraes Gimenes Bosco has tackled the large topic of endemic systemic mycoses of North and South America, covering histoplasmosis, blastomycosis, coccidiomycosis, sporotrichosis and paracocidiomycosis, with special emphasis on *Sporothrix brasiliensis*, an emerging infectious zoonotic disease of Brazil, with the cat being the pivotal amplifying host and cause of infections in man and dog. The same author tackles the fascinating pathogen *Pythium insidiousum*, an oomycete which causes 'swamp cancer' in horses and severe cutaneous and alimentary disease of dogs and serious disease in humans in certain geographical regions where the pathogen is endemic, such as Thailand, northern Australia and the southern states of the USA.

Pieckova from Slovakia addresses the risk of mycotoxins, whether they are present in foods (e.g. poorly stored grain) or inhaled from moist environments where fungal growth is favoured. The pathomechanism by which these potent fungal toxins derange function of the respiratory system are addressed, as is their role in food security in developing nations where storage of grain may be suboptimal. Kumari and Tirkey from India discuss tenuazonic acid, a potent mycotoxin elaborated by *Alternaria* spp. growing on cereals and fruits, including tomatoes. Raghuwanshi from India discuss the use of phytochemicals (medicinal plants) as potential drugs to treat candidiasis and other fungal infections, while Sharma and Katiyar from India look at re-tasking coumarin derivatives as antifungal agents.

Seyedmousavi, affiliated with the NIH (USA), provides an up-to-date summary of the comparative knowledge on *Aspergillus* infections in humans and various types of animals. A two-step molecular barcode, using ITS and β-tubulin, is suggested for taxonomic division of the genus, including new cryptic species, while the growing place of MALDI-TOF is mentioned, as well as aspects of azole resistance for this genus as a worrying concern especially in Europe.

Throughout this book, despite the multiauthor format, certain themes emerge and are reinforced through a repetition in different conceptual frameworks. Fungal disease is more complex and challenging because the pathogens are eukaryotic. The knowledge gap is greater for fungal diseases than those caused by bacteria or viruses, and there is an urgent need for new antifungal agents. Fungi are just as important plant pathogens as animal pathogens, and key insights can come from botany as well as human and veterinary medicine. Animals are oftentimes important sentinels for human disease. To understand any fungal infection, it's vital to appreciate its environmental niche and host: environment-pathogen interactions are key to understanding the aetiopathogenesis of disease.

I trust you will enjoy this book and find it helpful to achieve a greater understanding of fungal disease pathogenesis.

Adjunct Professor
Charles Sturt University
Bathurst Australia Veterinary Specialist,
Centre of Veterinary Education
University of Sydney
Camperdown Australia

Richard Malik

Preface

Fungi are vitally different from other pathogenic microbes. The pathogenicity of these eukaryotes depends on the immune status of the host. The frequency of mycotic diseases is increasing day by day, and the whole animal kingdom including humans is affected by them. The scenario has worsened with the emergence and re-emergence of fungal opportunists. Besides, endemic mycoses are now being evident from other parts of the world owing to travel across the globe. Although few pathogenic fungi are communicable from person to person, fungal infections transmitted through zoonotic and sapronotic sources are of main concerns.

Apart from mycoses, fungal pathogens indirectly harm the homeotherms by secreting toxins which are capable of exerting detrimental effects even in a very low concentration. These heat-resistant organic compounds enter through inhalation and/or ingestion and cause mycotoxicoses-state of ill health including occupational hazards. Moreover, the absence of specific clinical symptoms makes their diagnosis difficult.

Despite the use of effective antifungals, the morbidity of mycotic diseases remained persistently high. Inevitable increase in antifungal resistance in established pathogens, global warming, advent of new pathogens and more cases of impaired immunity jeopardize existing therapeutic options. The escalating demand of new antifungals could be fulfilled by adding new dimensions in the area of antifungal drug discovery, like identification of new drug targets, lead optimization, molecular validation and development of plant-based drugs.

This book is an effort to refine our knowledge about human and animal mycoses and provide a scalable platform to current trends of mycological research. This amalgam of chapters has three parts. Part I includes six chapters describing mycoses and their etiological agents. Part II contains two chapters addressing mycotoxins and their deleterious effects. Part III summarizes reviews on antifungal drug candidates.

Last but not the least, we would like to express our deep sense of gratitude to all the contributors for their benevolent contribution.

Varanasi, India

Karuna Singh
Neelabh Srivastava

Contents

Abbreviations

5FC	5-Flucytosine
3-NPA-3	Nitropropionic acid
AAL toxins	*Alternaria alternata lycopersici* toxins
AAL-TA	*Alternaria alternata* toxins TA
AAL-TB	*Alternaria alternata* toxins TB
ABPA	Allergic bronchopulmonary aspergillosis
AD	Autosomal dominant
Af	Aflatoxin
AFP	Antifungal peptide/protein
Ag	Silver
AGID	Agar gel immunodiffusion
ALT	Altenuene
AmB	Amphotericin B
AME	Alternariol monomethyl ether
AMP	Adenosine monophosphate
AMP	Antimicrobial peptides
ampB	Amphotericin B
AOH	Alternariol
AR	Autosomal recessive
AST	Aspartate aminotransferase
ATCC	American Type Culture Collection
ATX	Altertoxin
aw	Water activity
BBB	Blood-brain barrier
BEA	Beauvericin
BID	Bis in die (twice in a day)
CA	Californian
CAM	Complementary and alternative system of medicines
CANV	*Chrysosporium* anamorph of *Nannizziopsis vriesii*
CARD9	Caspase recruitment domain-containing protein 9
CAPD	Continuous ambulatory peritoneal dialysis
Cd	Cadmium
CDR	Candida drug resistance
CGD	Chronic granulomatous disease

CLSI	Clinical & Laboratory Standards Institute
CNS	Central Nervous System
Co	Cobalt
COPD	Chronic obstructive pulmonary disease
CT	Computed tomography
CTN	Citrinin
Cu	Copper
DMSO	Dimethyl sulfoxide
DNA	Deoxyribonucleic acid
DON	Deoxynivalenol
DTM	Dermatophyte test medium
EA	Ergot alkaloids
EC	Oesophageal cancer
EFSA	European Food Safety Authority
EIA	Enzyme immunoassay
ELEM	Equine leukoencephalomalacia
ELI025	Elicitin
ELISA	Enzyme-linked immunosorbent assay
ENN	Enniatins
ETP	Epipolythiodioxopiperazine
EUA	European University Association
FA	Fusaric Acid
FAD	Flea allergy dermatitis
FB1	Fumonisin B1
FdUMP	5-Fluorodeoxyuridine monophosphate
FeLV	Feline leukaemia virus
fg	Femtogram
FIC	Fractional inhibitory concentration index
FIV	Feline immunodeficiency virus
FUS	Fusaproliferin
FUTP	Fluorouridine triphosphate
GDP	Guanosine diphosphate
GGT	Gamma glutyltransaminase
GlcNAc	N-acetylglucosamine
GMS	Gomori methenamine silver
gp 43	Glycoprotein 43
HA	Hemagglutination assay
HACCP	Hazard analysis and critical control point
Hc 100 gene	*H. capsulatum* 100 gene
HE	Haematoxylin and Eosin
Hg	Mercury
HIES	Hyper-IgE syndrome
HIV	Human immunodeficiency virus
HSCT	Haematopoietic stem cell transplantation
Hsp	Heat shock protein
IC_{50}	Half maximal inhibitory concentration

ICT	Immunochromatography
ID	Immunodiffusion
IDH	Isoepoxydon dehydrogenase
IEC	Intestinal epithelial cell
IFN-γ	Interferon γ
IGS	Intergenic spacer
IgE	Immunoglobulin E
IgG	Immunoglobulin G
IgM	Immunoglobulin M
IL	Interleukin
ITS	Internal transcribed spacer
IUPAC	International Union of Pure and Applied Chemistry
kg	Kilogram
kDa	Kilodalton
LAD	Leukocyte adhesion deficiency
LLC-PK1 renal cells	Epithelial cell line originally derived from porcine kidneys
LRS	Lactated Ringer's solution
LSD	Lysergic acid diethylamide
MALDI-TOF	Matrix-assisted laser desorption/ionization-time of flight
MDR	Multidrug resistance
MFC	Minimum fungicidal concentration
μg	Microgram
μM	Micrometre
mg	Milligram
MIC	Minimum inhibitory concentration
ml	Millilitre
Mn	Manganese
MON	Moniliformin
MS	Mass spectrometry
mVOCs	Microbial volatile organic compounds
MW	Molecular weight
Na-K-ATPase	Sodium-potassium adenosine triphosphatase enzyme
Ni	Nickel
NIH/3T3	3-Day transfer, inoculums 3×10^5 cells from NIH mouse embryonic fibroblast cells
NIOSH	National Institute for Occupational Safety and Health
NRPS-PKS	Non-ribosomal peptide synthetase and polyketide synthase hybrid enzyme
OTA	Ochratoxin A
OTB	Ochratoxin B
OT-GSH	OTA-glutathione conjugate
OTHQ	OTA-hydroquinone
OTQ	OTA-quinone
PA	Penicillic acid
pK_a	Acid dissociation constant
PAMP	Pathogen-associated molecular patterns

PAS	Periodic Acid-Schiff
PAT	Patulin
PBS	Phosphate Buffered Saline
PCM	Paracoccidioidomycosis
PCR	Polymerase Chain Reaction
PCRF	Pathogenic *Chrysosporium*-related fungi
phen	1,10-phenanthroline
PIDs	Primary immunodeficiencies
PO	Per os
PPE	Personal protective equipment
PPE	Porcine pulmonary oedema
PS	Phylogenetic species
QuEChERS	Quick, easy, cheap, effective, rugged and safe
rDNA	Ribosomal DNA
RNA	Ribonucleic acid
RT	Room temperature
ROS	Reactive oxygen species
SAR	Structure-activity relationship
SCN	Severe congenital neutropenia syndrome
SFD	Snake fungal disease
SH	Sulfhydryl group
SID	Sem´el in di´e (once a day)
SIDA	Stable isotope dilution assay
SIV	Simian immunodeficiency virus
SLUDGE	Salivation, lacrimation, urination, defecation, gastrointestinal distress and emesis
SsCBF	*S. schenckii* Concanavalin A-binding fraction
SSKI	Saturated solution of potassium iodide
T-2	T-2 toxin
TAS1	TeA synthetase 1
TCs	Trichothecenes
TDI	Total daily intake
TeA	Tenuazonic acid
TEER	Transepithelial electrical resistance
TEN	Tentoxin
Th1	T helper 1
TID	Ter in die (three times a day)
Tl	Toll-deficient
TNF-α	Tumour necrosis factor-α
UBI	Ubiquicidin
VO	Vanadium
VOCs	Volatile organic compounds
WHO	World Health Organization
ZEN	Zearalenone
Zn	Zinc

Editors and Contributors

About the Editors

Karuna Singh is an Associate Professor at the Department of Zoology, Mahila Mahavidyalaya, Banaras Hindu University, Varanasi, India. She received her BSc, MSc (Zoology) and PhD degrees from Rani Durgavati University, Jabalpur, India, and has been actively engaged in teaching and research for the last 16 years. Chiefly working in the field of Animal Mycology, her current areas of research interest are molecular epidemiology of cryptococcosis, mycotoxicoses and drug design. She has published more than 40 original research papers and book chapters, filed 2 patents and edited 1 book. She has led two research projects funded by the CST-UP and Department of Science and Technology, New Delhi.

Neelabh Srivastava is a Senior Research Fellow (ICMR) at the Department of Zoology (MMV), Banaras Hindu University, India, where he works in the fields of drug design and antifungal peptides. He has published more than 30 research papers and filed 1 patent. He has won the prestigious Young ISHAM Fellowship twice and currently serves as a Reviewer and Editorial Board Member for several prominent national and international scientific journals.

Contributors

Eduardo Bagagli Laboratory of Fungal Biology, Department of Microbiology and Immunology, Institute of Biosciences, Universidade Estadual Paulista, UNESP/Botucatu, Botucatu, SP, Brazil

Jéssica Luana Chechi Laboratory of Medical Mycology, Department of Microbiology and Immunology, Institute of Biosciences, Universidade Estadual Paulista, UNESP/Botucatu, Botucatu, SP, Brazil

René Chermette Department of Parasitology, Mycology and Dermatology, Ecole nationale vétérinaire d'Alfort, Dynamyc Research Group, EnvA, UPEC, USC ANSES, Maisons-Alfort, France

Hans Garcia Garces Laboratory of Fungal Biology, Department of Microbiology and Immunology, Institute of Biosciences, Universidade Estadual Paulista, UNESP/Botucatu, Botucatu, SP, Brazil

Jacques Guillot Department of Parasitology, Mycology and Dermatology, Ecole nationale vétérinaire d'Alfort, Dynamyc Research Group, EnvA, UPEC, USC ANSES, Maisons-Alfort, France

Richa Gumasta Medical Mycology Laboratory, Department of Biological Science, RD University, Jabalpur, India

Fungal Disease Diagnostic and Research Centre, Jabalpur, India

Diksha Katiyar Department of Chemistry, MMV, Banaras Hindu University, Varanasi, India

Theerapong Krajaejun Department of Pathology, Faculty of Medicine, Ramathibodi Hospital, Mahidol University, Bangkok, Thailand

Ankita Kumari Animal Mycology Laboratory, Department of Zoology, Mahila Mahavidyalaya, Banaras Hindu University, Varanasi, India

Sandra de Moraes Gimenes Bosco Laboratory of Medical Mycology, Department of Microbiology and Immunology, Institute of Biosciences, Universidade Estadual Paulista, UNESP/Botucatu, Botucatu, SP, Brazil

Ramona Moraru Clinique vétérinaire, Le Châtelet-en-Brie, France

Alana Lucena Oliveira Laboratory of Medical Mycology, Department of Microbiology and Immunology, Institute of Biosciences, Universidade Estadual Paulista, UNESP/Botucatu, Botucatu, SP, Brazil

Roman S. Ovchinnikov National Research Center on Epidemiology and Microbiology n.a. N.F. Gamaleya, Moscow, Russia

Federal Research Center, All-Russian Scientific Research Institute of Experimental Veterenary Medicine n.a. K.I. Skryabin and Y.R. Kovalenko, Moscow, Russia

Giselle Souza da Paz Laboratory of Medical Mycology, Department of Microbiology and Immunology, Institute of Biosciences, Universidade Estadual Paulista, UNESP/Botucatu, Botucatu, SP, Brazil

Elena Piecková Faculty of Medicine, Slovak Medical University, Bratislava, Slovakia

Ana Carolina do Prado Laboratory of Medical Mycology, Department of Microbiology and Immunology, Institute of Biosciences, Universidade Estadual Paulista, UNESP/Botucatu, Botucatu, SP, Brazil

Richa Raghuwanshi Department of Botany, Mahila Mahavidyalaya, Banaras Hindu University, Varanasi, India

Seyedmojtaba Seyedmousavi National Institutes of Health (NIH), Bethesda, MD, USA

Rajesh Kumar Sharma Department of Chemistry, MMV, Banaras Hindu University, Varanasi, India

Shanker Mohan Singh Medical Mycology Laboratory, Department of Biological Science, RD University, Jabalpur, India

Fungal Disease Diagnostic and Research Centre, Jabalpur, India

Neha Nidhi Tirkey Animal Mycology Laboratory, Department of Zoology, Mahila Mahavidyalaya, Banaras Hindu University, Varanasi, India

Dmitry B. Vasyliev Department of Herpetology, Moscow State Zoo, Moscow, Russia

Danielle Hamae Yamauchi Laboratory of Fungal Biology, Department of Microbiology and Immunology, Institute of Biosciences, Universidade Estadual Paulista, UNESP/Botucatu, Botucatu, SP, Brazil

Part I

Human and Animal Mycoses

Pythiosis

1

Sandra de Moraes Gimenes Bosco, Jéssica Luana Chechi, Giselle Souza da Paz, and Theerapong Krajaejun

Abstract

Pythiosis is a granulomatous disease that affects subcutaneous, vascular, ocular, and gastrointestinal tissues of many humans and animals. The disease is caused by the fungus-like pathogen *Pythium insidiosum*, an oomycete found predominantly in tropical and subtropical areas of the world. Pythiosis in animals (horses and dogs) has been more prevalent in the American continent, while the disease in humans has been mostly reported from Thailand. Susceptibility according to age, sex, and breed of the animals is not considered as determinants for infection. Continuous stay in stagnant water with the presence of abundant plant material is the most important factor responsible for the infection, since the disease is acquired in aquatic environment due to the penetration of motile biflagellate zoospores into injured skin. Diagnosis of pythiosis is often difficult, delayed, and time-consuming due to the lack of clinical experience and diagnostic techniques. Treatment of choice is frequently relied upon extensive surgery, which, however, is not always possible due to the great extension of the lesions. Frequently, the disease has a poor prognosis, leading animals to death or euthanasia. In humans, removal of the infected organ or death is the common outcome. Therefore, new diagnosis methodologies and treatment approaches against pythiosis and its etiological agent would be among the most studied fields in the near future.

S. de Moraes Gimenes Bosco (✉) · J. L. Chechi · G. S. da Paz
Laboratory of Medical Mycology, Department of Microbiology and Immunology, Institute of Biosciences, Universidade Estadual Paulista, UNESP/Botucatu, Botucatu, SP, Brazil
e-mail: smgbosco@ibb.unespr.br

T. Krajaejun (✉)
Department of Pathology, Faculty of Medicine, Ramathibodi Hospital, Mahidol University, Bangkok, Thailand
e-mail: theerapong.kra@mahidol.edu

K. Singh, N. Srivastava (eds.), *Recent Trends in Human and Animal Mycology*,
https://doi.org/10.1007/978-981-13-9435-5_1

Keywords
Pythiosis · Oomycete · Animal · Human

1.1 Definition and Etiology

Pythiosis is a life-threatening infectious disease of humans and animals living in tropical and subtropical regions and exhibits granulomatous lesions mainly in subcutaneous, gastrointestinal, vascular, and ocular tissues. It is caused by fungus-like organism *Pythium insidiosum*, belonging to the kingdom Stramenopila, phylum Oomycota, class Oomycetes, order Peronosporales, and family Pythiaceae, in which genera *Pythium* and *Phytophthora* are included [1]. Both genera are phytopathogens. *Phytophthora infestans*, the causative agent of "late blight" or "potato blight" that caused the famous episode of "Irish potato famine," occurred in Ireland between 1845 and 1849 [2]. *P. insidiosum* is the main species among the pathogenic oomycetes that can infect animals, although recently two reports of pythiosis were described in humans due to *Pythium aphanidermatum* [3, 4].

There are some differences between oomycetes and true fungi (Kingdom Fungi). The mitochondrial crests in true fungi are flattened, while in oomycetes, they are tubular. The fungi of the phylum Chytridiomycota have uniflagellated zoospores, whereas the fungi of the phylum Oomycota have biflagellated spores. The cell wall is composed mainly of chitin, mannans, and α and β glucans in true fungi, while in oomycetes, it is composed of β glucans, cellulose, and hydroxyproline. The main difference is found in the cellular membrane which explains the failure of medical treatment of pythiosis: the absence of ergosterol in oomycetes, which is the main target for antifungal agents [5–7].

P. insidiosum is an oomycete that reproduces asexually in aquatic environments by producing biflagellate zoospores, associated with the infective form of the pathogen. The pathogen develops in plant material submerged in water. The sporangium is produced at the hyphal branches and their extremities in which the protoplasm of the vesicle begins progressive cleavage to form the biflagellated zoospores. The time elapsed between the formation of undifferentiated sporangia and the release of motile zoospores takes around 35 min, when in adequate climate conditions, with water, plant material, and temperatures around 37 °C–40 °C [8]. Once released, the zoospores penetrate into the plant material submerged in water (to continue its life cycle) or penetrate into injured skin of animals present in aquatic environments.

It can be cultivated aerobically in culture media which does not contain cycloheximide. Culturing on Sabouraud agar media at 37 °C shows a rapid growth (24 h) which can be attributed to the coenocytic hyphae [1, 8]. The colony has filamentous aspect with a short mycelium of white to beige color, and the texture may vary between membranous and velvety (Fig. 1.1). Large and sparsely septate hyphae, around 5–9 μm wide, branched predominantly in right angle without any spores were observed (mycelia sterilia) (Fig. 1.2). The spores are produced only in aquatic environments in nature or an appropriate induction medium in in vitro condition as shown in Fig. 1.3 [9].

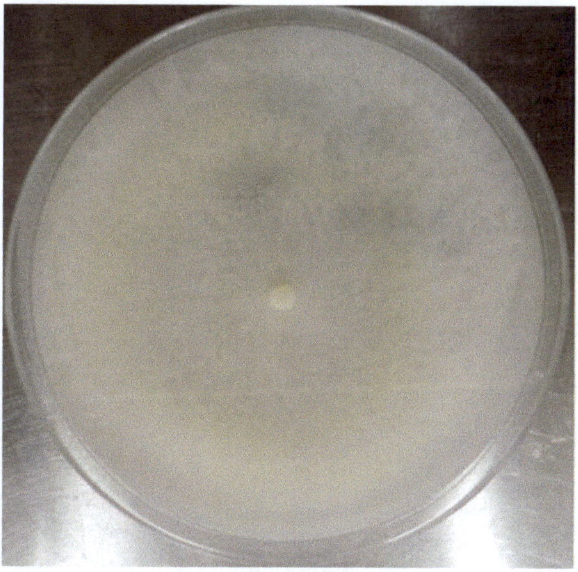

Fig. 1.1 Macroscopic aspect of 5 days of culture of *P. insidiosum* on 4% Sabouraud dextrose agar

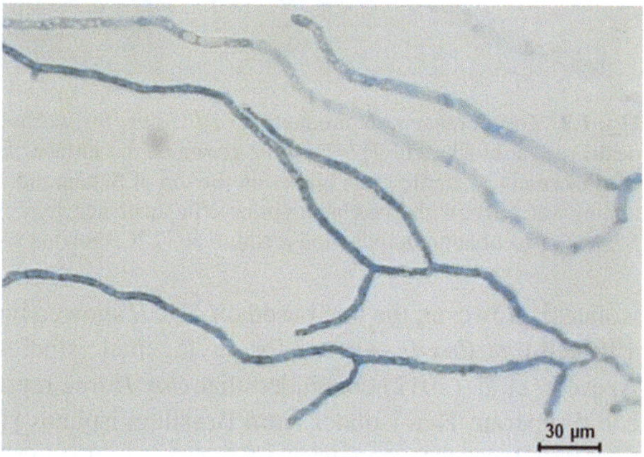

Fig. 1.2 Microscopic aspect of coenocytic hyphae of *P. insidiosum* with averaging 8 mm in diameter and right-angle branches. Mycelia sterilia (mycelium with absence of fruiting body) (lactophenol cotton blue, 200 ×)

Molecular studies based on phylogenetic analysis on rDNA (ITS and IGS regions) have shown that *P. insidiosum* comprises different genotypes, which were classified as clades I, II, and III. Clade I consists of isolates from American continent (Costa Rica, Brazil, Haiti, and the United States), whereas clade II consists of isolates from Asia (India, Thailand, Japan, and Papua New Guinea) and Australia. One isolate of this clade was found from a patient in the United States who might have acquired the infection from the Middle East. Clade III, which is most distantly related to the other two clusters, consists of many isolates from Thailand and a few isolates from the United States [10, 11]. Kammarnjesadakul et al. (2011) performed phylogenetic analysis based on ITS and *Cox II*, among isolates of *P. insidiosum* from Americas (n = 2) and Thailand (n = 33) [12]. Both analyses showed similar trees, clustering the isolates in clades A (American isolates), B, and C (Thai

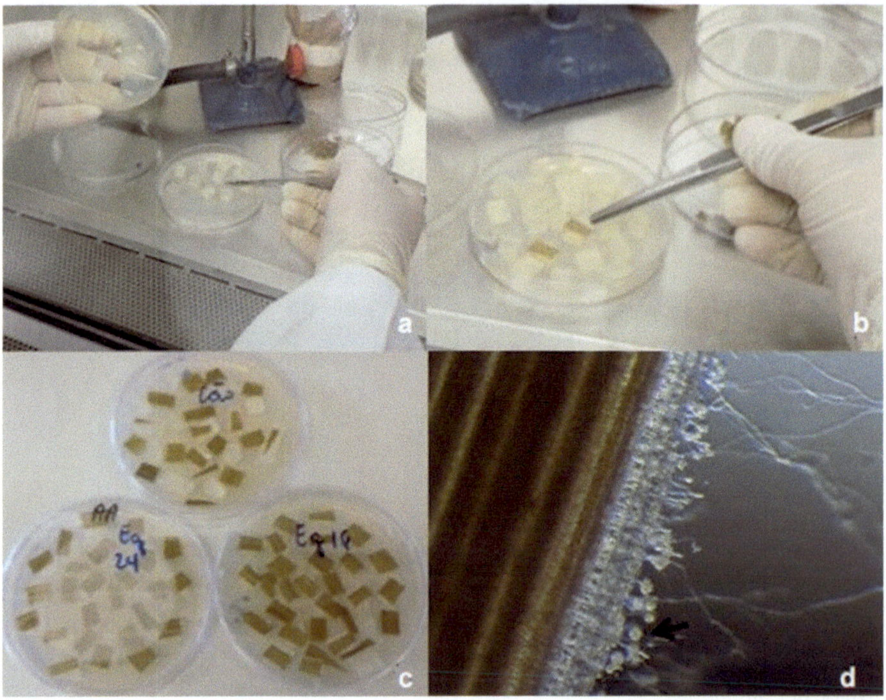

Fig. 1.3 Zoosporangenesis production of *Pythium insidiosum*. (**a**) Placement of 2% Sabouraud dextrose agar blocks with *P. insidiosum* grown on the surface of Petri dishes containing water agar. (**b**) Placement of sterile grass blades on the top of Sabouraud 2% blocks. (**c**) View of plates containing water agar with grass blades prior to be incubated at 37 °C for 2 days. (**d**) Grass blades after 5 hours of incubation in induction medium at 37 °C showing sporangium (arrow)

isolates). However, the tree based on *Cox II* showed high resolution, and the authors affirmed that *Cox II* is better for phylogenetic studies on *P. insidiosum*. Likewise, Azevedo et al. (2012) concluded that *Cox II* tree represented better phylogeny and could separate Thai isolates from Brazilian isolates [13]. The phylogenetic evaluation based on the sequences of exo-1,3-β-glucanase gene also showed a well-separated tree, in which Thai isolates clustered in clades I and II and Brazilian isolates clustered in clade III [14]. The authors also reinforce the genetic variation, mainly found in Thai isolates, and suggest an ancient population of *P. insidiosum* from Asian continent.

1.2 Epidemiology

Pythiosis is most frequently found in tropical and subtropical regions of the world. Different animal species are affected by disease, such as equines, dogs, cattle and small ruminants, cats, wild animals, and humans. However, the most affected species are horses, dogs, and humans [6].

Age, sex, and breed of the animal are not considered as determinants of infection. The most important factor in this respect is the continuous stay in stagnant water with abundant plant material, where the zoospores can penetrate into injured skin [1, 8]. Human pythiosis has been strongly associated to thalassemic patients from Thailand [15–19].

Historically, pythiosis was supposed to be firstly recognized by British veterinarians in horses from Indonesia presenting skin lesions; however, at that time, the disease was named hyphomycosis destruens equi [20]. The etiological agent was isolated in the early 1900s by Dutch scientists working with horses in Indonesia. The authors recognized the hyphal nature of the agent, but no identification was given to the etiological agent, since no sporulation of the fungus was observed [21, 22]. Equine pythiosis had different synonyms over time, such as hyphomycosis destruens equi (Indonesia), granular dermatitidis (Japan), equine phycomycosis, swamp cancer (Australia, United States), leeches (United States), espundia (Latin America), and summer sores (Australia, Latin America, United States) [6].

Equine pythiosis is mainly found in American continent, highlighting the Midwest region of Brazil, in Pantanal of Mato Grosso and Mato Grosso do Sul states [23, 24]. Owing to the great amount and variety of plant material, high temperatures, and abundance of water, Brazilian Pantanal promotes the maintenance of the pathogens in the environment. This geographic region of Brazil is economically important for cattle pasture, and the population of the animals here is estimated around eight million. During the rainy season (November to March), this region observes several flooded areas in which cattle are raised extensively in pastures. Horses are employed as work animals to assist in the removal of livestock from flooded areas to land (Fig. 1.4). Besides Brazil [23–25], equine pythiosis is also reported in Costa Rica [9], Venezuela [26], the United States [27–29], India [20], Egypt [30], and Australia [31, 32], and recently, the first case was reported in Thailand [33]. Pythiosis has also been reported in donkeys [34, 35].

Canine pythiosis is reported mainly in the United States [36–38] and Brazil [39–41]. A single case of pythiosis in dog was reported in Africa [42] and Venezuela [43]. Pythiosis in cats was reported in the United States [44, 45].

Fig. 1.4 Pantanal region of Mato Grosso do Sul state, Brazil, showing (**a**) cattle in flooded area and the (**b**) importance of equine for removing livestock from flooded areas to land

Pythiosis in ruminants has been reported in cattle, sheep, and goats. Among ruminants, bovines are undoubtedly the most affected species, especially calves [46, 47]. Outbreaks of cattle pythiosis have been reported in Venezuela and Brazil. Pérez et al. (2005) showed 63 Brahman beef calves, of both sexes, with subcutaneous and cutaneous lesions in the lower extremities of limbs [48]. In Brazil, Gabriel et al. (2008) reported lesions in 76 mixed breed cattle of both sexes and ages ranging from few months to 3 years old in the west region of Rio Grande do Sul state, during the months of January and February (rainy season) [49]. All animals showed nodular and ulcerated skin lesions, mainly in lower extremity of limbs, with different sizes, and some of them drained purulent exudate. Konradt et al. (2016) reported 23 calves with skin lesions, which began after 15 days of animals being exposed to extensive marshy regions [50]. The time of evolution of the lesions was around 20–30 days. It is important to emphasize that in these outbreaks the disease had a good prognosis, by the observation of remission of the lesions, a fact not observed in other species. Besides cattle, pythiosis has also been reported in goats [51] and sheep [52–54]. In wild animals, pythiosis has been reported in camel [55, 56], tiger [57], jaguar [58], and a single case in birds [59].

Human pythiosis was firstly reported in Thailand in the middle of the 1980s [60, 61]. Apart from the majority of cases of human pythiosis observed in the Thai patients, some human cases have also been reported in other countries, such as the United States [62, 63], Brazil [64, 65], Israel [66], and India [67]. Some imported cases of severe keratitis have been reported in patients after swimming in regions where pythiosis occurs, such as in Canada [68], France [69], and Spain [70]. These patients were reported to swim in Costa Rica, Thailand, Brazil, or Colombia, respectively.

1.3 Pathogeny

P. insidiosum produces the infective form (zoospores) in aquatic environment associated with plant material. It was demonstrated that zoospores have chemotaxis to wounds, damaged skin, hairs, and intestinal mucosa [1, 8]. Once the biflagellate zoospores attach to the injured skin of an animal or human exposed to contaminated water, they detach the flagella, encyst, and germinate as hyphae into host tissue. Ravishankar et al. (2001) evaluated the force exerted by the tip of the hyphae on different surfaces (human and animal skin) and observed that the forces of up to 6.9 mN is not enough to penetrate the healthy skin [71]. It is likely that the pathogen employs secreted proteolytic enzymes, in addition to hyphal tip exertion, to facilitate its penetration into host tissue [71, 72].

Little is known about virulence factors of *P. insidiosum*. Besides these data about temperature optima, proteolytic enzymes, and tissue pressure, Krajaejun et al. (2011) demonstrated that some genes are related to thermal adaptation, sterol binding, antioxidation, and immunomodulation as putative virulence factors [73]. Elicitins form a protein family that could function as a pathogen-associated molecular pattern (PAMP), sterol carrier, and immunomodulator. The elicitin (ELI025) is a secreted protein identified in *P. insidiosum*, and it can evade host antibody response, as described by Lerksuthirat et al. in 2015 [7]. Chechi et al. (2018) reported some putative virulence factors for *P. insidiosum*, such as enolase, heat-shock protein (Hsp) 70, and glucan 1,3-beta-glucosidase [74]. Enolase may be involved in degradation of extracellular matrix and adhesion; Hsp 70 helps during the process of heat stimulation and may also play a role in adhesion; glucan 1,3-beta-glucosidase is a hydrolytic enzyme that may favor the hyphal branching and cell wall extension, favoring the growth of hyphae in tissues. Krajaejun et al. (2006) demonstrated that human patients from different regions of Thailand recognized a protein of 74 KDa that was later identified as putative exo-1,3-beta-glucanase [19, 75, 76].

As hyphae grow on animal tissue, they release exo-antigens that act as chemotactic factors for antigen-presenting cells, especially dendritic cells. These cells secrete interleukin 4 (IL-4) that leads to immune response type 2, which in turn secretes more IL-4 and IL-5. The secretion of IL-4 stimulates B cells to produce IgM, IgG, and IgE. IL-5 and IgE mobilize eosinophils and mast cells to the site of infection, forming an eosinophilic granulomatous reaction leading to the development of a Splendore-Hoeppli-like material around hyphae, which is found in the center of eosinophilic micro-abscesses. The pathogenic process of pythiosis has been associated with the degranulation of eosinophils and mast cells (inflammatory cells) around the hyphae, and this fact contributes to tissue damage, leading to extensive lesions. In equines, the reaction is so pronounced that the eosinophils strongly degranulate around the hyphae of *P. insidiosum* forming the firm concretions called "kunkers." It is interesting to highlight that the pathogen is camouflaged inside the eosinophilic material, a fact that represents an evolutionary strategy that protects *P. insidiosum* to be fully presented to host defenses [77].

Experimental pythiosis can be established by inoculating zoospores subcutaneously into rabbits [78]. It has been observed that pythiosis in rabbits may behave differently from natural infection in animals, while naturally infected horses, for example, develop an extensive ulcerative lesion and rabbits form a well-encapsulated and fibrous abscess [79] (Fig. 1.5). Efforts for searching new experimental models for pythiosis have been undertaken in the recent years. Zanette et al. (2013) have developed a new model for pythiosis infection in a Toll-deficient (Tl) *Drosophila melanogaster* (fruit fly) [80]. It was observed that the infected Tl mutant flies had

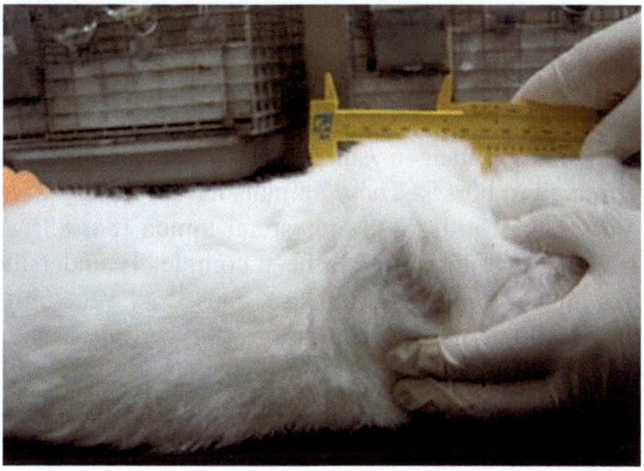

Fig. 1.5 Experimental pythiosis in rabbits showing large subcutaneous nodule of firm consistency, without ulceration. "Copyright of Gen National Publishing Group (Rio de Janeiro, Brazil), Book: Doenças Infecciosas Em Animais de Produção e Companhia, Chapter of Pitiose (page 949), reprinted with permission"

lower survival rates (73.7%), after 7 days postinoculation of 1 × 10^5 zoospores/ml, when compared with control group (98.6%) that received sterile PBS. The authors concluded that this model may represent a new alternative for studying the virulence of *P. insidiosum* but is still questionable for therapeutic studies, since medications may be toxic for these flies themselves and because host-pathogen relationship has not been taken into account. Tondolo et al. (2017) have proposed a murine model for evaluating the dissemination of pythiosis in immunosuppressed BALB/c mice [81]. The authors observed that when the control group (nonimmunosuppressed mice) were inoculated, the animals self-healed by increasing the levels of IL-2, IFN-γ, and IL-17A, cytokines characteristic of the Th1/Th17 response. On the other hand, immunosuppressed infected mice showed high levels of IL-10, IL-6, and TNF-α and showed dissemination of the pathogen through the lungs, kidneys, and liver, mimicking the vascular/disseminated pythiosis observed in humans [81]. A new approach for experimental infection was evaluated by the use of embryonated chicken eggs [82]. The authors evaluated different zoospore concentration and observed that the inoculation of 50 zoospores/egg produced mortality in 30% of the embryos after 48 hours and 95% embryos died within 72 h. Hyphae of *P. insidiosum* were histologically demonstrated in umbilical cords of 95% among the embryos, suggesting that this model might be suitable for evaluating pythiosis [82].

1.4 Clinical Aspects

Undoubtedly, horses are the most affected animals of pythiosis. In general, the lesions are characterized by pyogranulomatous irregular ulcerative wounds, which drain a serosanguinolent exudate with fetid odor, and the animal exhibits an intense pruritus. Inside the granulomatous tissue, irregular and yellowish-colored mass called "kunkers" only found in equines are noticed [77]. In most cases, it is observed as a single lesion in the ventral portion of the thoracic and abdominal walls and face. The great majority of lesions are found in the distal extremities of the limbs. The more distal is the lesion in the limbs, the worse is the prognosis in horses, as demonstrated by Watanabe et al. (2015) by observing a series of 28 cases of equine pythiosis in the central region of southeastern Brazil, São Paulo state [83]. In nine animals with lesions in the metacarpophalangeal joint, three of them were euthanized without therapy, five died after therapy (surgery and antifungal), and only one had remission of lesion after surgical debridement, potassium iodine, and amphotericin B therapy [83].

Besides cutaneous/subcutaneous lesions, intestinal pythiosis has also been observed in horses. These animals have a history of colic as a result of large tissue masses that obstruct the intestinal lumen [84, 85]. Although the pathogenesis of intestinal pythiosis has not been completely understood, it is suggested that previous injury to intestinal mucosa, caused by vegetal material or pathogenic agents, may facilitate the penetration of *P. insidiosum* in intestines. Another hypothesis is the active penetration of the agent [84]. Reis Jr. and coworkers in 2003 reported three cases of systemic pythiosis in equines through dissemination of *P. insidiosum* from chronic lesion in mammary glands, nasal cavity, and limbs. The pathogen was detected in the liver and lungs by immunohistochemistry and molecular methods [86]. Nasal pythiosis was also reported in two horses from semiarid region of Brazil. Both animals showed swelling in the rhinofacial region with a serosanguineous nasal discharge. The surface of granuloma had a yellow-gray granular aspect and cavitations of different sizes containing masses of necrotic tissue, "kunkers." One of the animals had dissemination of the disease to lungs [25]. Bone lesions have also been described in horses and are located mainly in limbs, probably by dissemination or contiguity of the pathogen from chronic subcutaneous lesions. Occasionally, equine pythiosis has also been reported affecting ligaments and tendons of limbs, joints, and bones, resulting in edema of the affected limb and laminitis [87]. Figure 1.6 shows some clinical aspects of equine pythiosis in Brazil. Pythiosis in dogs may be observed in two presentations: gastrointestinal and cutaneous/subcutaneous. The gastrointestinal form is most prevalent, and frequently, the diagnosis is only *postmortem* [36, 40]. Dogs become infected by drinking water, in lagoons or rivers, contaminated with zoospores. The main clinical sign is chronic anorexia, weight loss, vomiting, bloody diarrhea, and the presence of a firm nodular mass on the stomach or intestine that is easily palpable on physical examination. Besides the stomach and intestines, lesions in the oropharynx and cranial portion of the esophagus and colon infection with metastasis to the prostate have also been reported in dogs, resulting in prostatic hypertrophy and tenesmus [88]. Cutaneous/subcutaneous lesions in dogs may be found as granulomatous and ulcerated wounds with reactive and irregular borders in limbs and the ventral region of the neck, face, perineum, and thorax, as well as regional lymphadenopathy [37, 40, 89]. Other anatomical sites are the dorsal lumbar and tail base region or lumbosacral region [39]. Normally, dogs with cutaneous lesions in the lumbosacral region have a previous history of flea allergy dermatitis (FAD) that represents the entrance of zoospores into injured skin while these dogs swim in lagoons or rivers [37, 39, 42]. A rare and unique condition in which intestinal and cutaneous pythiosis occurred simultaneously in a dog was reported by Pereira et al. (2010) in Brazil [40]. The animal had an ulcerative lesion in the right thoracic region for 12 months of unresponsiveness to antifungal therapy. Two months prior to death and concurrent with skin lesion, the dog became anorexic with frequent vomiting and bloody stools. During necropsy, it was observed that the large intestine contained two lesions that caused luminal narrowing [40]. Coinfection with pythiosis in dog was reported by Connolly et al. (2012) in a male dog from the United States presenting anorexia, weight loss, and vomiting for 1 month. Prognosis was considered bad and the dog was euthanized [90]. During necropsy, the stomach and duodenum were found to be diffusely dark red with marked thickenings. Further gastric lymph nodes were found to be enlarged. The lungs of this animal showed some granulomatous lesions, and histopathology revealed the presence of yeasts that were identified as *Blastomyces dermatitidis* by immunohistochemistry and serology [90]. There is also a report of a primary pulmonary pythiosis in a dog, which presented enlarged cervical lymph node and nonproductive cough, which was unresponsive to antibiotic therapy. Pulmonary auscultation revealed harsh lung sounds more prominent in the right

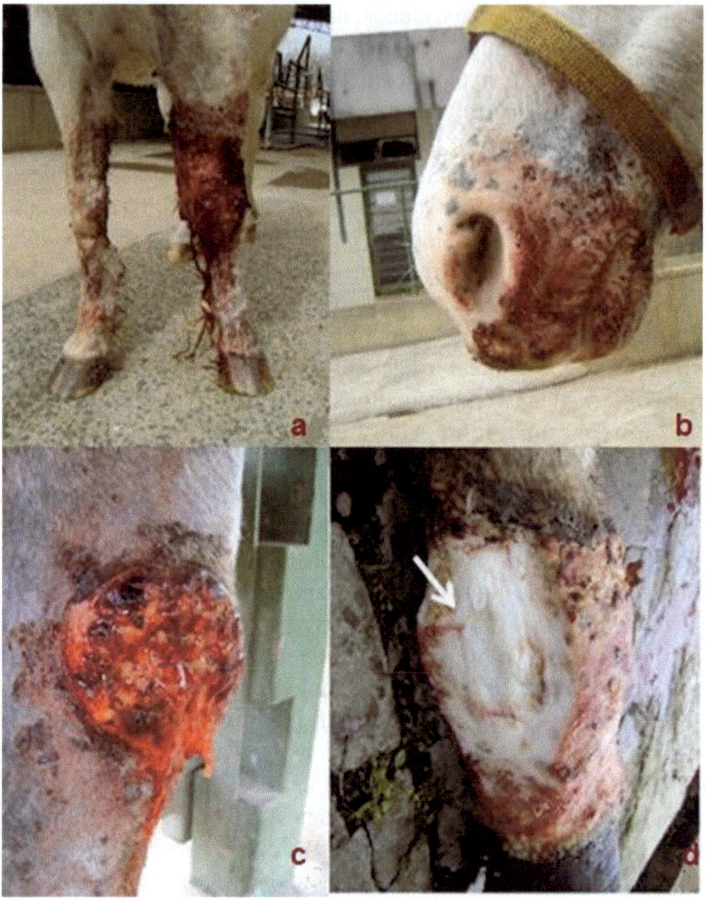

Fig. 1.6 Clinical manifestations of equine pythiosis in Brazil. (**a**) Granulomatous lesion in the anterior left limb showing large edema in the scapular region. (**b**) Same animal exposed in (a) presenting blood on the lips due to biting the lesion in an attempt to relieve the itching. (**c**) Granulation tissue and exudation of serosanguinolent viscous liquid. (**d**) Granulomatous lesion sectioned evidencing several "kunkers" (arrow). "Copyright of Gen National Publishing Group (Rio de Janeiro, Brazil), Book: DoençasInfecciosasemanimais de Produção e Companhia, Chapter of Pitiose (page 951) reprinted with permission"

hemithorax. The dog had no history of cutaneous or intestinal pythiosis [91]. Figure 1.7 illustrates clinical aspect of intestinal (Fig. 1.7a) and cutaneous/subcutaneous (Fig. 1.7b) lesions in dogs from Brazil.

Feline pythiosis has rarely been reported in literature, and the cases are restricted mainly to the retrobulbar, nasal, and intestinal lesions [44, 92]. In the first report on feline pythiosis, protrusion of the right nictitating membrane with conjunctivitis, soft swelling on the hard palate caudal to the last molar tooth on the right, mild respiratory stridor, and bilateral submandibular lymphadenopathy were observed [92]. Two other cats developed intestinal obstruction, and clinical presentations

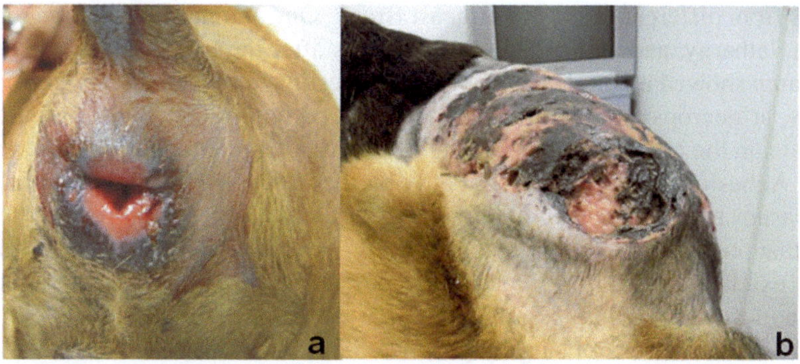

Fig. 1.7 Clinical manifestations of canine pythiosis in Brazil. (**a**) Intestinal pythiosis in a dog of Teckel breed. Note the swelling of the perianal region, with edema and exposure of the mucosa of the rectum. (**b**) Cutaneous/subcutaneous pythiosis in dog of German shepherd breed. Note ulcerated lesion with necrotic areas in the lumbosacral region. "Copyright of Gen National Publishing Group (Rio de Janeiro, Brazil), Book: Doenças Infecciosas Em Animais de Produção e Companhia, Chapter of Pitiose (page 952), reprinted with permission"

were anorexia and vomiting [44]. Fortin and coworkers in 2017 reported an unusual presentation of pythiosis in oral cavity in a cat which had multilobulated, sublingual mass since 3 months [45]. Nondomestic felids, such as jaguar (*Panthera onca*) and Bengal tiger (*Panthera tigris tigris*), have been reported with pulmonary and cutaneous pythiosis, respectively. The jaguar presented dyspnea and leukocytosis which was unresponsive to antibacterial therapy. By an exploratory thoracotomy, it was observed that multilobular masses with extensive fibrosis and numerous caseonecrotic foci were present in the left lung, which were removed and submitted to immunohistochemistry, confirming the diagnosis of pythiosis [58]. The Bengal tiger presented vomiting, diarrhea, and weight loss. Radiographic examination revealed the presence of an abdominal mass, which was removed during necropsy and was found to be presumptive for pythiosis. The diagnosis was confirmed by immunoblotting [57].

Among ruminants, bovine is the most affected species. Gabriel et al. (2008) described an outbreak in Pantanal region of Brazil in which 76 bovines were affected, and the animals showed some multifocal, nodular, and ulcerated lesions, lacking the presence of "kunkers," in the medial and lateral portions of the anterior and posterior limbs [49]. In this outbreak, a spontaneous healing of lesions was observed which reinforces the aspect of resistance of bovines to pythiosis [47]. Two other outbreaks of pythiosis have also been reported in sheep in the semiarid northeastern region of Brazil. The main clinical signs were local swellings with ulcerative lesions in the limbs, ventral abdomen, and prescapular regions. Limb ulcerated lesions were dry or wet and had a dark red or brown to black surface. Three sheep were necropsied, and two of them revealed dissemination to lungs, several multifocal nodules. One showed dissemination to prescapular lymph node, and another animal displayed cutaneous lesion that extended to the sesamoid bone [52, 53]. Two cases reported of gastroenteric pythiosis in lambs from different farms in northeast Brazil. One animal showed food regurgitation, lethargy, and anorexia and died 5 days after the first symptoms. The second lamb showed no clinical sign of gastrointestinal disorders before death. Both lambs, at necropsy, showed ulceration, covered by yellowish caseous granular exudate, in the mucosa of the esophagus, reticulum, rumen, omasum, and abomasum. Adhesions were observed between the serosa of the forestomach and abomasum to the liver and diaphragm [53]. Pythiosis has been reported in one goat that presented lameness and weight loss associated with an extensive, ulcerative, exudative, and pruriginous cutaneous lesion in the metatarsal-phalangeal region of the left hind limb with 1 month of evolution. The animal was clinically cured after two weeks of surgery [51].

P. insidiosum also causes pythiosis in humans [19, 93, 94]. The disease in humans has been reported mostly from Thailand [75], although some cases have been diagnosed in other countries as well [6, 19, 64]. Unlike the clinical features observed in the animals, human patients with pythiosis usually present with symptoms and signs associated with the *P. insidiosum* infection of eye (so-called ocular pythiosis) and arterial tissue of the upper or lower extremities (so-called vascular pythiosis) [19, 93]. Patients with the ocular infection seek for medical care due to the presence of corneal ulcer, keratitis, ocular pain, conjunctivitis, impaired visual acuity, tissue swelling, and hypopyon [75, 93]. Patients with the vascular infection seek for medical attention because they suffer from intermittent claudication or gangrenous ulcer of an arm or a leg (due to arterial insufficiency), alongside with some other clinical features, such as fever, itching, cellulitis, tissue swelling, unpalpable arterial pulse, and groin or abdominal mass [75, 94]. Average duration of the clinical symptoms of the patients with vascular pythiosis from first notice to seeking medical care (~2 months) is markedly longer than that of patients with ocular pythiosis (~2 weeks) [75]. This can be attributed to the eye being a highly concerned organ, and ocular tissue is an immune privilege site, which could promote the infection caused by *P. insidiosum*.

The *P. insidiosum* infection of other organs is relatively rare, compared to the ocular and vascular infections. Infection of the gastrointestinal tract has been reported in one Thai patient who presented with upper gastrointestinal bleeding and bloody mucous stool [75]. Seven reported cases suffered from the *P. insidiosum* infection of the head and neck region [95]. Their clinical presentations (i.e., severe headache, cellulitis, sinusitis, facial palsy, toothache, and seizure) are associated with internal and external carotid arteritis and inflammation of head and neck soft tissues along with brain abscess. Although the skin is a common site of pythiosis in animals (i.e., horses, dogs, cattle, sheep, and goats), only some human patients had the *P. insidiosum* infection of cutaneous/subcutaneous tissues. These patients presented with prolonged (up to several months) painful subcutaneous nodule and ulcer, or with acute onset (up to several days) of cellulitis, on the arm or leg [75].

In general, human pythiosis is commonly observed in patients aged 20–60 years and male workers with an agriculture-related occupation (i.e., farmer and

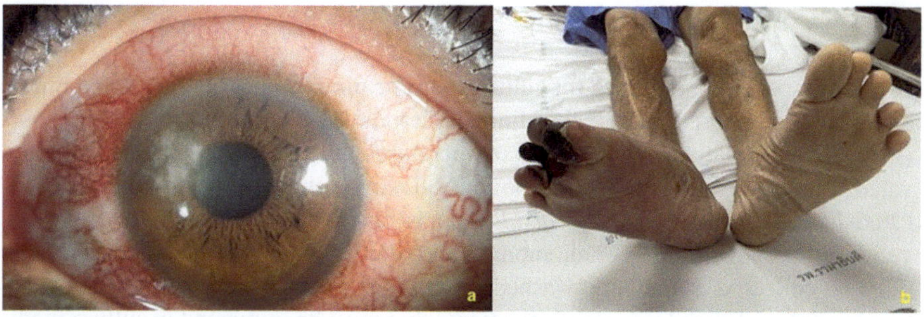

Fig. 1.8 Human pythiosis in Thailand. (**a**) Ocular pythiosis presents with keratitis. Using slit-lamp biomicroscope, inhomogeneous dense infiltration was observed at anterior to mid-corneal stroma and multiple small dots spreading with tentacle-liked opacities in superficial layer. Photograph by Dr. Passara Jongkhajornpong and Dr. Kaevalin Lekhanont. (**b**) Vascular pythiosis presents with gangrenous ulcers of the 1st, 2nd, and 3rd toes of the right foot. Photograph by Dr. Sirawat Srichatrapimuk and Dr. Maria Nina Chitasombat

fisherman) [75]. Interestingly, almost all patients with non-ocular pythiosis (vascular, cutaneous, gastrointestinal tract, and head and neck tissue) had an underlying condition, mostly hematological disorders (thalassemia, paroxysmal nocturnal hemoglobinuria, anemia of chronic disease, idiopathic thrombocytopenic purpura, leukemia, and hemophilia A) [75]. In contrast, most patients with ocular pythiosis reported no such underlying diseases [75, 93]. Figure 1.8 shows clinical aspects of keratitis (Fig. 1.8a) and cutaneous lesions (Fig. 1.8b) of pythiosis from Thai patients.

In Brazil, only a single case of human disease was reported. The patient reported fishing activity 1 week prior to the onset of the lesion that was initially suspected as a bacterial infection and showed no improvement on treatment with antibacterials. The lesion worsened and became ulcerative. Additionally, a skin biopsy revealed broad coenocytic hyphae, leading to the erroneous diagnosis of zygomycosis by physicians. Antifungal therapy (amphotericin B, oral itraconazole, and potassium iodine) was unsuccessful, and the patient was submitted to an extensive surgical procedure to remove the entire lesion, with margins of border and depth (amputation of the leg was considered prior to this surgical procedure). This case report is considered didactic and reflects well upon the difficulty in establishing a correct and early diagnosis of pythiosis [64, 65].

1.5 Diagnosis

Diagnosis of pythiosis is difficult and time-consuming, requires equipment, and depends on skilled healthcare personnel. Definitive diagnosis of the disease relies on clinical and epidemiological analysis and information from laboratory investigations, which include culture identification, serodiagnosis, hispathological examination, and molecular-based assays (PCR and sequence homology analysis) as briefly described below.

1.5.1 Culture Identification

A clinical specimen, obtained for culture identification, should be transferred immediately at room temperature to the clinical microbiology laboratory, as *P. insidiosum* is sensitive to relatively low or high temperatures [76]. The organism can be cultivated, in aerobic condition, in Sabouraud dextrose medium preferably at 37 °C (98.6 F), and the growth of *P. insidiosum* is observed in less than 24 h, as flat, yellow-to-brownish, non-aerial, submerged colony. The border of infected tissues must be collected aseptically, washed three times in sterile saline solution or distilled water, and sliced into small fragments that must be immersed in culture media (Sabouraud, blood agar, potato dextrose agar, and corn meal agar). Only in horses the culture must be performed with "kunkers" (Fig. 1.9); otherwise, the isolation is not possible. *P. insidiosum* may be easily recovered in culture; however, since the culture media above are not selective and allow the rapid growth of fungal contaminants, such as *Zygomycetes*, its recognition becomes difficult for technicians who have no experience in cultivating this pathogen. Microscopically, *P. insidiosum* is observed as sparsely septate, right angle-branching, relatively broad hyphae, which resemble filamentous fungi, such as *Aspergillus* spp., *Fusarium* spp., and the *Zygomycetes*. Once the pathogen is cultivated, it is necessary to demonstrate the zoospore production, as suggested by Mendoza and Prendas in 1988 [9].

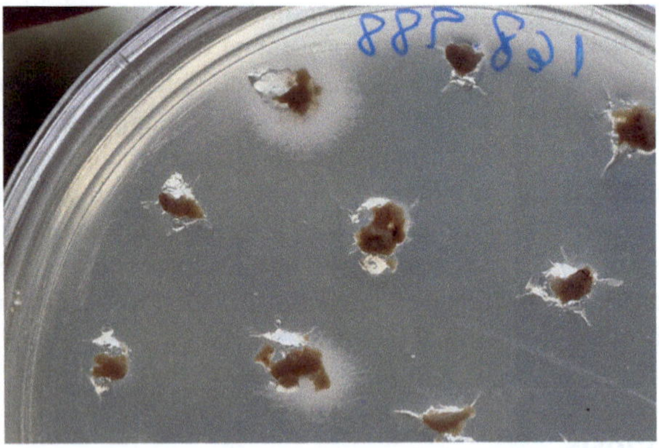

Fig. 1.9 Isolation of *P. insidiosum* from cultured fragments of "kunkers," removed from equine pythiosis, on 4% Sabouraud agar with 24 h of incubation at 37 °C. Note that the hyphae of *P. insidiosum* emerge from the "kunkers" and present hyaline color, with low aerial mycelium. "Copyright of Gen National Publishing Group (Rio de Janeiro, Brazil), Book: Doenças Infecciosas Em Animais de Produção e Companhia, Chapter of Pitiose (page 953), reprinted with permission"

1.5.2 Serodiagnosis

Detection of the anti-*P. insidiosum* antibodies in serum samples could aid efficient diagnosis of pythiosis. Several immunological techniques have been employed to accomplish this. Initially, an immunodiffusion (ID) assay has been developed using the crude protein extract of *P. insidiosum* [96–98]. Although ID is a relatively inexpensive and highly specific assay, its diagnostic performance is compromised by reportedly low sensitivity and thus leads to false-negative result. The disadvantages of the ID test have been overcome by the introduction of the enzyme-linked immunosorbent assay (ELISA) [99–101], which improves detection sensitivity while retaining its specificity. Western blot analysis, an assay with high sensitivity and specificity, has been reported for detecting the anti-*P. insidiosum* antibodies [19, 102, 103]. However, clinical use of ELISA and Western blot is cumbersome due to their multistep procedures, relatively long turnaround time, and requirements of routinely unavailable equipment. To serve the needs for a user-friendly and rapid serodiagnostic assay for pythiosis, hemagglutination assay (HA), which employs the *P. insidiosum* crude protein extract-coated sheep red cells, has been developed which provides the results within an hour [104, 105]. However, due to the limited detection sensitivity (84–88%) and specificity (82–99%) of HA, an immunochromatography (ICT), another user-friendly and rapid diagnostic format, has been invented for convenient, efficient, and rapid detection of the antibodies of *P. insidiosum* [19, 106].

Chareonsirisuthigul et al. (2013) compared these serodiagnostic assays, using a set of 37 pythiosis sera and 248 control sera, concluding that both ELISA and ICT exhibited equivalent diagnostic performance (100% sensitivity and specificity) [105]. An improved version of ICT has been developed by Intaramat et al. (2016) using the protein A/G conjugated with colloidal gold, which allow detection of the serum anti-*P. insidiosum* antibodies in both animals and humans [107]. They compared the protein A/G-based ICT and ELISA, using 85 sera from humans and animals (horses, dogs, cattle, and rabbits) having pythiosis and 143 sera from those with healthy condition or other diseases, and demonstrated that ELISA had 99% sensitivity and 100% specificity, while the protein A/G-based ICT had 91% sensitivity and 100% specificity. In their study, most of the sera with false-negative reads by ICT are the sera with weakly positive reads by ELISA. It should be noted that sera samples from patients with the ocular infection are likely to be read negative by all of the serodiagnostic assays. Thus, serodiagnostic interpretation of such samples should be done with caution.

1.5.3 Hispathological Examination

Use of the non-specific histological staining assays, such as Grocott's methenamine silver stain or periodic acid-Schiff (Fig. 1.10), is not effective for discrimination of *P. insidiosum* from some other filamentous fungi (*Zygomycetes*, *Aspergillus* spp., and *Fusarium* spp.) because these organisms share microscopic morphologies.

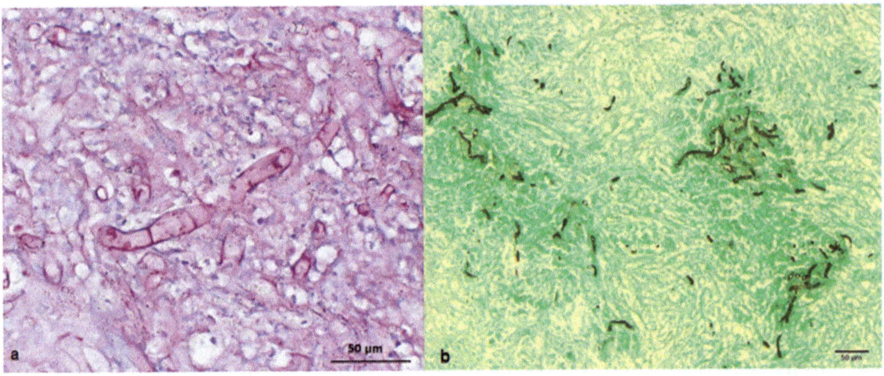

Fig. 1.10 *P. insidiosum* hyphae in histological sections. (**a**) PAS staining of cutaneous/subcutaneous lesion in dog, evidencing hyphae of pink coloration (400 ×). (**b**) Gomori-Grocott staining in granulomatous areas of the subcutaneous tissue in experimental pythiosis in rabbits, evidencing hyphae of blackish color (200 ×). "Copyright of Gen National Publishing Group (Rio de Janeiro, Brazil), Book: DoençasInfecciosasemanimais de Produção e Companhia, Chapter of Pitiose (page 954), reprinted with permission"

Thus, misinterpretation of the histological examination of *P. insidiosum*-infected tissue is likely to occur [64, 65, 108]. A more specific histological staining assay, employing the specific antibodies to anti-*P. insidiosum*, is then needed. Keeratijarut et al. (2009) developed the in-house immunoperoxidase assay for histodiagnosis of pythiosis, using the rabbit antiserum raised against the *P. insidiosum* crude protein extract [109]. In their evaluation of the tissue sections from 19 patients with pythiosis immunohistostaining assay demonstrated 100% sensitivity and 94% specificity for detection of *P. insidiosum* in the infected tissues. The detection specificity of the assay was affected by cross-reactivity of the rabbit antiserum with *Fusarium* species. To improve the detection specificity, Inkomlue et al. (2016) developed an immunohistochemical assay by using the rabbit anti-ELI025 antibody [109].

Elicitins are the proteins found only in the oomycetes (especially the genera *Pythium* and *Phytophthora*) and not in any other fungi. It could, therefore, serve as a better target for a more specific histological detection of *P. insidiosum*. The authors compared the sensitivity and specificity of the immunohistochemical assays using the rabbit antiserum raised against the elicitin protein (ELI025) and crude protein extract of *P. insidiosum* [110]. While both assays showed equivalently high sensitivity (100%), only the rabbit anti-ELI025 antibody failed to stain the fungus *Fusarium* and, thus, was more specific than the rabbit anti-*P. insidiosum* crude protein extract antibody (100% vs. 98%).

1.5.4 Molecular-Based Assays

Use of the molecular-based assays for detection of pathogens has become increasingly popular, due to its ease to perform and high diagnostic efficiency. The ribosomal DNA (rDNA) region (which is a multicopy gene that contains 18S rRNA, ITS1, 5.8S rRNA, ITS2, 28S rRNA, and the intergenic spacer) is the most popular target sequence used to detect many microorganisms, including *P. insidiosum* [111]. The molecular-based assays for detection of *P. insidiosum* are mostly relied on sequence homology analysis and PCR detection. For sequence homology, Badenoch et al. (2009) used a set of fungal universal primers (NS1 and NS2) to amplify and sequence the rDNA gene from the isolated organism. BLAST search of the obtained sequence indicates that the organism is *P. insidiosum* [112]. Similar sequence-based approaches targeting the rDNA gene have been successfully used by other investigators to identify *P. insidiosum* from pure cultures and infected tissues [42, 64, 86, 113].

The PCR-based assays rely on one or two sets of the rDNA gene-targeting primers for amplification of a defined size of amplicon, without sequence homology analysis. Grooters and Gee (2002) designed two sets of primers in their nested PCR for specific identification of *P. insidiosum* [113]. Because the primers of Grooters and Gee failed to amplify one of the Thai strains of *P. insidiosum*, Vanittanakom et al. (2004) designed a new set of the rDNA-specific primers and successfully identified all Thai strains tested in their study [114]. Some advances in the PCR-based assays include the use of a different target gene (glucanase-encoding gene) and a rapid amplification platform (real-time PCR) to increase the assay efficiency and decrease the assay turnaround time [115]. Rujirawat et al. (2017) developed the single nucleotide polymorphism-based multiplex PCR (targeting the rDNA gene) to identify and genotype *P. insidiosum* simultaneously [116].

1.6 Treatment

1.6.1 Animal Pythiosis

Treatment of animal pythiosis is difficult, since the prognosis is frequently worsened due to the delay in diagnosis. Currently available antifungals do not act on *P. insidiosum* due to lack of ergosterol in the cell membrane [117]. This is the reason why extensive surgical debridement is the main choice in the majority of the cases, when such approach is feasible. It is important to consider that surgery should contain pathogen-free surgical margin; otherwise, recurrences certainly occur.

There are antifungals routinely used in agriculture (plant fungicide) for the treatment of phytopathogens, including the genus *Pythium* and *Phytophthora.* However, such antifungals are toxic for animals, which make it impracticable to be used in the treatment of animal pythiosis. Among these plant fungicides, Brown et al. (2008) tested mefenoxam, as well as several other antifungals (itraconazole, voriconazole, posaconazole, terbinafine, and caspofungin), and found that only mefenoxam inhibited radial growth of *P. insidiosum* in vitro [118]. Recently, the mixture of mefenoxam with itraconazole and terbinafine was used for the treatment of canine gastrointestinal pythiosis with satisfactory results. The dose of mefenoxam employed was 4 mg/kg twice daily during a period of 18 months with no significant clinical, hematologic, or biochemical abnormalities. Despite the tolerability of mefenoxam by this dog, the authors' advice cautions in its prescription until the safety is well evaluated in a larger population of dogs [119].

Equine pythiosis has been treated in Brazil, in the majority of the cases, by surgery procedures to remove the entire lesion with pathogen-free margins (if the lesion is small) or the excess of granulation tissue (if the lesion is large). Other possibilities of treatment include immunotherapy, intravenous regional perfusion with amphotericin B, and oral administration of potassium iodide, in association with/without surgery. Immunotherapy consists of administering a macerate of sonicated hyphae conditioned in bottles (resuspended in 2 ml of sterile water at the moment of application). It is commercially available in Brazil as Pitium-Vac™, with a medium cost per dose of US$16.00. The recommendation for using immunotherapy is that the lesion is unique and is not more than 60 days old. If more than one dose is required, the interval between them is 14 days [120]. The main mechanism of action of immunotherapy for healing wound is the switch of Th2 response to Th1, which is mediated by mononuclear cells, mainly monocytes and lymphocytes [121, 122]. Immunotherapy, when associated with surgical removal in horses, reaches up to 75% cure [24].

Intravenous regional perfusion with amphotericin B has been employed for the treatment of lesions in the middle-third and extremities of limbs of horses. It was demonstrated that if such procedure was performed after surgery, around 58% of the horses had the wounds healed 60 days after one administration of intravenous regional perfusion; and in 42% of the horses, a second administration was necessary 14 days after the first one [123, 124].

Photodynamic therapy was evaluated experimentally as a new and promising therapeutic approach [125, 126].

1.6.2 Human Pythiosis

Treatment of human pythiosis is difficult and challenging. Therapeutic options for the disease include the use of antimicrobial drugs, administration of an immunotherapeutic vaccine, and surgical removal of an infected organ. Only a few patients with pythiosis had a favorable response to the medical treatment using a combination of antifungal drugs (itraconazole and terbinafine) guided by in vitro susceptibility analysis [62]. In contrast, most patients underwent a combination of medical treatment and radical surgery of an affected organ to control the *P. insidiosum* infection [75, 94]. Conventional antifungal drugs (itraconazole and terbinafine) are generally ineffective against *P. insidiosum* because the organism lacks antifungal drug targets in the ergosterol biosynthetic pathway [117]. Potassium iodide oral solution (SSKI) has been used in some patients, but only the patients with cutaneous pythiosis showed good responses [75, 94]. Use of the immunotherapeutic vaccine prepared from the *P. insidiosum* crude protein extract showed a limited efficiency in the treatment of pythiosis [75, 94, 127, 128].

1.7 Prophylaxis, Control, and Public Health Concerns

There is no vaccine as a prophylactic measure aiming to avoid new cases of the disease. As mentioned before, vaccine is employed as immunotherapy for switching the immune response from Th2 to Th1 profile.

It is strongly recommended to avoid the permanence of animals on stagnant water with vegetation, since infective zoospores are found in this environment. However, such measure is unpractical in many farms, since pasture is more available in such areas and attracts the attention of animals for food. Reviewing management practices during rainy seasons, such as keeping animals in a drier environment, may be a good prophylactic practice to prevent new infections.

The prevention in humans is equally difficult, since it is quite impossible to prohibit swimming in rivers and lagoons as well as stopping agricultural practices. The use of personal protective equipment (PPE), such as adequate swimming goggles for swimming in rivers/lagoons, rubber gum boots, gloves, and glasses to enter flooded fields (such as rice paddies), should be recommended. Another possibility is the advertisement for advice for the risk of infection in rivers/lagoons in areas endemic for pythiosis.

Direct zoonotic transmission between animals to humans is not reported, since infection is acquired in water environment through penetration of zoospores in injured skin. It is, therefore, important to note that the concept of zoonosis also includes the environment as a source of infection, and in this sense, pythiosis may be classified as saprozoonosis. Accidental laboratory infection has never been reported; however, care should be taken while dealing with in vitro methods for zoospore production [129].

References

1. De Cock AW, Mendoza L, Padhye AA et al (1987) *Pythium insidiosum* sp. nov., the etiologic agent of pythiosis. J Clin Microbiol 25:344–349
2. Ristaino JB (2002) Tracking historic migrations of the Irish potato famine pathogen, *Phytophthora infestans*. Microb Infect 4:1369–1377
3. Calvano TP, Blatz PJ, Vento TJ et al (2011) *Pythium aphanidermatum* infection following combat trauma. J Clin Microbiol 49:3710–3713
4. Farmer AR, Murray CK, Driscoll IR et al (2015) Combat-related *Pythium aphanidermatum* invasive wound infection: case report and discussion of utility of molecular diagnostics. J Clin Microbiol 53:1968–1975
5. Sekhon AS, Padhye AA, Garg AK (1992) *In vitro* sensitivity of *Penicillium marneffei* and *Pythium insidiosum* to various antifungal agents. Eur J Epidemiol 8:427–432
6. Gaastra W, Lipman LJA, De Cock AW et al (2010) *Pythium insidiosum*: an overview. Vet Microbiol 146:1–16
7. Lerksuthirat T, Lohnoo T, Inkomlue R et al (2015) The elicitin-like glycoprotein, ELI025, is secreted by the pathogenic oomycete *Pythium insidiosum* and evades host antibody responses. PLoS One 10:e0118547
8. Mendoza L, Hernandez F, Ajello L (1993) Life cycle of the human and animal oomycete pathogen *Pythium insidiosum*. J Clin Microb 31:2967–2973

9. Mendoza L, Prendas J (1988) A method to obtain rapid zoosporogenesis of *Pythium insidiosum*. Mycopathologia 104:59–62
10. Schurcko A, Mendoza L, De Cock AW et al (2003a) Evidence for geographic clusters: Molecular genetic differences among strains of *Pythium insidiosum* from Asia, Australia and Americas are explored. Mycologia 95:200–208
11. Schurko A, Mendoza L, Levesque CA et al (2003b) A molecular phylogeny of *Pythium insidiosum*. Mycol Res 107:537–544
12. Kammarnjesadakul P, Palaga T, Sritunyalucksana K et al (2011) Phylogenetic analysis of *Pythium insidiosum* Thai strains using cytochrome oxidase II (*COX II*) DNA coding sequences and internal transcribed spacer regions (ITS). Med Mycol 49:289–295
13. Azevedo MI, Botton SA, Pereira DI et al (2012) Phylogenetic relationships of Brazilian isolates of *Pythium insidiosum* based on ITS rDNA and cytochrome oxidase II gene sequences. Vet Microbiol 159:141–148
14. Ribeiro TC, Weiblen C, Azevedo MI et al (2017) Microevolutionary analyses of *Pythium insidiosum* isolates of Brazil and Thailand based on exo-1,3-β-glucanase gene. Infect Genet Evol 48:58–63
15. Sathapatayavongs B, Leelachaikul P, Prachaktam R et al (1989) Human pythiosis associated with Thalassemia Hemoglobinopathy Syndrome. J Infect Dis 159:274–280
16. Wanachiwanawin W, Thianprasit M, Fucharoen S et al (1993) Fatal arteritis due to *Pythium insidiosum* infection in patients with thalassaemia. Trans R Soc Trop Med Hyg 87:296–298
17. Imwidthaya P (1994) Human pythiosis in Thailand. Postgrad Med J 70:558–560
18. Thianprasit M, Chaiprasert A, Imwidthaya P (1996) Human pythiosis. Curr Top Med Mycol 7:43–54
19. Krajaejun T, Kunakorn M, Pracharktam R et al (2006) Identification of a novel 74-kilo dalton immunodominant antigen of *Pythium insidiosum* recognized by sera from human patients with pythiosis. J Clin Microbiol 44:1674–1680
20. Smith F (1884) The pathology of bursattee. Vet J 19:16–17
21. DeHaan J, Hoogkamer LJ (1901) Hyphomycosisdestruens. Veeartsennijkundige Bld voor Ned Indië 13:350–374
22. Witkamp J (1924) Bijdrage tot de kennis van de Hyphomycosis destruens. Ned Ind Blad voor Diergeneeskd en Dierenteelt 36:229–245
23. Santurio JM, Alves SH, Pereira DB et al (2006) Pythiosis: an emergent mycosis. Acta Scientiae Veter 34:1–14
24. Santos CEP, Ubiali DG, Pescador CA et al (2014) Epidemiological survey of equine pythiosis in the Brazilian Pantanal and nearby areas: results of 76 cases. J Eq Vet Sci 34:270–274
25. Souto EPF, Maia LA, Olinda RG et al (2016) Pythiosis in the nasal cavity of horses. J Comp Path 155:126–129
26. Salas Y, Márquez A, Canelón J et al (2012) Equine pythiosis: report in crossed bred (criole venezuelan) horses. Mycopathologia 174:511–517
27. Bridges CH, Emmons CW (1961) A phycomycosis of horses caused by *Hyphomyces destruens*. J Am Vet Med Assoc 138:579–589
28. Goad MEP (1984) Pulmonary pythiosis in a horse. Vet Pathol 21:261–262
29. White SD, Ghoddusi M, Grooters AM et al (2008) Cutaneous pythiosis in a nontravelled California horse. Vet Dermatol 19:391–394
30. Mosbah E, Karrouf GIA, Younis EA et al (2012) Diagnosis and surgical management of pythiosis in draft horses: report of 33 cases in Egypt. J Eq Vet Sci 32:164–169
31. Miller RI (1983) Investigations into the biology of three 'phycomycotic' agents pathogenic for horses in Australia. Mycopathologia 81:23–28
32. Shipton WA (1987) *Pythium destruens* sp. nov., an agent of equine pythiosis. J Med Vet Mycol 25:137–151
33. Tonpitak W, Pathomsakulwong W, Sornklien C et al (2018) First confirmed case of nasal pythiosis in a horse in Thailand. JMM Case Reports 5:e005136
34. Álvarez JAC, Viloria MIV, Ayola SCP (2013) Evaluación clínica e histopatológica de la pitiosis cutánea en burros (*Equus asinus*). Rev Med Vet 25:9–19

35. Maia LA, Olinda RG, Araújo TF et al (2016) Cutaneous pythiosis in a donkey (*Equus asinus*) in Brazil. J Vet Diag Invest 28:436–439
36. Fischer JR, Pace LW, Turk JR et al (1994) Gastrointestinal pythiosis in Missouri dogs: eleven cases. J Vet Diagn Invest 6:380–382
37. Dykstra M, Sharp NJ, Olivry T et al (1999) A description of cutaneous-subcutaneous pythiosis in fifteen dogs. Med Mycol 37:427–433
38. Berryessa NA, Marks SL, Pesavento PA et al (2008) Gastrointestinal pythiosis in 10 dogs from California. J Vet Intern Med 22:1065–1069
39. Torres-Neto R, Bosco SMG, Amorin RL et al (2010) Cutaneous pythiosis in a dog from Brazil. Vet Dermatol 21:202–204
40. Pereira DI, Schild AL, Motta MA et al (2010) Cutaneous and gastrointestinal pythiosis in a dog in Brazil. Vet Res Commun 34:301–306
41. Fernandes CPM, Giordani C, Grecco FB et al (2012) Gastric pythiosis in a dog. Rev Iberoam Micol 29:235–237
42. Rivierre C, Laprie C, Guiard-Marigny O et al (2005) Pythiosis in Africa. Emerg Infect Dis 11:479–481
43. Mendoza L, Arias M, Colmenarez V et al (2005) Intestinal canine pythiosis in Venezuela confirmed by serological and sequencing analysis. Mycopathologia 159:219–222
44. Rakich PM, Grooters AM, Tang KN (2005) Gastrointestinal pythiosis in two cats. J Vet Diagn Invest 17:262–269
45. Fortin JS, Calcutt MJ, Kim DY (2017) Sublingual pythiosis in a cat. Acta Vet Scand 59:63
46. Miller RI, Olcott BM, Archer M (1985) Cutaneous pythiosis in beef calves. J Am Vet Med Assoc 186:984–986
47. Santurio JM, Monteiro AB, Leal AT et al (1998) Cutaneous Pythiosis insidiosi in calves from the Pantanal region of Brazil. Mycopathologia 141:123–125
48. Pérez RC, Luis-León JJ, Vivas JL et al (2005) Epizootic cutaneous pythiosis in beef calves. Vet Microbiol 109:121–128
49. Gabriel AL, Kommers GD, Trost ME et al (2008) Outbreak of cutaneous pythiosis in cattle. Pesq Vet Bras 28:583–587
50. Konradt G, Bassuino DM, Bianchi MV et al (2016) Cutaneous pythiosis in calves: an epidemiologic, pathologic, serologic and molecular characterization. Med Mycol Case Rep 14:24–26
51. Carmo PMS, Portela RA, Silva TR et al (2015) Cutaneous pythiosis in a goat. J Comp Path 152:103–105
52. Tabosa IM, Riet-Correa F, Nobre VM et al (2004) Outbreaks of pythiosis in two flocks of sheep in northeastern Brazil. Vet Pathol 41:412–415
53. Pessoa CRM, Riet-Correa F, Pimentel LA et al (2012) Pythiosis of the digestive tract in sheep. J Vet Diagn Invest 24:1133–1136
54. Bernard FD, Conhizak C, Ambrosini F et al (2015) Pythiosis in sheep from Paraná, southern Brazil. Pesq Vet Bras 35:513–517
55. Wellehan JF, Farina LL, Keoughan CG et al (2004) Pythiosis in a dromedary camel (*Camelus dromedarius*). J Zoo Wildl Med 35:564–568
56. Videla R, Amstel SV, O'Neill SH et al (2012) Vulvar pythiosis in two captive camels (*Camelus dromedarius*). Med Mycol 50:219–224
57. Buergelt C, Powe J, White T (2006) Abdominal pythiosis in a bengal tiger (*Panthera tigris tigris*). J Zoo Wildl Med 37:186–189
58. Camus AC, Grooters AM, Aquilar RF (2004) Granulomatous pneumonia caused by *Pythium insidiosum* in a central American jaguar, *Panthera onca*. J Vet Diagn Invest 16:567–571
59. Pesavento PA, Barr B, Riggs SM et al (2008) Cutaneous pythiosis in a nestling white-faced Ibis. Vet Pathol 45:538–541
60. Thianprasit M (1986) Fungal infection in Thailand. Jpn J Dermatol 96:1343–1345
61. Thianprasit M (1990) Human pythiosis. Trop Dermatol 4:1–4
62. Shenep JL, English BK, Kaufman L et al (1998) Successful medical therapy for deeply invasive facial infection due to *Pythium insidiosum* in a child. Clin Infect Dis 27:1388–1393

63. Hilton RE, Tepedino K, Glenn CJ et al (2016) Swamp cancer: a case of human pythiosis and review of the literature. Br J Dermatol 175:394–397
64. Bosco SMG, Bagagli E, Araujo JP Jr et al (2005) Human pythiosis, Brazil. Emerg Infect Dis 11:715–717
65. Marques SA, Bagagli E, Bosco SMG et al (2006) *Pythium insidiosum*: relato do primeiro caso de infecção humana no Brasil. An Bras Dermatol 81:483–485
66. Tanhehco TY, Stacy RC, Mendoza L et al (2011) *Pythium insidiosum* keratitis in Israel. Eye Contact Lens 37:96–98
67. Rathi A, Chakrabarti A, Agarwal T et al (2018) *Pythium* keratitis leading to fatal cavernous sinus thrombophlebitis. Cornea 37:518–522
68. Hung C, Leddin D (2014) Keratitis caused by *Pythium insidiosum* in an immunosuppressed patient with Crohn's disease. Clin Gastroenterol Hepatol 12:xxi–xxii
69. Lelievre L, Borderie V, Garcia-Hermoso D et al (2015) Case Report: Imported *Pythium insidiosum* keratitis after a swim in Thailand by a contact lens-wearing traveler. Am J Trop Med Hyg 92:270–273
70. Castellar FR, Jiménez CS, Zarzuelo AH et al (2017) Intraocular minocycline for the treatment of ocular pythiosis. Am J Health-Syst Pharm 74:821–825
71. Ravishankar JP, Davis CM, Dacis DJ et al (2001) Mechanics of solid tissue invasion by the mammalian pathogen *Pythium insidiosum*. Fung Genet Biol 34:167–175
72. Davis DJ, Lanter K, Makselan S et al (2006) Relationship between temperature optima and secreted protease activities of three *Pythium* species and pathogenicity toward plant and animal hosts. Mycol Res 110:96–103
73. Krajaejun T, Khositnithikul R, Lerksuthirat T et al (2011) Expressed sequence tags reveal genetic diversity and putative virulence factors of the pathogenic oomycete *Pythium insidiosum*. Fung Biol 115:683–696
74. Chechi JL, Franckin T, Barbosa LN et al (2018) Inferring putative virulence factors for *Pythium insidiosum* by proteomic approach. Med Mycol 57:92–100
75. Krajaejun T, Sathapatayavongs B, Pracharktam R et al (2006) Clinical and epidemiological analyses of human pythiosis in Thailand. Clin Infect Dis 43:569–576
76. Krajaejun T, Chongtrakool P, Angkananukul K et al (2010) Effect of temperature on growth of the pathogenic oomycete *Pythium insidiosum*. Southeast Asian J Trop Med Public Health 41:1462–1466
77. Mendoza L, Newton JC (2005) Immunology and immunotherapy of the infections caused by *Pythium insidiosum*. Med Mycol 43:477–486
78. Miller RI, Campbell SF (1983) Experimental pythiosis in rabbits. Sabouraudia 21:331–341
79. Trolezi R, Azanha JM, Paschoal NR et al (2017) *Stryphnodendron adstringens* and purified tannin on *Pythium insidiosum*: *in vitro* and *in vivo* studies. Ann Clin Microbiol Antimicrob 16:7
80. Zanette RA, Santurio JM, Loreto ES et al (2013) Toll-deficient *Drosophila* is susceptible to *Pythium insidiosum* infection. Microbiol Immunol 57:732–735
81. Tondolo JSM, Loreto ES, Ledur PC et al (2017) Chemically induced disseminated pythiosis in BALB/c mice: a new experimental model for *Pythium insidiosum* infection. PLoS One 12:e0177868
82. Verdi CM, Jesus FPK, Kommers G et al (2018) Embryonated chicken eggs: an experimental model for *Pythium insidiosum* infection. Mycoses 61:104–110
83. Watanabe MJ, Alonso J, Alves AL et al (2015) Equine pythiosis: report of 28 cases from São Paulo State, Brazil. Semina: Ciências Agrárias 36:909–916
84. Brown CC, Roberts ED (1988) Intestinal pythiosis in a horse. Aust Vet J 65:88–89
85. Bezerra PS Jr, Pedroso PMO, Pavarini SP et al (2010) Equine intestinal pythiosis in Southern Brazil. Arq Bras Med Vet Zootec 62:481–483
86. Reis JL Jr, de Carvalho EC, Nogueira RH et al (2003) Disseminated pythiosis in three horses. Vet Microbiol 96:289–295
87. Mendoza L, Alfaro AA, Villalobos J (1988) Bone lesions caused by *Pythium insidiosum* in a horse. J Med Vet Mycol 26:5–12

88. Jaeger GH, Rotstein DS, Law JM (2002) Prostatic pythiosis in a dog. J Vet Intern Med 16:598–602
89. Oldenhoff W, Grooters A, Pinkerton ME et al (2014) Cutaneous pythiosis in two dogs from Wisconsin, USA. Vet Dermatol 25:52–e21
90. Connolly SL, Frank C, Thompson CA et al (2012) Dual infection with *Pythium insidiosum* and *Blastomyces dermatitidis* in a dog. Vet Clin Pathol 41:419–423
91. Kepler D, Cole R, Lee-Fowler T et al (2017) Pulmonary pythiosis in a canine patient. Vet Radiol Ultrasound 60:E20–E23
92. Bissonnette KW, Sharp NJ, Dykstra MH et al (1991) Nasal and retrobulbar mass in a cat caused by *Pythium insidiosum*. J Med Vet Mycol 29:39–44
93. Krajaejun T, Pracharktam R, Wongwaisayawan S et al (2004) Ocular pythiosis: is it underdiagnosed? Am J Ophthalmol 137:370–372
94. Chitasombat MN, Larbcharoensub N, Chindamporn A et al (2018a) Clinicopathological features and outcomes of pythiosis. Int J Infect Dis 71:33–41
95. Chitasombat MN, Petchkum P, Horsirimanont S et al (2018b) Vascular pythiosis of carotid artery with meningitis and cerebral septic emboli: a case report and literature review. Med Mycol Case Rep 21:57–62
96. Mendoza L, Kaufman L, Standard PG (1986) Immunodiffusion test for diagnosing and monitoring pythiosis in horses. J Clin Microbiol 23:813–816
97. Imwidthaya P, Srimuang S (1989) Immunodiffusion test for diagnosing human pythiosis. Mycopathologia 106:109–112
98. Pracharktam R, Changtrakool P, Sathapatayavongs B et al (1991) Immunodiffusion test for diagnosis and monitoring of human pythiosis insidiosi. J Clin Microbiol 29:2661–2662
99. Mendoza L, Kaufman L, Mandy W et al (1997) Serodiagnosis of human and animal pythiosis using an enzyme-linked immunosorbent assay. Clin Diagn Lab Immunol 4:715–718
100. Krajaejun T, Kunakorn M, Niemhom S et al (2002) Development and evaluation of an in-house enzyme-linked immunosorbent assay for early diagnosis and monitoring of human pythiosis. Clin Diagn Lab Immunol 9:378–382
101. Grooters AM, Leise BS, Lopez MK et al (2002) Development and evaluation of an enzyme-linked immunosorbent assay for the serodiagnosis of pythiosis in dogs. J Vet Intern Med 16:142–146
102. Mendoza L, Nicholson V, Prescott JF (1992) Immunoblot analysis of the humoral immune response to *Pythium insidiosum* in horses with pythiosis. J Clin Microbiol 30:2980–2983
103. Supabandhu J, Vanittanakom P, Laohapensang K et al (2009) Application of immunoblot assay for rapid diagnosis of human pythiosis. J Med Assoc Thai 92:1063–1071
104. Jindayok T, Piromsontikorn S, Srimuang S et al (2009) Hemagglutination test for rapid serodiagnosis of human pythiosis. Clin Vaccine Immunol 16:1047–1051
105. Chareonsirisuthigul T, Khositnithikul R, Intaramat A et al (2013) Performance comparison of immunodiffusion, enzyme-linked immunosorbent assay, immunochromatography and hemagglutination for serodiagnosis of human pythiosis. Diagn Microbiol Infect Dis 76:42–45
106. Krajaejun T, Imkhieo S, Intaramat A et al (2009) Development of an immunochromatographic test for rapid serodiagnosis of human pythiosis. Clin Vaccine Immunol 16:506–509
107. Intaramat A, Sornprachum T, Chantrathonkul B et al (2016) Protein A/G-based immunochromatographic test for serodiagnosis of pythiosis in human and animal subjects from Asia and Americas. Med Mycol 54:641–647
108. Kaufman L (1998) Penicilliosis marneffei and pythiosis: emerging tropical diseases. Mycopathologia 143:3–7
109. Keeratijarut A, Karnsombut P, Aroonroch R et al (2009) Evaluation of an in-house immunoperoxidase staining assay for histodiagnosis of human pythiosis. Southeast Asian J Trop Med Public Health 40:1298–1305
110. Inkomlue R, Larbcharoensub N, Karnsombut P et al (2016) Development of an anti-elicitin antibody-based immunohistochemical assay for diagnosis of pythiosis. J Clin Microbiol 54:43–48

111. Krajaejun T, Sathapatayavongs B, Sullivan TD (2011) *Pythium*. In: Liu D (ed) Molecular detection of human fungal pathogens, 1st edn. CRC press, Florida, pp 851–863
112. Badenoch PR, Mills RA, Chang JH et al (2009) *Pythium insidiosum* keratitis in an Australian child. Clin Exp Ophthalmol 37:806–809
113. Grooters AM, Gee MK (2002) Development of a nested Polymerase Chain Reaction assay for the detection and identification of *Pythium insidiosum*. J Vet Intern Med 16:147–152
114. Vanittanakom N, Supabandhu J, Khamwan C et al (2004) Identification of emerging human-pathogenic *Pythium insidiosum* by serological and molecular assay-based methods. J Clin Microbiol 42:3970–3974
115. Keeratijarut A, Lohnoo T, Yingyong W et al (2015) Detection of the oomycete *Pythium insidiosum* by real-time PCR targeting the gene coding for exo-1,3-β-glucanase. J Med Microbiol 64:971–977
116. Rujirawat T, Sridapan T, Lohnoo T et al (2017) Single nucleotide polymorphism-based multiplex PCR for identification and genotyping of the oomycete *Pythium insidiosum* from humans, animals and the environment. Infect Genet Evol. 54:429–436
117. Lerksuthirat T, Sangcakul A, Lohnoo T et al (2017) Evolution of the sterol biosynthetic pathway of *Pythium insidiosum* and related oomycetes contributes to antifungal drug resistance. Antimicrob Agents Chemother 61:e02352–e02316
118. Brown TA, Grooters AM, Hosgood GL (2008) *In vitro* susceptibility of *Pythium insidiosum* and a *Lagenidium* sp. to itraconazole, posaconazole, voriconazole, terbinafine, caspofungin and mefenoxam. Am J Vet Res 69:1463–1468
119. Hummel J, Grooters A, Davidson G et al (2011) Successful management of gastrointestinal pythiosis in a dog using itraconazole, terbinafine and mefenoxam. Med Mycol 49:539–542
120. http://pitiose.com.br/loja-virtual/pitium-vac
121. Mendoza L, Mandy W, Glass R (2003) An improved *Pythium insidiosum*-vaccine formulation with enhanced immunotherapeutic properties in horses and dogs with pythiosis. Vaccine 21:2797–2804
122. Santurio JM, Leal AT, Leal AB et al (2003) Three types of immunotherapies against pythiosis insidiosi developed and evaluated. Vaccine 21:2535–2540
123. Dória RG, Freitas SH, Linardi RL et al (2012) Treatment of pythiosis in equine limbs using intravenous regional perfusion of amphotericin B. Vet Surg 41:759–765
124. Dória RG, Carvalho MB, Freitas SH et al (2015) Evaluation of intravenous regional perfusion with amphotericin B and dimethylsulfoxide to treat horses for pythiosis of a limb. BMC Vet Res 11:152
125. Pires L, Bosco SMG, Silva Junior NF et al (2013) Photodynamic therapy for pythiosis. Vet Dermatol 24:130–136
126. Pires L, Bosco SMG, Baptista MS et al (2014) Photodynamic therapy in *Pythium insidiosum* - an *in vitro* study of the correlation of sensitizer localization and cell death. PLoS One 9:e85431
127. Wanachiwanawin W, Mendoza L, Visuthisakchai S et al (2004) Efficacy of immunotherapy using antigens of *Pythium insidiosum* in the treatment of vascular pythiosis in humans. Vaccine 22:3613–3621
128. Permpalung N, Worasilchai N, Plongla R et al (2015) Treatment outcomes of surgery, antifungal therapy and immunotherapy in ocular and vascular human pythiosis: a retrospective study of 18 patients. J Antimicrob Chemother 70:1885–1892
129. Bosco SMG, Hussni CA, Santurio JM et al (2016) Pitiose. In: Megid J, Ribeiro MG, Paes AC (eds) Doenças infecciosas em animais de produção e companhia, 1st edn. Roca, Rio de Janeiro, pp 946–957

Superficial Mycoses in Dogs and Cats

2

Ramona Moraru, René Chermette, and Jacques Guillot

Abstract

Dermatophytosis, *Malassezia* otitis and dermatitis represent the superficial mycoses with greatest significance in dogs and cats. The causative agents, dermatophytes and *Malassezia* yeasts, have evolved independently in order to adapt to survival and development within the cutaneous ecosystem. The most prevalent dermatophyte species, *Microsporum canis*, may cause outbreaks at least in its principal hosts and at the same time have the ability to infect a wide range of mammals, including humans. The non-lipid-dependent species *Malassezia pachydermatis* is a common cause of otitis externa and pruritic dermatitis in dogs and sometimes in cats. Dermatophytosis should be considered in the differential diagnosis of many skin diseases, and diagnostic tests are systematically required in dogs and cats. The diagnosis of *Malassezia* dermatitis is based on clinical signs and the presence of high number of yeasts in lesional skin together with clinical and mycological responses to a specific therapy. The treatment of superficial mycoses includes the use of topical and systemic antifungals. In case of dermatophytosis, the disinfection of the environment may be required.

Keywords

Dermatophyte · Ringworm · *Malassezia* · Dog · Cat

R. Moraru
Clinique vétérinaire, Le Châtelet en Brie, France

R. Chermette · J. Guillot (✉)
Department of Parasitology, Mycology and Dermatology, Ecole nationale vétérinaire d'Alfort, Dynamyc Research Group, EnvA, UPEC, USC ANSES, Maisons-Alfort, France
e-mail: jacques.guillot@vet-alfort.fr

K. Singh, N. Srivastava (eds.), *Recent Trends in Human and Animal Mycology*,
https://doi.org/10.1007/978-981-13-9435-5_2

2.1 Introduction

Superficial mycoses of the skin are fungal infections that affect mostly the superficial layers of the skin, hairs and any other skin appendages, such as claws in carnivores [1]. Dermatophytoses and *Malassezia* infections represent the superficial mycoses with greatest impact in dogs and cats [2]. The causative agents, dermatophytes and *Malassezia* yeasts, respectively, are placed in distant branches of the fungal tree of life and are likely to have evolved independently to adapt to survival and development within a cutaneous ecosystem.

Dermatophytes are filamentous fungi, which are able to use hard keratin, a fundamental component of the skin and skin appendages, as a growth substrate. According to their affinity, three categories of dermatophytes are usually defined, namely, (i) the zoophilic ones, which develop as parasites in mammals, but more rarely in birds, (ii) the anthropophilic dermatophytes in man and (iii) the geophilic dermatophytes, which develop in the soil, but some of them are able to become parasite in animals and man. In all cases, dermatophytic spores can survive in the environment leading to resistance and spreading the fungus. In dogs and cats, the most prevalent dermatophyte species, *Microsporum canis* and *Trichophyton mentagrophytes*, are true parasites; when in contact with the skin, they infect stratum corneum and hairs causing cutaneous lesions. Dermatophytes are also significant due to their zoonotic potential and thc concern of companion animals' owners confronted with severe and chronic inflammatory skin diseases [3]. *Malassezia* yeasts are normal commensals and occasional pathogens of the skin for many warm-blooded animal species [4]. They have different enzymes including lipases, phospholipases and aspartyl protease, which allow them to survive in the cutaneous environment. The non-lipid-dependent species *M. pachydermatis* is a very common cause of otitis externa and pruritic dermatitis in dogs. It is also regularly recovered from the skin of cats along with other *Malassezia* species. The genetic basis of the keratinolytic activity of dermatophytes or the lipophilic nature of *Malassezia* yeasts and corresponding adaptation to animal skin (putatively starting from an ancestral state as a plant or soil resident) are still in discussion [5, 6].

This chapter provides an overview of dermatophytes and *Malassezia* yeasts, their significance in companion animals, different steps to diagnose superficial mycoses, rational control measures in order to treat dogs and cats and also prevention of animal and/or human against these infections.

2.2 Causative Agents

2.2.1 Dermatophyte Species

Dermatophytes are members of the phylum Ascomycota. They are closely related to *Coccidioides immitis* within the order Onygenales [7]. More than 20 different dermatophyte species have been isolated from companion animals, but the most commonly isolated pathogens are *Microsporum canis*, *Trichophyton mentagrophytes*

Table 2.1 Major dermatophyte species infecting dogs and cats

Dermatophyte species	Major hosts	Frequent sources of contamination	Risk of human contamination
Microsporum canis	Cats, dogs and many other mammals (including humans)	Young cats	Yes
Trichophyton mentagrophytes	Small rodents (mice, rats), rabbits, dogs, cats	Small rodents, rabbits, dogs	Yes
Microsporum (Nannizzia) gypseum	Dogs, cats, horses	Soil	Yes (but very seldomly)
Microsporum (Nannizzia) persicolor	Small rodents (moles and voles), dogs and cats	Small rodents, soil	Yes (but very seldomly)

and to a lesser extent *Microsporum (Nannizzia) gypseum* and *Microsporum (Nannizzia) persicolor* (Table 2.1) [1, 3]. *Microsporum canis* is by far the most predominant in cats with over 90% of the feline isolates in most of the surveys conducted worldwide. *Trichophyton mentagrophytes* remains a frequent pathogen of small rodents like mice, rats and chinchillas but can also be found in dogs and cats. *Microsporum (Nannizzia) persicolor* is hosted normally by microtid rodents, by dogs and sometimes by cats that frequently borrow and capture preys. The hedgehog-associated *Trichophyton erinacei* can also be found in dogs. *T. simii* is sparsely recorded from birds, carnivores and primates [3]. *Microsporum (Nannizzia) gypseum* is a geophilic species regularly isolated from lesions of dermatophytosis in a wide variety of hosts, particularly dogs, horses and occasionally cats. Other geophilic dermatophytes such as *M. cookei*, *M. praecox*, *T. ajelloi* or *T. terrestre* may be isolated in culture from the coat or skin lesions in companion animals, but they are supposed to be nonpathogenic. The major dermatophyte species that may be isolated from dogs and cats are listed in Table 2.1. The classification of dermatophytes fungi has been recently revised [8], and the names used in Table 2.1 have been modified accordingly. Studies about the fungal flora of dogs and cats without any cutaneous lesions clearly demonstrated that dermatophytes are not part of the normal skin microbiome of companion animals [9–11].

A retrospective epidemiological survey concerning mycological analyses in 2055 animals within a 3-year period was recently conducted by the mycology laboratory of the Veterinary College of Alfort (ENVA), France [12]. The prevalence rate of positive culture for dermatophytes was 13.8% in dogs with a majority of *M. canis* identified (71.2%). The prevalence seemed to be associated with the way of life, hunting or outdoor dogs being more frequently infected. Among cats, the prevalence rate of positive culture was 22.3% with *M. canis* found in 90.8% of the cases. In purebred cats, cultures were significantly more frequently positive than in crossbred or household pet cats. There was no association between the sex of the animals and the prevalence of positive culture.

Isolation of anthropophilic dermatophytes such as *M. audouinii*, *T. rubrum*, *T. tonsurans*, *T. violaceum* and *Epidermophyton floccosum* from companion animals has also been reported but only on rare occasions. This should be carefully considered when observed in mycological culture from an animal, as it does not necessarily indicate a pathogenic role of the isolated fungus. This could simply reflect the dermatophytic flora of humans in the concerned area rather than a real infection of the animals. Nevertheless, some authentic cases due to *T. rubrum* [13, 14], *T. tonsurans* [15] or *E. floccosum* [16, 17] have been documented in dogs. In such cases, close contact with an infected person and predisposing factors explaining a higher susceptibility of the animal such as a tumoural disease, immunosuppressive chemotherapy or ageing have been mentioned.

2.2.2 *Malassezia* Species

Malassezia yeasts are included in the phylum *Basidiomycota* and subphylum *Ustilaginomycotina* in which they constitute a specific order *Malasseziales* [18]. *Malassezia* yeasts belong to normal cutaneous or mucosal microbiota of most (probably all) warm-blooded vertebrates [4]. These atypical fungal organisms have an affinity for lipids as substrates, and the term "lipophilic yeasts" is frequently used to characterise them. Most of the species exhibit an absolute requirement for long fatty acid chains, and they cannot be isolated by culture unless specific nutrients are provided in the medium. To date, lipid-dependent yeasts include 15 species: *M. sympodialis*, *M. globosa*, *M. restricta*, *M. furfur*, *M. obtusa*, *M. slooffiae*, *M. dermatis*, *M. japonica*, *M. yamatoensis*, *M. nana*, *M. equina*, *M. caprae*, *M. cuniculi*, *M. brasiliensis* and *M. psittaci* [18]. Of these, three species (*M. dermatis*, *M. japonica* and *M. yamatoensis*) have been isolated exclusively from human skin so far. *Malassezia pachydermatis* is the only non-lipid-dependent species; it grows on Sabouraud glucose agar without any special additional requirements [19]. Table 2.2 gives the list of the *Malassezia* species that may be isolated from dogs and cats.

Table 2.2 *Malassezia* species recovered from the skin of dogs and cats

Species	Major animal hosts	Related diseases	Risk of human contamination
Malassezia pachydermatis[a]	Dogs, cats (and many other mammals), birds	Otitis and dermatitis in dogs and cats	Yes
Malassezia sympodialis[b]	Cats and other mammals (including humans)	Otitis in cats	Not reported
Malassezia globosa[b]	Cats and other mammals (including humans)	Otitis in cats	Not reported
Malassezia slooffiae[b]	Cats, pigs and other mammals (including humans)	Otitis in cats, dermatitis	Not reported
Malassezia nana[b]	Cats and cattle	Otitis in cats	No

[a]Non-lipid-dependent *Malassezia* yeasts, which are able to grow on routine mycological media (like Sabouraud dextrose agar) without lipid supplementation

[b]Lipid-dependent yeasts, which require lipid-supplemented media (like Dixon's medium). Fifteen lipid-dependent species are currently described: *M. furfur*, *M. sympodialis*, *M. globosa*, *M. obtusa*, *M. restricta*, *M. slooffiae*, *M. dermatis*, *M. japonica*, *M. yamatoensis*, *M. nana*, *M. caprae*, *M. equina*, *M. cuniculi*, *M. brasiliensis* and *M. psittaci* [18]

2.3 Transmission and Predisposing Factors

The occurrence of dermatophytosis or *Malassezia* dermatitis is not only influenced by a vast number of intrinsic factors relating to the animals themselves but also to extrinsic factors such as climatic conditions or environmental issues including overcrowding. Some factors may need more intensive monitoring and/or treatment, while others may suggest a less aggressive approach.

2.3.1 Dermatophytosis as a Contagious Disease

Dermatophytosis is easily transmitted through direct contact with infected animals or indirectly from contaminated fomites, and all the circumstances that favour those contacts should be considered as predisposing factors. This accounts for a higher occurrence of dermatophytosis when animals are confined in catteries or shelters. Moreover, the great resistance of the dermatophyte spores during months or years increases the role of reservoir played by the environment, and the use of material that may be shared between companion animals for grooming or transportation favours the contamination [3]. Enzootic situation regularly occurs in catteries with *M. canis*, and eradication of dermatophytosis is particularly difficult in that case due to a large number of animals (including kittens) in a confined environment or to the diffusion of *M. canis* through exchanges of cats for reproduction and pet exhibitions. Occurrence of infection is also high in stray dogs and cats, and there is a high risk of human contamination when a puppy or kitten from such a population is adopted.

Any breed is susceptible to dermatophyte infection. However, Yorkshire terrier may be at increased risk for generalised dermatophytosis. Short-haired Pointers, Labrador, fox terrier, Groenendael, beagle, Pointer, Jack Russell terrier and Jagdterrier also appear to be predisposed to dermatophytosis, caused by *M. persicolor* and *M. gypseum*, probably due to increased contact with contaminated soil or preys. Persian and Angora cats have also been found to be frequently infected. In fact, no particular racial factors have been evidenced at present, but long-haired breeds seem to be more susceptible. Familial predisposition has also been seen in cats.

Susceptibility to dermatophytosis depends also on the general health status of animals. The parallel evolution of diseases such as hyperadrenocorticism, or the use of some treatments, mainly corticotherapy, may favour appearance and severity of fungal lesions through impairment of immunity. In cats, the prevalence of fungal flora was investigated in regard to the potential immunosuppressive effect of retroviruses such as the feline immunodeficiency virus (FIV) and the feline leukaemia virus (FeLV). Only one study reported high prevalence of *M. canis* in FIV-infected cats compared to FIV-negative animals [20]. On the contrary, another study [21] suggested that the association may be related to differences in the environment rather than to the retroviral infection status of the cats [22].

2.3.2 *Malassezia* Dermatitis as an Opportunistic Disease

Malassezia yeasts are considered to be opportunistic pathogens that play a significant role in the development of different animal diseases such as otitis externa or seborrhoeic dermatitis [2]. Pathogenicity is clearly associated with yeast overgrowth or sensitisation to the yeasts. Several investigations clearly indicated that some breeds are predisposed to the development of abnormally high populations of *Malassezia* yeasts. In dogs, the list includes basset hounds, dachshunds, cocker spaniels, Shar Pei, poodles, bulldogs and West Highland white terriers [23]. In cats, Devon Rex, Peterbald and Sphinx (Fig. 2.2f) seem to be more frequently colonised by *Malassezia* yeasts [24].

Atopic dermatitis is the most frequently diagnosed concurrent disease in dogs with *Malassezia* dermatitis. However, not all dogs with atopic dermatitis have *Malassezia* dermatitis and vice versa [23]. Ectoparasitoses such as ear mite or flea infestations or pruritus from secondary infections may be responsible for yeast overgrowth. *Malassezia* yeasts are frequently isolated from cats with head and neck pruritus syndrome. Any debilitating disease may play a role by making dogs and cats more susceptible to *Malassezia* dermatitis. In cats, the isolation of *Malassezia* has been associated with retroviral infections [25], paraneoplastic syndromes [26], thymoma [27] and diabetes mellitus [28]. Based on these findings, *Malassezia* overgrowth may be considered as a marker of life-threatening, underlying diseases in some cats.

2.4 Clinical Signs and Lesions

2.4.1 Dermatophytosis in Dogs and Cats

Dermatophytes invade hair shafts and cornified epithelium. As a consequence, dermatophytosis usually presents as patchy areas of alopecia [3, 23]. Lesions can be detected on any part of the body. However, the head and the forelimbs are more frequently affected. Multiple lesions may coalesce, while a spontaneous healing at the centre may occur. Dermatophytosis is typically considered as non-pruritic, but some animals, especially adult cats, may be moderate to intensely pruritic. Uncommon clinical manifestations including folliculitis, feline miliary dermatitis, feline acne, pemphigus-like syndromes and pseudomycetoma may occur.

In cats, dermatophytosis includes a large range of clinical presentations (Fig. 2.1a and b). In kittens, lesions of alopecia are usually localised on the nose, the edges of the ears, the fingers and the tail. Lesions are sometimes very small, and their detection requires an accurate observation and the use of the Wood's lamp. Miliary dermatitis is frequently observed in cats mainly in hypersensitivity states such as flea allergy or atopy, but also in some dermatophyte infections. Miliary dermatitis includes inflammation, pruritus and the presence of small crusts on the back and around the neck. In case of chronic infection in a debilitated animal, extensive lesions may be found. Among the other clinical presentations of dermatophytosis,

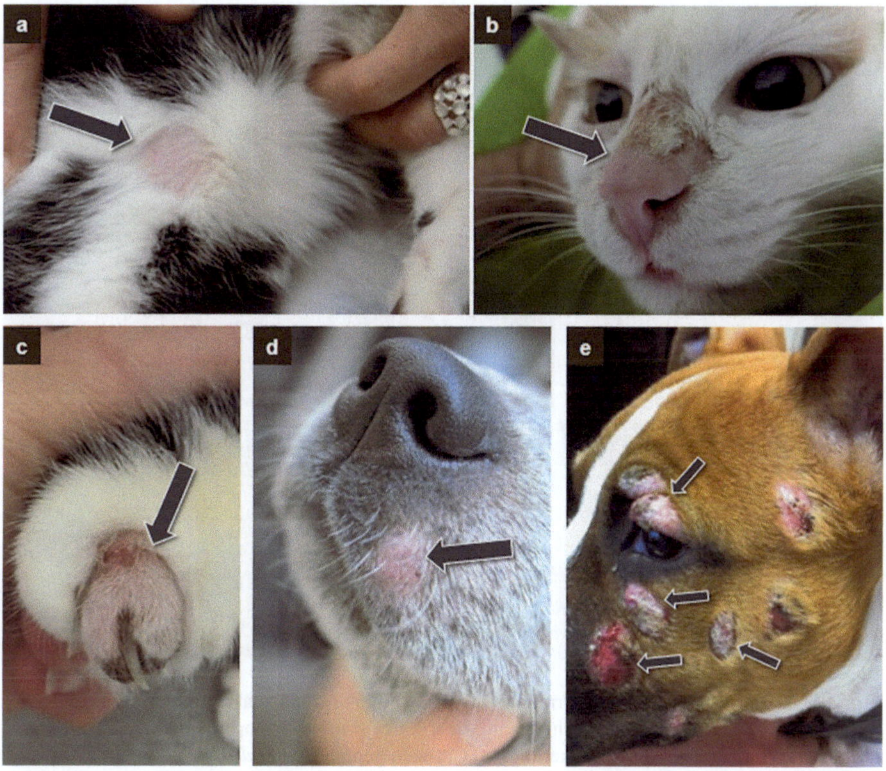

Fig. 2.1 Lesions of dermatophytosis in cats (**a**, **b** and **c**) and dogs (**d** and **e**) (black arrows). Typical lesions present as patchy areas of alopecia. In case of kerion (**e**), there is a severe inflammation with suppurative folliculitis

onyxis and perionyxis due to *M. canis* (Fig. 2.1c), associated with or without typical ringworm lesions, are difficult to detect and to treat. Exceptionally, as described in humans, mycetoma-like lesions have been reported, mostly in Persian cats, in which the dermatophyte develops into the dermis and the subcutis leading to nodulous lesions [29, 30].

In dogs, dermatophytosis is usually associated with isolated or multiple well-circumscribed areas of alopecia without any pruritus (Fig. 2.2d) [3, 23]. Other clinical presentations are less frequently reported. In case of kerion, there is a severe inflammation with suppurative folliculitis, and the palpation of the lesions may express some droplets of pus (Fig. 2.1e). In hunting dogs or animals with a burrowing instinct, contamination from rodents, from insectivores or from the soil with a particular dermatophyte (*M. persicolor*, *M. gypseum*, *T. mentagrophytes*, *T. erinacei*) is frequent and may be responsible for lesions on the face, especially the bridge of the nose. In this case, dermatophytosis can be misdiagnosed with an autoimmune disease especially pemphigus foliaceus. Pseudomycetoma due to *M. canis* is possible in dogs [31] but less frequently than in cats.

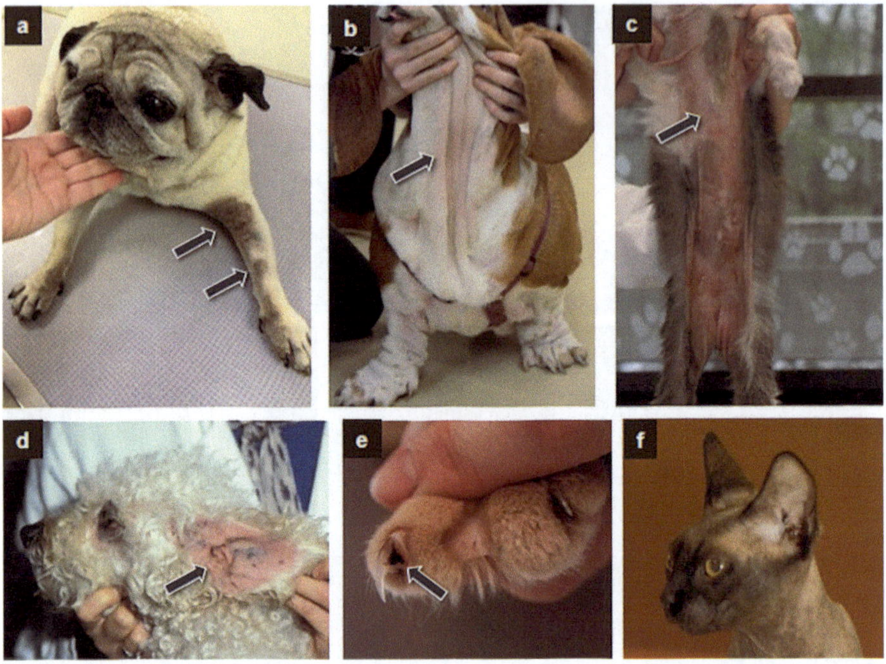

Fig. 2.2 Lesions of *Malassezia* dermatitis in dogs (**a**, **b** and **d**) and cats (**c** and **e**) (black arrows). Skin lesions are erythematous, with varying degrees of traumatic alopecia. (**a**) Dog with severe limb lesions have marked skin thickening. (**c**) Cat with generalised *Malassezia* dermatitis. Lesions are characterised by alopecia, erythema, greasy adherent brownish scales and hyperpigmentation on the head, the ventral face of the neck and the abdomen. The presence of brown, greasy material in the nailfolds may be associated with the presence of *Malassezia* yeasts. (**e**) Sphinx cats (**f**) seem to be more frequently colonised by *Malassezia* yeasts

2.4.2 *Malassezia* Dermatitis in Dogs and Cats

Malassezia yeasts are frequently detected in case of otitis externa in dogs and cats [2, 23]. Several studies demonstrated that such cases responded to antifungal therapy, supporting the current opinion that *Malassezia* yeasts act as opportunistic secondary pathogens within the external ear canal in dogs and to a lesser extent in cats. Dufait is credited for the first report of *Malassezia* yeasts as a cause of more widespread dermatitis in dogs [32]. He described a series of 50 dogs with pruritic dermatitis from which the yeasts could be readily recovered by cytology or culture and which responded to antifungal therapy. These first observations were confirmed by other studies [33, 34], and the veterinary dermatology community progressively considered the potential role of *Malassezia* yeasts as a cause of canine skin disease. Skin lesions associated with *Malassezia* yeasts in dogs are erythematous, with varying degrees of traumatic alopecia [23]. Scaling is often prominent, and a greasy exudate is a feature of lesions in intertriginous and interdigital areas in some dogs (Fig. 2.2a and b); significant exudation is normally accompanied by nauseous smell.

Interdigital lesions are common, and in more severe cases, erythema and alopecia extend to affect the accessory carpal areas and medial aspects of the limbs. Some dogs with severe limb lesions have marked skin thickening, resulting in the formation of erythematous and alopecic ridges. Pedal skin disease may also progress to involve the claw folds with red-brown staining of the claw and exudation in the claw fold. Hyperpigmentation and lichenification are frequently observed in dogs with chronic disease and is particularly common in West Highland white terriers. Dogs with concurrent otitis externa show erythematous vertical ear canals and pinnae with varying degrees of lichenification and scaling, accompanied by a yellow or brownish ceruminous discharge (Fig. 2.2d). Although skin lesions may be confined to one area, multiple regions are usually affected, especially the limbs, the ventrum, the ears and the face.

Cases of *Malassezia* dermatitis have also been observed in cats. In atopic animals, cutaneous lesions related to *Malassezia* overgrowth commonly occur on the face, ventral neck, abdomen and ear canals. These lesions are characterised by alopecia, erythema, greasy adherent brownish scales, hyperpigmentation, easily plucked hair and follicular casts [25] (Fig. 2.2c). Exfoliative erythroderma, greasy exudation and varying degrees of pruritus may be seen in cases secondary to a severe systemic disease [26–28]. The presence of brown, greasy material in the nailfolds may be associated with the presence of *Malassezia* yeasts but not with pruritus (even in Devon Rex cats) (Fig. 2.2e) [35].

2.5 Diagnosis

2.5.1 Diagnosis of Dermatophytosis

Dermatophytosis should be considered in the differential diagnosis of many skin diseases, and diagnostic tests are systematically required [3, 23]. Examination of the hair coat with an ultraviolet lamp (Wood's lamp) is a good screening method for dermatophytosis in cats and to a lesser extent in dogs. When exposed to the light, hairs – but not the scales – invaded by *M. canis* glow fluorescent yellow green (Fig. 2.3a). However, some topical medications may destroy the fluorescence. Moreover, hairs infected by other dermatophyte species such as *T. mentagrophytes, M. persicolor* or *M. gypseum* never fluoresce. As a consequence, when Wood's lamp examination is negative, it does not necessarily mean that dermatophytosis is not occurring. The observation of fluorescence should systematically be confirmed by microscopic examination of hairs, which remains the gold standard diagnostic tool even though the recognition of infected hairs is not always easy and may require an experienced eye.

Hairs should be collected through skin scrapings or under Wood's lamp examination. After digestion with a clearing solution, such as potassium hydroxide or chlorolactophenol, infected hairs become enlarged with a swollen structure with a rough and irregular surface (Fig. 2.3c). Clusters or chains of fungal spores (2–4 μm for *M. canis*) are typically present on the surface of infected hairs. The infected

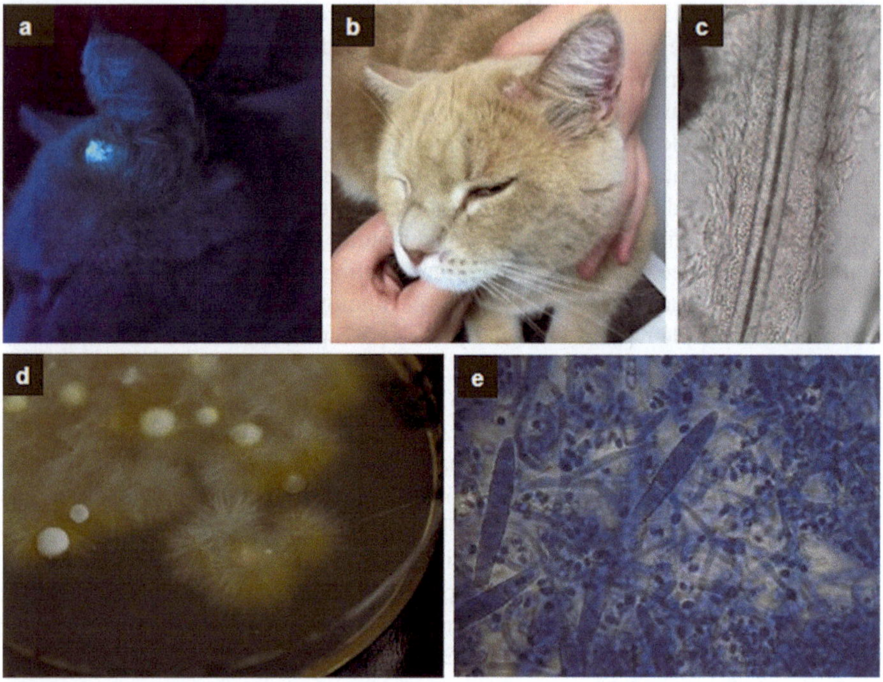

Fig. 2.3 Experimental diagnosis of dermatophytosis in companion animals. (**a**, **b**) Wood's lamp examination in a cat with a lesion on the head. (**c**) A hair infected by *Microsporum canis*. (**d**) Colonies of *Microsporum canis* on Sabouraud dextrose agar. (**e**) Microscopic aspect of a 7-day-old culture of *Microsporum (Nannizzia) persicolor*

scales show thin, segmented hyaline hyphae and spores. Dermoscopy may be an alternative for the detection and examination of infected hairs in cats. This is a non-invasive technique that allows for illuminated magnification of the skin. Scarampella et al. [36] published dermoscopic observations in 12 cats with dermatophytosis and in control animals with another cause of hair loss. In most of the infected cats, opaque, curved or broken hairs with a homogenous thickness ("comma hairs") were evidenced.

Mycological culture remains the most reliable technique for confirming dermatophytosis in dogs and cats [3]. Samples may be collected by scraping the cutaneous lesions, plucking hairs under Wood's lamp examination or brushing the hair coat with a sterile toothbrush or a little piece of sterile carpet. Several media are suitable for mycological cultures, e.g. Sabouraud dextrose agar supplemented with antibiotics (Fig. 2.3d). Dermatophyte Test Media (DTM) are regularly used in veterinary medicine. However, only a very few attempts have been made to evaluate the performance of such media with material obtained from animals, and the use of DTM is not recommended for the diagnosis of animal dermatophytoses because they may give a high number of false-positive results [37]. The samples collected from the animals should be sent to a laboratory with an expertise in veterinary mycology.

Fungal colonies are usually identified by microscopic examination (Fig. 2.3e). The number of fungal colonies may vary between mechanical carriers and infected animals. Mechanical carriage is due to the contamination of the environment and is usually associated with a limited number of dermatophyte colonies in culture. Infection leads to a massive production of spores (arthroconidia) and is usually associated with a very high number of dermatophyte colonies in culture.

Polymerase chain reaction (PCR) for the diagnosis of dermatophytosis in dogs and cats was used in only two studies [38, 39]. Cafarchia et al. [38] collected hair samples from dogs and cats with a clinical suspicion of dermatophytosis. Mycological culture was found positive for 59 out of 183 (32.2%) samples. Infected hairs were detected by direct microscopic examination in 22 out of 183 samples (12.0%). One-step PCR and nested PCR were reported to be positive for 49 of 183 (26.8%) samples and 63 of 183 (34.4%) samples, respectively. Over the past few years, PCR test is commercially available for the diagnosis of dermatophytosis in dogs and cats in European countries and the USA.

2.5.2 Diagnosis of *Malassezia* Dermatitis

Malassezia dermatitis should be suspected in animals with inflammatory skin diseases characterised by erythematous and/or greasy lesions, especially when lesions involve folded areas. In dogs, it may mimic or complicate atopic disease and food allergy. Hyperpigmentation and lichenification are frequently observed in animals with chronic disease and are particularly common in West Highland white terriers. Dogs with concurrent otitis externa show erythematous vertical ear canals and pinnae with varying degrees of lichenification and scaling accompanied by a yellow or brownish ceruminous discharge. Although skin lesions may be confined to one area, multiple regions are usually affected, especially the limbs, ventral neck, abdomen, ears and the face. The diagnosis of *Malassezia* dermatitis is based on clinical signs, the presence of high number of yeasts in lesional skin (Fig. 2.4a) and a clinical and

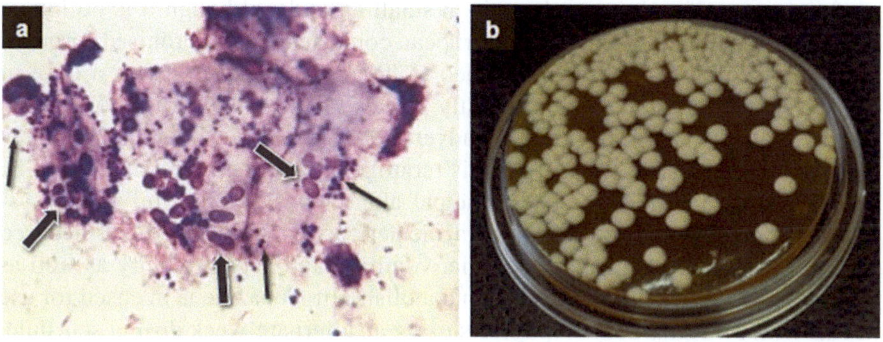

Fig. 2.4 Experimental diagnosis of *Malassezia* dermatitis in companion animals. (**a**) *Malassezia* yeasts on cytological examination (large arrows). Many cocci are also visible (thin arrows). (**b**) Colonies of *Malassezia pachydermatis* after 7 days of culture on Dixon's medium

mycological response to antifungal therapy [23]. The tape strip technique is convenient and reliable: clear adhesive tape is pressed on the surface of the skin, thus collecting the stratum corneum cells and superficial microbial organisms. Because a small population of yeasts might create disease in sensitised animals, and in view of the variations in population sizes between different dog breeds and anatomic sites, trial therapy should be given whenever *Malassezia* yeasts are readily identified in cytologic specimens obtained from compatible lesions (Fig. 2.4a). Definitive identification of *Malassezia* species is not simple and requires subculturing on specific culture media (Fig. 2.4b) as well as molecular characterisation.

2.6 Treatment

2.6.1 Treatment of Dermatophytosis

Recently, Moriello et al. (2017) reviewed the existing literature and provided consensus recommendations for veterinary clinicians about the diagnosis and treatment of dermatophytosis in cats and dogs [40].

Antifungal treatment should be systematically used to shorten the course of the infection and to reduce dissemination of infective material which is composed of fragments of hair covered by microscopic fungal spores and of infected scales into the environment. Infective material is easily spread and can remain viable in the environment for up to 18 months under optimal temperature and humidity. Infected animals, with or without clinical signs, and contaminated environments represent long-term reservoirs for infection of other animals and owners.

The combination of systemic and topical treatments is required. Systemic antifungals are supposed to contribute to speed up the resolution of the infection, whereas topical antifungals are required to reduce the risk of transmission and environmental contamination. Conventional systemic treatment relies on oral antifungal drugs: griseofulvin, ketoconazole and more recently itraconazole or terbinafine (Table 2.3). Griseofulvin was the most commonly used systemic treatment for dermatophytosis in small animals, although it is no longer licensed for animal use in several European countries. The micronised formulation of griseofulvin was recommended to be administered orally at a dose of 25 mg/kg twice daily with a fatty meal to promote drug absorption. Haematological and gastrointestinal adverse effects may occur and are probably more common in cats. Griseofulvin is teratogenic and therefore should not be given to pregnant animals. The principal alternatives to griseofulvin for systemic therapy of dermatophytosis are azole derivatives. Ketoconazole is licensed for use in dogs. It may cause anorexia, vomiting and hepatotoxicity as well as may interfere with steroid hormone metabolism. Itraconazole is licensed for use in cats with *M. canis* dermatophytosis using an alternate-week dosing schedule, reflecting its incorporation rate into stratum corneum and hair. For concurrent topical treatment, many products have been proposed (Table 2.4). The decision to use topical therapy should be based upon the owner's ability and willingness

Table 2.3 Systemic antifungal drugs for the treatment of superficial mycoses in dogs and cats

Antifungal drugs[a] (chemical family)	Indications, dosage and frequency of administration	Comments on use	Adverse effects
Itraconazole (azoles)	For the treatment of dermatophytosis or *Malassezia* dermatitis	In most countries, the drug is registered for use in cats but not in dogs	Itraconazole has lower toxicity than ketoconazole, and at regular dosages, adverse effects are very seldom observed
	5 mg/kg administered every 24 h	Because of its high lipophilicity, the drug has proved to be effective in an alternate-week regimen (one week off and one week on) for the treatment of dermatophytosis	The drug should not be administered to pregnant bitches and queens (even if teratogenic effects have been reported only in rodents and at very high doses)
Ketoconazole (azoles)	For the treatment of dermatophytosis or *Malassezia* dermatitis	In some European countries, the drug is registered for use in dogs (but not in cats)	The drug is teratogenic and must not be administered to pregnant bitches and queens
	5 mg/kg administered every 12 h	The absorption is improved when the drug is given with food	Anorexia, vomiting and diarrhoea are sometimes observed
			Ketoconazole has hepatotoxic effects, including elevated serum alanine transaminase activity
			Ketoconazole interferes with the metabolism of other drugs and with steroid hormone metabolism
Griseofulvin (polyenes)	For the treatment of dermatophytosis only	In many countries, the drug is no longer used and is not registered for use in dogs and cats	The drug is highly teratogenic and must not be administered to pregnant bitches and queens
	25 mg/kg administered every 12 h (micronised form)	The drug should be administered with a fatty meal (the fat enhances absorption)	Gastrointestinal disorders are sometimes observed
	5 mg/kg administered every 12 h (ultramicronised form)		Myelosuppression has been documented in FIV-infected cats

(continued)

Table 2.3 (continued)

Antifungal drugs[a] (chemical family)	Indications, dosage and frequency of administration	Comments on use	Adverse effects
Terbinafine (allylamines)	For the treatment of dermatophytosis or *Malassezia* otitis or dermatitis	The drug is commonly used for the treatment of dermatophytosis (especially onychomycosis) in humans	No teratogenicity has been reported in rodents or rabbits
	20–40 mg/kg administered every 24 h	Terbinafine is registered for use in dogs for the treatment of otitis externa	Vomiting may sometimes be observed in cats

[a]Lufenuron is a chitin synthesis inhibitor commonly used for the prevention of flea infestations in dogs and cats. Since chitin is a component of fungal cell walls, several studies have investigated whether lufenuron has useful antifungal activity. The first retrospective study was conducted in Israel and suggested that lufenuron treatment was strongly associated with recovery in many dogs and cats infected by diverse fungal infections, including dermatophytosis and *Malassezia* dermatitis [41]. However, the results of other investigations were contradictory, and increasing scepticism about efficacy of lufenuron rapidly occurred [42]. To date, the use of lufenuron is not recommended for the treatment of superficial mycoses in dogs and cats [40]

to pour or sponge the product over the entire hair coat of the infected animal. Spot treatment of lesions is not recommended. The frequency of topical treatment should be at least twice a week.

This combined treatment should be continued for at least 10 weeks. The general recommendation is to stop antifungal administration after two consecutive negative cultures. Three negative results are preferred when multiple cats are involved. If lesions persist after 8 weeks of treatment, veterinarians should suspect (i) that the treatment is not being administered correctly by the owner, (ii) that an underlying disorder is interfering with the normal action of the immune system or (iii) that the animal has a genetic background that makes it more susceptible to dermatophyte infection. The presence of resistant strains is regularly suspected, but resistance of dermatophytes to antifungal drugs has been proved only in a very few instances and thus should not be considered as most likely in cases of treatment failure. Lack of environmental control is possibly a reason for the recurrence.

Clipping of the hair coat may be recommended, especially in severely infected animals, in long-haired cats or in multi-animal households. Clipping makes topical therapy application easy and allows better penetration of the drug. In households with one or two pets, spot clipping of lesions may be enough. All whiskers should be clipped. In cats, clipping the coat may require sedation. Clipping must be performed carefully and in an area that can be easily disinfected. Infected hairs should be burned or placed in a plastic biohazard bag and autoclaved. Disposable clothing should be used in order to limit the risk of zoonotic transmission of the disease.

Dermatophyte species may not have the same susceptibility to currently available antifungals. As a consequence, the specific identification of the dermatophyte is required for the choice of the specific treatment for a better understanding of the

Table 2.4 Topical antifungal drugs for the treatment of dermatomycoses in dogs and cats

Antifungal drugs	Indications, dosage and frequency of administration	Comments on use	Adverse effects
Shampoos			
Miconazole + chlorhexidine	For the treatment of dermatophytosis or *Malassezia* dermatitis	Lathering or rubbing process may macerate fragile hairs and increase the release and dispersal of spores	No adverse effect has been documented
	2% miconazole and 2% chlorhexidine twice weekly		
Rinses			
Enilconazole	For the treatment of dermatophytosis or *Malassezia* dermatitis	The entire body must be treated and the antifungal agent left to dry on the skin	Topical application of enilconazole is well tolerated (including by cats)
	0.2% solution twice weekly	Careful application (using sponges and by patting rather than rubbing) is recommended	
		After application, the coat and skin can be dried with a hairdryer	
Lime sulphur	For the treatment of dermatophytosis	Lime sulphur is commonly used in the USA but is not available in all European countries	Lime sulphur has an offensive odour and may stain light-coloured hair
	1:32 or 1:16 twice weekly	The entire body must be treated and the antifungal agent left to dry on the skin	Oral ulceration has sometimes been observed in cats. As a consequence, cats should be collared to prevent them from licking the solution
		Careful application (using sponges and by patting rather than rubbing) is recommended	
Creams, gels, ointments and suspensions			
Several compounds available (e.g. miconazole)		The efficacy of these products has not been demonstrated specifically in dogs and cats with dermatophytosis or *Malassezia* dermatitis	The products may be messy or easily groomed off by the animals

Captan, povidone-iodine and chlorhexidine (alone and at a concentration lower than 3%) have been found to be ineffective against dermatophytes in both in vitro and in vivo studies [40]
Sodium hypochlorite solution has been used as topical treatment of dermatophytosis in cats. However, it dries and irritates the skin and bleaches the hair coat. The use of this product is not recommended

epidemiology of the infection and also for preventing new contamination. Moriello et al. suggested the objectives of environmental disinfection which are (i) to minimise the risk of dermatophyte transmission to other animals and owners and (ii) to minimise fomite carriage on the hair coat of animals that can complicate monitoring of dermatophytosis [40]. Among the different disinfectants, which have been tested for removing infective material from the environment, the following proved to be the most effective: sodium hypochlorite (household bleach) at concentrations from 1:10 to 1:100, enilconazole as a spray and environmental fogger both at a concentration of 20 mg/mL, accelerated hydrogen peroxide and potassium peroxymonosulfate [43–45].

In catteries and animal shelters, dermatophytosis is very difficult to eradicate and creates a significant health hazard for people in contact with the animals. The cost of antifungal drugs and the reluctance of the breeders to admit that their colony is infected usually account for lack of compliance with treatment. Most recommendations for the control of dermatophytosis in catteries are based on the concept of a total treatment programme, which is associated with the use of reliable diagnostic tools, both topical and systemic treatment of all the cats and strong environmental decontamination procedures.

In dog and cat breeding units as well as in animal shelters, the main risk is represented by the introduction of an infected animal. Management plans usually include screening, monitoring and treatment procedures. At the point of entry, animals should be carefully examined, vaccinated against major infectious disease and treated for ectoparasites and intestinal worms. The animals should also be screened for dermatophytosis via Wood's lamp examination and fungal culture. Animals should then be transferred to a quarantine ward until the results of the tests are known. The provision of a separate area for the treatment of animals with dermatophytosis is preferable. Treatment decisions should be made according to the results of fungal culture. Colony-forming unit count combined with clinical examination can help to differentiate mechanical carriers from infected animals. Mechanical carriers should be treated with a single topical application of an antifungal drug before introduction within the colony. Infected animals are kept in quarantine and treated using a combination of systemic and topical antifungal drugs. These animals should not be reintroduced to the colony before two consecutive negative fungal cultures have been obtained.

2.6.2 Treatment of *Malassezia* Dermatitis

Topical treatments licensed for canine *Malassezia* otitis externa in veterinary medicine generally contain either azole antifungal drugs (principally miconazole or posaconazole) or nystatin [2, 23]. These are normally combined with antibiotics and a glucocorticoid, reflecting the need to control concurrent bacterial infection and reduce inflammation and proliferative pathological changes (e.g. stenosis) of the ear canal. Combined antibacterial and antifungal drug administration may also prevent the switch from bacterial to yeast infection, or vice versa, that may be encountered when antibacterial or antifungal monotherapy is used in dogs with otitis externa or otitis media. Concurrent use of ear cleaners is recommended when cerumen is excessive. Animals with *Malassezia* otitis should receive a complete dermatologic evaluation, because failure to identify and correct predisposing, primary and other perpetuating factors may result in persistent or recurrent disease.

Because *Malassezia* yeasts are located within the stratum corneum, topical therapy alone may be sufficient to resolve the clinical signs of infection, provided the owner and pet are compliant. An evidence-based review about the treatment of *Malassezia* dermatitis in dogs concluded that there was clear evidence for the effectiveness of the twice-weekly use of a 2% miconazole/2% chlorhexidine shampoo [46]. The use of oral ketoconazole (10 mg/kg once daily) and oral itraconazole (5 mg/kg once daily) for 3 weeks was also evident. Itraconazole might be preferred to ketoconazole because it is better tolerated. As in dermatophytosis, the keratinophilic and lipophilic properties of this drug enable intermittent administration, with the advantage of reducing costs and the risk for adverse effects and potentially improving compliance. Severe claw fold infections may require longer treatment or higher doses, and cases of otitis externa may not respond adequately. Concerning otitis externa, identification and correction of primary causes and predisposing factors are essential for successful management, although many dogs with *Malassezia* dermatitis require regular maintenance therapy to prevent relapse. Clinical and cytologic assessments should be repeated to determine the efficacy of antifungal therapy and to establish the evidence of concurrent diseases. Relapsing infection is common when primary causes and predisposing factors are not identified and corrected.

2.7 Prevention

Contact with infected animals or contaminated environments represents the major risk of dermatophyte infection; therefore, it should be prevented. This prophylactic strategy is very simple but not always possible because infected animals do not necessarily show obvious clinical signs. Asymptomatic carriers are frequently observed in feline populations.

To protect animals, the use of antifungal drugs has been proposed as a preventive measure. Oral antifungal drugs are not proved to be effective, whereas topical treatments are more expensive. Rinses or shampoos containing enilconazole or miconazole are licensed for dogs and cats in most European countries. The general recommendation is to apply an antifungal shampoo or rinse to the entire body of any dog or cat, which has been in contact with an infected animal or a contaminated area. Under optimum conditions, infective fungal spores germinate within 6 hours on the skin of pet carnivores, so the preventive application of antifungal drug should be performed in the day following the presumptive contamination.

Efforts to develop vaccines to prevent dermatophytosis in dogs and cats still continue. There are only a few products, which are currently commercialised in some Central or Eastern European countries. These are live vaccines that may contain different dermatophyte species (e.g. *Microsporum canis* and *Trichophyton mentagrophytes*). Investigations proving that these vaccines are protective against challenge exposure are still lacking. As a consequence, use of these vaccines should not be recommended for prevention of dermatophytosis in dogs and cats [41].

References

1. Miller WH, Griffin CE, Campbell KL (2013) Fungal and algal skin diseases. In: Muller and Kirk's small animal dermatology, 7th edn. Elsevier, New York, pp 223–283
2. Bond R (2010) Superficial veterinary mycoses. Clin Dermatol 28:226–236
3. Chermette R, Ferreiro L, Guillot J (2008) Dermatophytoses in animals. Mycopathologia 166:385–405
4. Sugita T, Boekhout T, Velegraki A et al (2010) Epidemiology of *Malassezia*-related skin diseases. In: *Malassezia* and the skin, 1st edn. Springer, Berlin, Heidelberg, pp 65–119
5. White TC, Findley K, Dawson TL Jr et al (2014) Fungi on the skin: dermatophytes and *Malassezia*. Cold Spring Harb Perspect Med 1:1–16
6. Wu G, Zhao H, Li C et al (2015) Genus-wide comparative genomics of *Malassezia* delineates its phylogeny, physiology, and niche adaptation on human skin. PLoS Genet 11:e1005614
7. Gräser Y, Scott J, Summerbell R (2008) The new species concept in dermatophytes-a polyphasic approach. Mycopathologia 166:239–256
8. de Hoog GS, Dukik K, Monod M et al (2017) Toward a novel multilocus phylogenetic taxonomy for the dermatophytes. Mycopathologia 182:5–31
9. Hoffmann AR, Patterson AP, Diesel A et al (2014) The skin microbiome in healthy and allergic dogs. PLoS One 9:e83197
10. Meason-Smith C, Diesel A, Patterson AP et al (2016) Characterization of the cutaneous mycobiota in healthy and allergic cats using next generation sequencing. Vet Dermatol 28:71–e17
11. Moriello KA, DeBoer DJ (1991) Fungal flora of the coat of pet cats. Am J Vet Res 52:602–606
12. Johannsen C, Guechi R, Hourigat S et al (2015) Dermatophytes species from companion animals in France; results of a 3-year period activity (2010–2012) of the Mycology laboratory at the Veterinary College of Alfort. In: Paper presented at the 19th Congress of the International Society for Human and Animal Mycology (ISHAM), University of Melbourne, Australia, 4–8 May 2015
13. Kano R, Hirai A, Yoshiike M et al (2002) Molecular identification of *Trichophyton rubrum* isolate from a dog by chitin synthase 1 (CHS1) gene analysis. Med Mycol 40:439–442
14. Kushida T, Watanabe S (1975) Canine ringworm caused by *Trichophyton rubrum*; probable transmission from man to animal. Sabouraudia 13:30–32
15. Brilhante RSN, Cordeiro RA, Gomes JMF et al (2006) Canine dermatophytosis caused by an anthropophilic species: molecular and phenotypical characterization of *Trichophyton tonsurans*. J Med Microbiol 55:1583–1586
16. Stenwig H, Taksdal T (1984) Isolation of *Epidermophyton floccosum* from a dog in Norway. Sabouraudia 22:171–172
17. Terreni AA, Gregg WB Jr, Morris PR et al (1985) *Epidermophyton floccosum* infection in a dog from the United States. Sabouraudia 23:141–142
18. Gaitanis G, Magiatis P, Hantschke M et al (2012) The *Malassezia* genus in skin and systemic diseases. Clin Microbiol Rev 25:106–141
19. Guillot J, Bond R (1999) *Malassezia pachydermatis* a review. Med Mycol 37:295–306
20. Mancianti F, Giannelli C, Bendinelli M et al (1992) Mycological findings in feline immunodeficiency virus-infected cats. J Med Vet Mycol 30:257–259
21. Sierra P, Guillot J, Jacob H et al (2000) Fungal flora on cutaneous and mucosal surfaces of cats infected with feline immunodeficiency virus or feline leukemia virus. Am J Vet Res 61:158–161
22. Mignon BR, Losson BJ (1998) Prevalence and characterisation of *Microsporum canis* carriage in cats. J Med Vet Mycol 35:15–19

23. Bond R, Guillot J, Cabanes J (2010) *Malassezia* yeasts in animal diseases. In: *Malassezia* and the skin. Springer, Berlin, pp 271–299
24. Ahman S, Perrins N, Bond R (2007) Carriage of *Malassezia* spp. yeasts in healthy and seborrhoeic Devon Rex cats. Med Mycol 45:449–455
25. Crosaz O, Legras A, Vilaplana-Grosso F et al (2013) Generalized dermatitis associated with *Malassezia* overgrowth in cats: a report of six cases in France. Med Mycol Case Rep 2:59–62
26. Godfrey DR (1998) A case of feline paraneoplastic alopecia with secondary *Malassezia*-associated dermatitis. J Small Anim Pract 39:394–396
27. Forster-Van Hijfte MA, Curtis CF, White RN (1997) Resolution of exfoliative dermatitis and *Malassezia pachydermatis* overgrowth in a cat after surgical thymoma resection. J Small Anim Pract 38:451–454
28. Perrins N, Gaudiano F, Bond R (2007) Carriage of *Malassezia* spp. yeasts in cats with diabetes mellitus, hyperthyroidism and neoplasia. Med Mycol 45:541–546
29. Nardoni S, Franceschi A, Mancianti F (2007) Identification of *Microsporum canis* from dermatophytic pseudomycetoma in paraffin-embedded veterinary specimens using a common PCR protocol. Mycoses 50:215–217
30. Yager JA, Wilcock BP, Lynch JA et al (1986) Mycetoma-like granuloma in a cat caused by *Microsporum canis*. J Comp Pathol 96:171–176
31. Abramo F, Vercelli A, Mancianti F (2001) Two cases of dermatophytic pseudomycetoma in the dog: an immunohistochemical study. Vet Dermatol 12:203–207
32. Dufait R (1983) *Pityrosporum canis* as the cause of canine chronic dermatitis. Vet Med Small Anim Clin 78:1055–1057
33. Larsson CE, Gandra CRP, Larsson M et al (1988) Dermatitis in dogs caused by *Malassezia (Pityrosporum) pachydermatis*. Ars Vet 4:63–68
34. Mason KV, Evans AG (1991) Dermatitis associated with *Malassezia pachydermatis* in 11 dogs. J Am Anim Hosp Assoc 27:13–20
35. Colombo S, Nardoni S, Cornegliani L et al (2007) Prevalence of *Malassezia* spp. yeasts in feline nail folds: a cytological and mycological study. Vet Dermatol 18:278–283
36. Scarampella F, Zanna G, Peano A et al (2015) Dermoscopic features in 12 cats with dermatophytosis and in 12 cats with self-induced alopecia due to other causes: an observational descriptive study. Vet Dermatol 26:282–e63
37. Guillot J, Latié L, Deville M et al (2001) Evaluation of the dermatophyte test medium Rapid Vet-D. Vet Dermatol 12:123–127
38. Cafarchia C, Gasser RB, Figueredo LA et al (2013) An improved molecular diagnostic assay for canine and feline dermatophytosis. Med Mycol 51:136–143
39. Daz browska I, Dworecka-Kaszak B, Brillowska-Daz browska A (2014) The use of a one-step PCR method for the identification of *Microsporum canis* and *Trichophyton mentagrophytes* infection of pets. Acta Biochim Pol 61:375–378
40. Moriello KA, Coyner K, Paterson S et al (2017) Diagnosis and treatment of dermatophytosis in dogs and cats: clinical consensus guidelines of the World Association for Veterinary Dermatology. Vet Dermatol 28:266–e68
41. Ben-Ziony Y, Arzi B (1999) Use of lufenuron for treating fungal infections of dogs and cats: 297 cases (1997–1999). J Am Vet Med Assoc 217:1510–1513
42. Zur G, Elad D (2006) *In vitro* and *in vivo* effects of lufenuron on dermatophytes isolated from cases of canine and feline dermatophytoses. J Vet Med B Infect Dis Vet Public Health 53:122–125
43. Moriello KA (2015) Kennel disinfectants for *Microsporum canis* and *Trichophyton* sp. Vet Med Int. https://doi.org/10.1155/2015/853937
44. Moriello KA (2016) Decontamination of laundry exposed to *Microsporum canis* hairs and spores. J Fel Med Surg 18:457–461
45. Moriello KA (2017) Decontamination of carpet exposed to *Microsporum canis* hairs and spores. J Fel Med Surg 19:435–439
46. Nègre A, Bensignor E, Guillot J (2009) Evidence-based veterinary dermatology: a systematic review of interventions for *Malassezia* dermatitis in dogs. Vet Dermatol 20:1–12

Pathogenic *Chrysosporium*-Related Fungi in Reptiles and Other Animals

3

Roman S. Ovchinnikov and Dmitry B. Vasyliev

Abstract

Pathogenic *Chrysosporium*-related fungi (PCRF) have manifested themselves in recent decades as the serious causative agents of mycoses in captive and free-living reptiles. The anamorphic (asexual) genus *Chrysosporium* Corda comprises a number of species including *Chrysosporium* anamorph of *Nannizziopsis vriesii* (CANV), which is considered as a main fungal pathogen in reptiles in many countries of the world. Due to increased popularity of exotic reptiles as pets, these infections have become widespread around the world in recent decades. Taxonomy and nomenclature of *Chrysosporium*-related fungi have been revised radically. The present chapter puts together the recent advances in classification, physiology, etiological significance, epidemiology, and occurrence of PCRF-induced mycoses in different species of reptiles. Mycoses in mammals including humans associated with PCRF are also colligated together with our published and unpublished experiences in clinical and laboratory diagnosis, antifungal susceptibility, therapy, and prevention of reptile mycoses caused by *Chrysosporium*-related fungi. The data demonstrate that PCRF-associated mycoses are important aspects of veterinary mycology and herpetology and thus should be explored further.

Keywords

Chrysosporium · *Nannizziopsis* spp. · Pathogenic fungi · Mycosis · Dermatomycosis · Reptile

R. S. Ovchinnikov (✉)
National Research Center on Epidemiology and Microbiology n.a. N.F. Gamaleya, Moscow, Russia

Federal Research Center, All-Russian Scientific Research Institute of Experimental Veterenary Medicine n.a. K.I. Skryabin and Y.R. Kovalenko, Moscow, Russia

D. B. Vasyliev
Department of Herpetology, Moscow State Zoo, Moscow, Russia

K. Singh, N. Srivastava (eds.), *Recent Trends in Human and Animal Mycology*,
https://doi.org/10.1007/978-981-13-9435-5_3

3.1 The Current Significance of the Fungal Infections in Reptiles

Exotic reptiles took up a niche of popular companion animals in recent decades. However, inadequate maintenance conditions predispose animals to various infectious diseases including mycoses. Due to their biological features (exothermy, etc.), reptiles are considered to be naturally predisposed for fungal infections [1].

However, the importance of fungal infections in reptiles is still underestimated in many cases. Fungi tend to occupy the lowest position in the differential diagnosis list of veterinary clinicians. The primary reason for this is that fungi are generally considered to be opportunistic pathogens in reptiles, rather than obligate pathogens such as viruses, parasites, and bacteria [2]. The anamorphic (asexual) genus *Chrysosporium* Corda includes mostly keratinophilic species that live on the remains of hair and feathers in soil. Except reptiles, fungi are rarely reported as animal pathogens. Pathogenic *Chrysosporium*-related fungi (PCRF) are able to cause superficial and deep mycoses that affect both captive and wild reptiles [3].

In cold-blooded animals including reptiles, fungal infections can be caused by a variety of fungal species. But in last 15 years, PCRF gained etiological importance, and now they are evaluated as emerging pathogens in reptiles [4]. In many publications, the most important *Chrysosporium*-related pathogen is designated as *Chrysosporium* anamorph of *Nannizziopsis vriesii* (abbreviated as CANV in many reports). However, there is some inconsistency in the traditional and modern nomenclature of *Chrysosporium*-related fungal pathogens.

Nannizziopsis spp. was isolated for the first time in the 1990s from sick captive day geckos imported from Madagascar to Germany and from chameleons in Canada [5, 6]. Outbreaks of CANV in Australia occurred on two separate occasions in 1994 and 1997 in crocodiles sourced from the same crocodile farm [7]. The disease was known as “yellow fungus” because of the characteristic color of skin lesions of affected animals. The most demonstrative yellow fungus manifestations can be seen in bearded dragons (*Pogona* spp.)

The reptile trade, which occurs on a worldwide scale, has obscured the provenance of CANV isolates recovered from sick captive reptiles [8]. Abarca et al. described the first isolation of CANV in Spain from green iguana in the year 2008 and in bearded dragon (*Pogona vitticeps*) in 2009 [9, 10]. In 2010, Hellebuyck et al. diagnosed CANV in girdled lizard (*Cordylus giganteus*) in Belgium [11], whereas in Australia, it was detected in *Pogona barbata* by Johnson et al. (2011) [12]. In Russia, CANV was mycologically detected for the first time in green iguana [13]. In subsequent years, a trend has been observed toward the spread of CANV among captive reptiles. Three cases of CANV in pet reptiles were detected in the year 2008 [14], and up to 2014, CANV became the dominant pathogen of reptilian fungal infections in Russia with a share of 37% [15]. Till date, CANV has been isolated from captive reptiles of Asia, Australia, Europe, and North America [2].

3.2 *Chrysosporium*-Related Fungal Species and Their Nomenclature and Host Specificity

The genus *Chrysosporium* is polyphyletic, having affiliation with at least two orders of the *Ascomycota*. About 65 *Chrysosporium* species are currently accepted, and their sexual morphs (teleomorphs) are found in a variety of genera such as *Aphanoascus*, *Arthroderma*, or *Nannizziopsis* [16].

As mentioned earlier, the main *Chrysosporium*-related pathogen in reptiles is identified in most reports as *Chrysosporium* anamorph *of Nannizziopsis vriesii* (CANV). The species *Nannizziopsis vriesii* (Apinis) Currah (*Ascomycota*, *Onygenales*, *Onygenaceae*) has white ascomata, asperulate peridial hyphae constricted at septa, hyaline and globose ascospores, and a *Chrysosporium* asexual morph.

In routine laboratory practice, the identification of *Chrysosporium*-like fungal isolates is performed on the basis of morphological features. However, several years ago from preliminary molecular phylogenetic analysis, it was suggested that the *Chrysosporium* anamorph of *N. vriesii* actually represented a species complex rather than a single species, containing members that could be allied to specific hosts [17]. Recently, Sigler et al. [8] and Stchigel et al. [18] published independently the latest taxonomic revisions of *Chrysosporium*-related fungi together with the relationships between specific fungal species with their hosts [8, 18]. According to Sigler et al. (2013), one lineage of *Chrysosporium*-related fungi represents the genus *Nannizziopsis* and comprises *N. vriesii*, *N. guarroi*, and six newly described species *N. dermatitidis*, *N. crocodili*, *N. barbata*, *N. infrequens*, *N. hominis*, and *N. obscura* isolated from chameleons and geckos, crocodiles, agamid and iguanid lizards, and humans.

N. guarroi and *N. dermatitidis* were found to be major pathogens of lizards. *N. guarroi* was described originally from captive green iguanas in Spain. They possibly acquire the fungus from pet trade. Interestingly, this case is the first case of *N. guarroi* in green iguanas coincided temporally with the first documented European cases of yellow fungus disease in bearded dragons [10, 19]. Till then, it has been isolated repeatedly from pet inland bearded dragons with yellow fungus disease in North America.

Human-related species of *Nannizziopsis* are *N. infrequens*, *N. hominis*, and *N. obscura*. Other two lineages comprise the genus *Ophidiomyces*, with the species *O. ophiodiicola* (occurring only in snakes), and *Paranannizziopsis* gen. nov., with three new species *P. australasiensis*, *P. californiensis*, and *P. crustacea* infecting squamates and tuataras [8].

Moreover, based on ribosomal ITS region, actin and β-tubulin gene sequence four new species of *Nannizziopsis*, viz., *N. chlamydospora*, *N. draconii*, *N. arthrosporioides*, and *N. pluriseptata* were described by Stchigel et al. (2013) [18], which differ from those described by Sigler et al. (2013) [18]. They also described *Chrysosporium longisporum*, which was renamed *Paranannizziopsis* by Sigler et al. (2013). However, Stchigel and coauthors were unable to define clear-cut host specificity in newly described *Chrysosporium*-related fungi. Thus, the modern nomenclature of *Chrysosporium*-related fungi is not completely established and requires further study.

Table 3.1 summarizes the current data on the nomenclature and host specificity of 15 *Chrysosporium*-related fungi pathogenic for reptiles, as well as their specific morphological features.

Table 3.1 Current nomenclature of *Chrysosporium*-related fungi pathogenic for reptiles

Fungus species	Reptile species (source of isolation)	Morphological features
Genus *Nannizziopsis* Currah (1985)		
N. arthrosporioides [18]	Water dragon (*Physignathus* sp.)	Colonies on peptone yeast extract (PYE) at 30 °C attaining a diameter of 34.0–37.0 mm after 14 days, yellowish white, zonate, felted, slightly cottony at center, with lobate margins; reverse yellowish white. Hyphae hyaline, septate, smooth walled, 1.0–4.0 μm wide, straight or twisted. Conidia 1(−2) celled, mostly sessile, also produced short protrusions or terminal, hyaline, thin and smooth walled, subglobose, pyriform, obovate, or claviform to cylindrical, 2.5–7.0 × 1.5–3.0 μm; intercalary conidia present, similar to the arthroconidia in shape and size; arthroconidia arranged in short terminal and intercalary chains, doliiform to cylindrical or irregularly shaped, 5.0–15.0 × 1.5–4.0 μm. Chlamydospores absent. Sexual morph not observed. Fetid (skunk-like) odor present on all the culture media tested
N. barbata [18]	Coastal bearded dragon (*Pogona barbata*)	Colonies on potato dextrose agar (PDA) were 5.5–6.0 cm in diameter, powdery, flat to slightly raised and cottony at the center, but otherwise zonate after 21 days. There was no growth at 35 °C. Aleurioconidia pyriform to clavate, measured 3.0–6.5 μm long and 1.8–2.5 μm wide, and sessile or borne on slightly swollen cells. Fission arthroconidia measuring 4.4–8.5 μm long and 1.7–3.5 μm wide, as well as undulate hyphae, are commonly produced. Moist colonies on PDA demonstrated budding
N. chlamydospora [18]	Inland bearded dragon (*Pogona vitticeps*)	Colonies on peptone yeast extract (PYE) at 30 °C attaining a diameter of 41.0–48.0 mm after 14 days, yellowish white elevated at the center and radially folded, compact, with an irregular margin; reverse yellowish white. Hyphae hyaline, septate, smooth walled, straight or twisted, 1–3(−4) μm wide. Conidia unicellular, sessile, on short protrusions or on side branches, less frequently terminal, hyaline, thin and smooth walled, pyriform, claviform, or cylindrical, 3.0–9.0 × 1.5–2.0 μm; intercalary conidia, cylindrical to doliiform, 6.0–10.0 × 1.5–2.0 μm; arthroconidia catenate, cylindrical to doliiform, 4.0–10.0 × 2.0–4.0 μm. Chlamydospores globose, broadly ellipsoidal or irregular, smooth and thick walled, 5–15(−20) μm in diameter. Sexual morph not observed. Fetid (skunk-like) odor produced on all the culture media tested

(continued)

Table 3.1 (continued)

Fungus species	Reptile species (source of isolation)	Morphological features
N. crocodili [8]	Saltwater crocodile (*Crocodylus porosus*)	Colonies on PDA were 4.0–5.5 cm in diameter, velvety to powdery, slightly to strongly zonate, and sometimes with exudate droplets after 21 days. Growth was slow at 35 °C (1.0–2.7 cm in diameter after 21 days). Aleurioconidia subglobose, measuring 1.5–2.5 μm long and 1.3–2.4 μm wide, and sessile or borne on swollen cells either on the vegetative mycelium or within ascomata-like structures (pseudogymnothecia). Arthroconidia measuring 3.7–7.5 μm long and 2.0–3.0 μm wide are produced at low frequency and often show germination. Undulate hyphae are formed
N. dermatitidis [8]	Chameleons, geckos	Colonies on PDA attained 3.8–4.7 cm in diameter after 21 days and were strongly zonate and powdery with a thin margin. Most isolates failed to grow at 35 °C. Aleurioconidia were clavate to pyriform and measured 2.8–7.5 μm long if single celled, up to 9.0 μm long if two celled, and 1.2–3.0 μm wide. Undulate branches and cylindrical to slightly barrel-shaped fission arthroconidia were formed with arthroconidia measuring 2.8–9.0 μm long and 1.5–3.0 μm wide. Transitory yeast-like colonies grown on PDA at 30 °C were composed of ovoid to cylindrical yeast-like cells and arthroconidia
N. draconii [18]	Inland bearded dragon	Colonies on PYE at 30 °C attaining a diameter of 32.0–38.0 mm after 14 days, yellowish white, felted, slightly elevated at center, with regular margin; reverse yellowish white to pale yellow at center. Hyphae hyaline, septate, smooth walled, 1–3(−5) μm wide. Conidia unicellular, mostly sessile, also produced on short protrusions or on side branches, or terminal, hyaline, thin and smooth walled, claviform or cylindrical, 4.0–7.0 × 1.5–2.0(−2.5) μm; intercalary conidia scarce, cylindrical, 4.0–9.0 × 1.5–2.0 μm; arthroconidia catenate, mostly cylindrical or doliiform, scarcely produced, 5.0–9.0 × 1.5–2.5 μm. Chlamydospores absent. Sexual morph not observed. Fetid (skunk-like) odor produced on all the culture media tested

(continued)

Table 3.1 (continued)

Fungus species	Reptile species (source of isolation)	Morphological features
N. guarroi [8]	Green iguana (*Iguana iguana*), inland bearded dragon, lizard (*Agama agama*)	Colonies on PDA 2.7–4.7 cm in diameter, powdery, sometimes sectoring to cottony, often strongly zonate, sometimes with exudate droplets. Growth at 35 °C was similar, with colonies attaining 2.3–4.0 cm in diameter. Aleurioconidia clavate to pyriform and measured 3.2–6.5 μm long and 1.5–2.5 μm wide. Undulate hyphae were common. Arthroconidia in chains measured 2.8–7.0 μm long and 2.0–3.7 μm wide and sometimes showed budding in young cultures. *N. guarroi* is distinguished from other reptile-associated *Nannizziopsis* species by its slightly lower growth rate at 30 °C and good growth at 35 °C. This species is considered the etiologic agent of yellow fungus disease in inland bearded dragons, a contagious and progressive necrogranulomatous dermatomycosis first observed about 15 years ago
N. pluriseptata [18]	Skink lizard (*Eumeces inexpectatus*)	Colonies on PYE at 30 °C attaining a diameter of 38.0–40.0 mm after 14 days, white to orange white, zonate, felted, slightly cottony at the center, with regular margins; reverse orange white. Hyphae hyaline, septate, smooth walled, 1.0–5.0 μm wide, straight. Conidia 1(−5) celled, mostly sessile, also produced on short protrusions or on side branches, or terminal, hyaline, thin and smooth walled, pyriform, obovate, claviform to cylindrical, 2.5–8.0(−15.0) × 1.5–2.5 μm; intercalary conidia occasionally present, cylindrical to doliiform or irregularly shaped, 2.5–5.0 × 2.0–2.5 μm; arthroconidia, disposed in lateral or terminal short chains, cylindrical to doliiform, 4.0–7.0 × 2.5–3.5 μm, usually bearing sessile conidia. Chlamydospores and sexual morph absent. Fetid (skunk-like) odor present on all culture media tested
N. vriesii Currah (1985) [8]	Lizard (*Ameiva* sp.)	*Nannizziopsis vriesii* is distinguished from all the other *Nannizziopsis* species described here by the production of ascomata (gymnothecia) produced on oatmeal salts agar (OAT) at 30 °C. Colonies on PDA were 4.5–5.5 cm in diameter after 21 days, velvety to slightly cottony, and furrowed. Growth was inhibited at 35 °C, with colonies attaining 2.5 cm in diameter. Aleurioconidia 2.5–6.0 μm long (in rare cases up to 8 μm long) and 1.5–2.7 μm wide. Isolates produced undulate hyphae and cylindrical fission arthroconidia measuring 2.7–7.3 μm long and 1.7–2.7 μm wide and sometimes showing yeast-like budding

(continued)

Table 3.1 (continued)

Fungus species	Reptile species (source of isolation)	Morphological features
Genus *Paranannizziopsis* [8]		Colonies were pale and moderately fast growing. Vegetative hyphae were narrow, branched, and septate, sometimes with racquet mycelia. Conidia (aleurioconidia) sessile or produced on slightly swollen cells or on short stalks and released by rhexolytic dehiscence. They are hyaline, smooth, pyriform, and clavate to obovate. Arthroconidia absent, intercalary, or produced in adjacent chains. Undulate lateral branches were produced. No teleomorph was produced. *Paranannizziopsis* species are distinguished from *Nannizziopsis* and *Ophidiomyces* species by the uncommon occurrence or absence of fission arthroconidia
P. australiensis [8]	Northern tuatara (*Sphenodon punctatus punctatus*), coastal bearded dragon, aquatic file snake (*Acrochordus* sp.)	Colonies on PDA attained 4.5–5.0 cm in diameter and were powdery or sometimes cottony, flat, or faintly zonate. There was no growth at 35 °C. Aleurioconidia sessile or subtended by slightly swollen cells from which one or two conidia were produced. Pyriform to clavate conidia of 3.5–8.0 μm long and 1.5–2.7 μm wide. Occasional intercalary arthroconidia and undulate hyphae were produced. Ascomatal initials occurred in cottony sectors and appeared as inflated cells with secondary proliferations
P. californiensis [8]	Tentacled snake (*Erpeton tentaculatum*)	Colonies on PDA attained 4.5–5.2 cm in diameter and were powdery and flat to slightly zonate. Growth at 35 °C was strongly inhibited. Aleurioconidia were clavate to pyriform or obovate, measured 4.0–8.5 μm long and 1.8–2.6 μm wide, and were sessile or borne on a slightly swollen cell. Undulate hyphae were uncommon, and arthroconidia were not observed. Ascomatal initials occurred in cottony sectors and were associated with large irregularly shaped cells. The latter measured 10.0–36.0 μm long and 3.5–9.5 μm wide
P. crustacea [8]	Tentacled snake	Colonies on PDA attained 5.8–6.5 cm in diameter and were powdery, flat, and occasionally with dense downy overgrowth. There was no growth at 35 °C. Aleurioconidia were clavate to pyriform or obovate, sessile or formed on short stalks, and measured 4.0–7.5 μm long and 2.0–3.5 μm wide. Undulate hyphae, fission arthroconidia, and occasional intercalary arthroconidia were produced. Arthroconidia measured 3.8–9.2 μm long and 1.9–2.7 μm wide

(continued)

Table 3.1 (continued)

Fungus species	Reptile species (source of isolation)	Morphological features
P. longispora [8]	Tentacled snake	The sequence of *C. longisporum* groups closest to *P. crustacea* but differs at 9 positions in the ITS region. This level of sequence difference, combined with morphological differences, including the absence of growth at 30 °C, absence of fission arthroconidia in chains, and longer conidia (3.0–13.0 μm long), provides support for the retention of both species
Genus *Ophidiomyces* [8]		Colonies were yellowish white and moderately fast growing. Vegetative hyphae were narrow, branched, and septate, occasionally with racquet mycelia. Conidia sessile or borne on short stalks and released by rhexolytic dehiscence (aleurioconidia). Aleurioconidia hyaline, smooth, and cylindrical to clavate. Arthroconidia were formed in chains by schizolytic fragmentation of hyphae or were sometimes intercalary. Short, undulate, and sparsely septate lateral branches were common. No teleomorph is known
O. ophiodiicola [8]	Captive and wild snakes (black rat snake, brown tree snakes, garter snake, green anacondas, broad-headed snake, carpet snakes, *Boa constrictor*, *Nerodia* species, timber rattlesnakes, eastern massasauga rattlesnakes)	Colonies on PDA were 4–6 cm in diameter and were velvety to powdery, dense, flat, frequently zonate, and sometimes with cottony sectors. Clear exudate droplets were often present. Most isolates failed to grow at 35 °C. Aleurioconidia were sessile or borne at the ends of short stalks, cylindrical to clavate, and 2.5–7.5 μm long and 1.5–2.5 μm wide. Arthroconidia were 3.0–12.5 μm long (in rare cases up to 15.0 μm long) by 1.5–3.5 μm wide. In young cultures on PDA, arthroconidia sometimes showed budding or germination. Undulate hyphae were commonly produced. Some isolates produced ascomatal initials in cottony sectors. Most isolates produced a strong to weak mercaptan-like odor
Genus *Chrysosporium* Corda (1833)		
C. longisporum [18]	Tentacled snake	Colonies on PYE at 25 °C attaining a diameter of 40.0–46.0 mm after 14 days, white to pale orange (M. 6A3), zonate, felted, slightly cottony at center, with regular margins; reverse pale orange (M. 5A2). Hyphae hyaline, septate, smooth walled, 1.0–5.0 μm wide, straight. Conidia 1(−2) celled, mostly sessile, or produced on short protrusions or on side branches or terminal, hyaline thin and smooth walled, pyriform, obovate, claviform to cylindrical, 3.0–13.0 × 2.0–3.5 μm; intercalary conidia present, cylindrical to doliiform, 3.0–6.0 × 2.0–3.0 μm, usually bearing sessile conidia; arthroconidia in chains absent. Chlamydospores and sexual morph absent. Fetid (skunk-like) odor present on all the culture media tested

3.3 Physiological and Morphological Features of Pathogenic *Chrysosporium*-Related Fungi

The morphological differences between genera and species of *Chrysosporium*-related fungi are not so clear as to differentiate these species without DNA sequencing. All isolates are moderately fast growing on PDA (potato dextrose agar) at 30 °C and have yellowish white, velvety to powdery, dense, and sometimes zonate colonies with uncolored to yellowish reverse [8]. Similar colonial morphology can be observed on SDA (Sabouraud's dextrose agar) (Fig. 3.1). In some cases, the heterogenous colonial morphology can be seen (Fig. 3.2). It is noteworthy that PCRF can actively grow on media enriched with the sheep blood (Fig. 3.3). In a number of isolates of *N. guarroi*, we observed the presence of incomplete (partial) hemolysis (Fig. 3.4). This indicates that some of PCRF fungi are able to synthesize the hemolysins.

All isolates of PCRF produce aleurioconidia which are solitary conidia released by lytic dehiscence. The aleurioconidia are commonly sessile, sometimes subtended by slightly swollen cells, or formed at the ends of short stalks. They are clavate or pyriform with truncate bases, occasionally subglobose or obovate, mostly single celled, and occasionally two celled (Fig. 3.5). *Nannizziopsis* and *Ophidiomyces* species commonly have chains of adjacent cylindrical arthroconidia that are produced by schizolytic fragmentation of the hyphae. Arthroconidia sometimes demonstrate budding and are found especially in the moist yeast-like colonies. In tissues, arthroconidia occur in the stratum corneum or deeper in the epidermis or in characteristic aggregates or tufts at the surface of lesions (Fig. 3.6). Aleurioconidia also can be seen in affected tissues. According to Sigler et al. (2013), the notable characteristic, occurring in members of all three genera (*Nannizziopsis*, *Paranannizziopsis*, and *Ophidiomyces*) but not in *Chrysosporium* species or dermatophytes, was the formation of short, solitary, undulate, lateral branches that were occasionally sparsely septate [8]. They may play a role in pathogenicity, at least in reptiles, by possibly

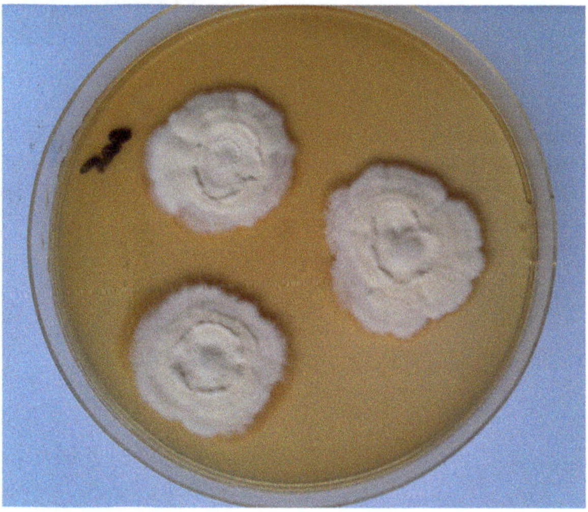

Fig. 3.1 Colonies of *Nannizziopsis guarroi* on SDA

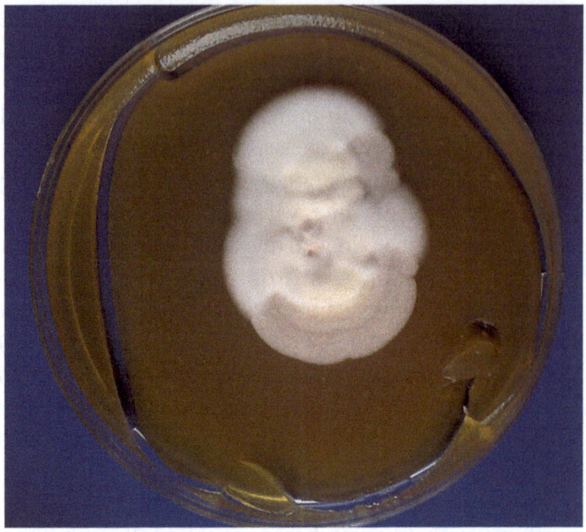

Fig. 3.2 Heterogenous colonial morphology of *N. guarroi*

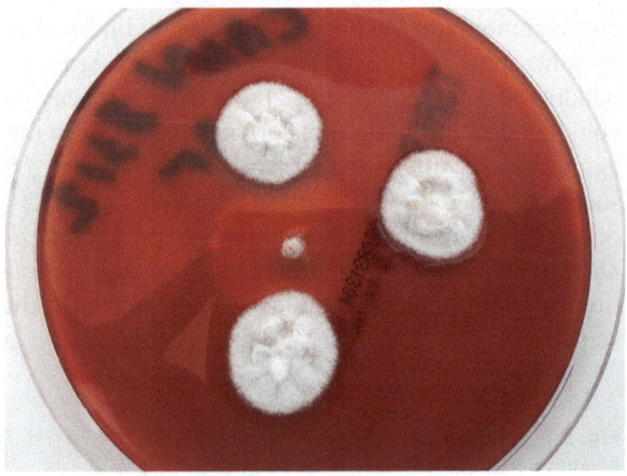

Fig. 3.3 Colonies of *N. guarroi* on agar with sheep blood

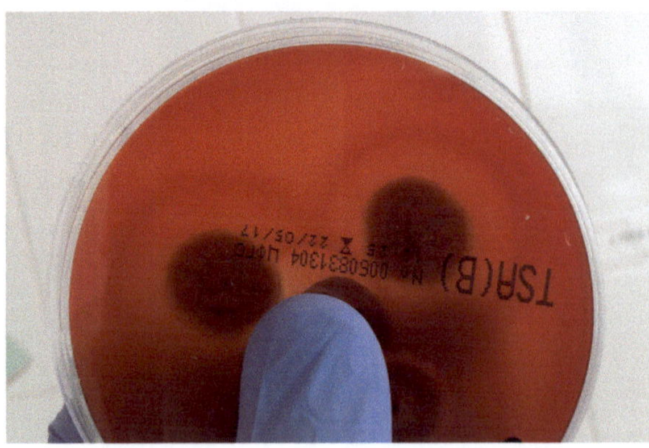

Fig. 3.4 Partial hemolysis caused by *N. guarroi* on agar with sheep blood

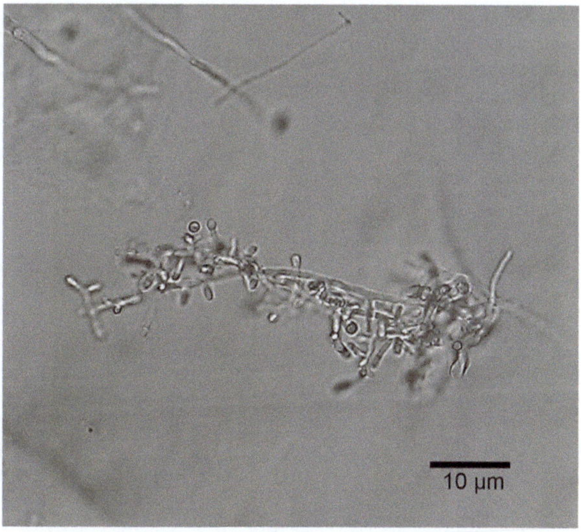

Fig. 3.5 Sessile aleurioconidia and hyphae of *N. guarroi* (400×)

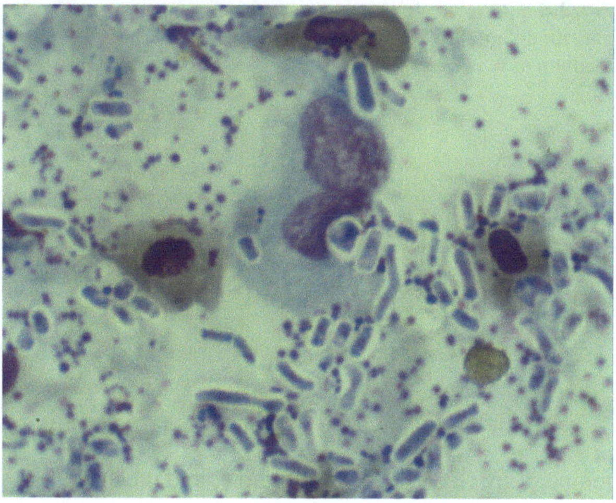

Fig. 3.6 Arthroconidia of *N. guarroi* in affected reptile skin. (Diff-Quik-stained cytological smear, 1000×)

aiding in attachment. PCRF isolates failed to produce ascospores with the exception of *N. vriesii* isolates, that were incubated on culture media for several months. Cultures often have a specific unpleasant odor which is compared with the smell of a skunk or the smell of a mercaptan.

The keratinolytic activity of PCRF is attributed to the keratinase which relates them with dermatophytic fungi. Keratinases are the most important virulence factors of dermatotropic fungi (*Microsporum* spp., *Trichophyton* spp.), ensuring the penetration of pathogen into the stratum corneum. These enzymes stipulate the positive results in in vitro hair perforation test for both dermatophytes and PCRF. The presence of keratinolytic activity indirectly confirms that PCRF are primary pathogens not opportunists. The temperature optimum for PCRF growth is 28–30°C. The

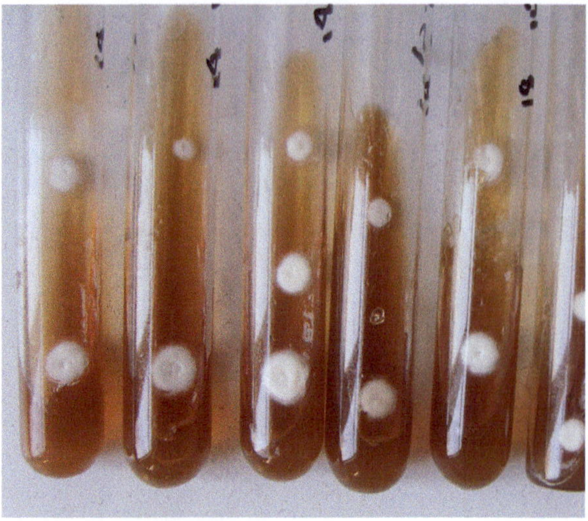

Fig. 3.7 Colonies of *N. guarroi* isolated from clinical samples on SDA supplemented with chloramphenicol and cycloheximide

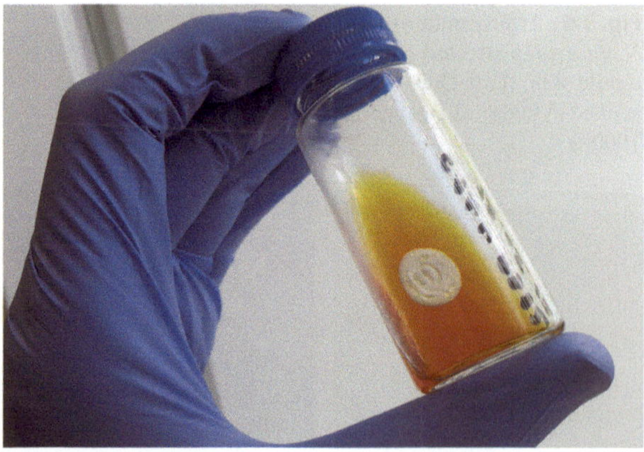

Fig. 3.8 Colonies of *N. guarroi* do not turn DTM medium red

temperature of 37 °C is unfavorable, and the growth of fungal cultures was found to be restricted. Thus, these fungi, unlike dermatophytes, are not thermotolerant and affect warm-blooded animals rarely except *Nannizziopsis guarroi*, which shows optimal growth at 35 °C [20].

This feature can be used to treat reptilian mycosis, and exposure of high temperature (37–39 °C) may be helpful for the treatment; however, it should be tolerable to the patient. The PCRF fungi are naturally resistant to cycloheximides, which are protein biosynthesis inhibitors in eukaryotes. Due to this feature, culture media with cycloheximide can be successfully used for the selective isolation of these fungi in the laboratory (Fig. 3.7).

In our practice, for PCRF isolation, we use media with selective supplement for dermatophytes containing cycloheximide and an antibiotic. PCRF fungi can also be isolated on DTM (dermatophyte test medium). Unlike dermatophytes, these fungi do not cause the alkalinization and redness of the medium (Fig. 3.8).

3.4 Clinical Manifestations of PCRF-Associated Mycoses in a Range of Reptile Species

PCRF mycoses are diagnosed in various representatives of the class Reptilia, including order Squamata which comprise lizards (Lacertilia), chameleons (Chameleontes), snakes (Ophidia), and order Crocodilia. In tortoises (Chelonia), there are no cases of PCRF mycoses reported in literature. But in the author's practice, rare cases of these infections have been found. Reptile species susceptible to PCRF mycoses are listed in Table 3.2.

Reptiles infected with PCRF can present with a range of clinical signs, from focal skin lesions to systemic disease. The most common clinical signs are associated with the integument. Crust formation, color change, and necrosis are commonly seen. Like any other cutaneous fungal lesions, PCRF infections tend to start as focal lesions that spread from a central point. Because of the invasive nature of this fungus, it is common to observe pyogranulomatous disease as it invades through the epidermis and dermis. Once the fungus invades through the integument, it can spread locally or systemically. However, the fungus is reported to be locally invasive [2].

Table 3.2 Species of reptiles found to be infected with pathogenic *Chrysosporium*-related fungi [2]

Common name	Scientific name	References
Lizards		
Ameiva	*Ameiva chaitzarni*	[21]
Day geckos	*Phelsuma* spp.	[22]
Green iguanas	*Iguana iguana*	[9]
Central inland bearded dragon	*Pogona vitticeps*	[23]
Coastal bearded dragon	*Pogona barbata*	[12]
Panther chameleon	*Furcifer pardalis*	[6]
Jackson's chameleon	*Chamaeleo jacksoni*	[6]
Jeweled chameleon	*Chamaeleo lateralis*	[6]
Parson's chameleon	*Chamaeleo parsonii*	[6]
Veiled chameleon	*Chamaeleo calyptratus*	[24]
Girdled lizard	*Cordylus giganteus*	[11]
Leopard geckos	*Eublepharis macularius*	[28]
Snakes		
Boa constrictor	*Boa constrictor*	[29]
Ball pythons	*Python regius*	[22]
Garter snakes	*Thamnophis* spp.	[22]
Brown tree snake	*Boiga irregularis*	[26]
Milk snake	*Lampropeltis triangulum*	[22]
Corn snake	*Pantherophis guttatus*	[22]
Tentacle snakes	*Erpeton tentaculatum*	[27]
Eastern massasauga rattlesnakes	*Sistrurus catenatus catenatus*	[3]
File snakes	*Acrochordus* spp.	[30]
Crocodilians		
Saltwater crocodiles	*Crocodylus porosus*	[7]

The clinical manifestations of PCRF-associated mycoses are supposed to vary among different species of reptiles. Thus, bearded dragons affected with PCRF (CANV) typically present with dermatitis characterized by crusts, ulcers, and pyogranulomatous disease [12, 23]. Originally described as yellow fungus disease, the crusts found on bearded dragons tend to have a yellow coloration. Lesions in bearded dragons are often multifocal and may include the head, oral cavity, limbs, ventrum, or dorsum. Infection tends to be aggressive and disseminate into the subcutaneous tissues which is usually followed by necrosis, sloughing, and ulceration involving muscle and bone. The infection can disseminate with a fatal outcome [19, 23, 30].

Chameleons are the other group of lizards that seem to be highly susceptible to PCRF (CANV) infections. Additionally, these infections are unlikely to be limited to a group from a single native origin. Affected chameleons often present with focal to multifocal necrotic (black) areas of skin surrounded by crusts. The lesions may be found on the body, limbs, and tail [6]. Other species of lizards that have been reported to develop PCRF (CANV) infections in captivity include day geckos (*Phelsuma* spp.) [22], a wild-caught girdled lizard (*Cordylus giganteus*) [11], green iguanas (*Iguana iguana*) [9, 14], and an ameiva (*Ameiva chaitzarni*) [21]. The lesions in these animals were similar to those described for bearded dragons and chameleons and included crusting and ulcerative dermatitis lesions on the head and body. An outbreak of CANV mycosis in colony of leopard geckos (*Eublepharis macularius*) was reported by Toplon et al. (2013) [28]. Histopathology of the affected animal revealed multifocal to coalescing dermal and subcutaneous heterophilic granulomas that contained septate fungal hyphae. The multifocal epidermal hyperplasia with hyperkeratosis was also observed. Moreover, hyphae of causative fungus, occasionally with terminal chains of arthroconidia found within the stratum corneum, were consistent with the CANV. In one case, focal extension of granulomatous inflammation into the underlying masseter muscle was also seen, and *Chrysosporium* anamorph of *Nannizziopsis vriesii* (CANV) was identified by sequencing of the internal transcribed spacer region of the rRNA gene.

According to our observations, sloughing yellowish scales and decolorized skin foci are often observed in green iguanas affected by PCRF (Figs. 3.9 and 3.10). In chameleon (*Chamaeleo calyptratus*), the mycosis was characterized by necrosis and parakeratosis of the epidermis, edema and infiltration of subepidermal tissues, and the formation of granulomas and severe scabby crusts (Figs. 3.11 and 3.12), whereas in monitor lizard (*Varanus exanthematicus*), a severe peripheral edema with muscle infiltration was manifested along with depigmentation and desquamation of the epidermis (Fig. 3.13). In above mentioned cases, CANV was culturally diagnosed as etiologic agent. The authors noticed that superficial infections in reptiles are clinically different from bacterial dermatoses and characterized by pronounced hyper- and parakeratose reactions accompanied by huge scabby crusts with hydrophobic properties resembling the structure of the squama (Fig. 3.14). The thickness of the crust correlates with the aggressiveness of the infection (Fig. 3.15).

In our opinion, the signs of muscle infiltration are unique for PCRF infection. The fungus is especially actively disseminating before molting, during the accumulation of lymph in the zona intermedia, where there is a demarcation of new and old epidermal skin generation. Therefore, the use of multivitamin preparations and

Fig. 3.9 Yellowish skin scales in green iguana infected by PCRF

Fig. 3.10 Decolorization of the skin infected by PCRF in green iguana

retinol for stimulation of molting in the case of fungal infections is not recommended. In other cases, molting usually leads to a reduction of lesions (in case of the concomitant etiotropic therapy).

According to the published data, bearded dragons and chameleons tend to be the most common case presentations of PCRF mycoses in reptiles. Based on the data obtained from Moscow Zoo, the predisposed species are green iguana (*Iguana*

Figs. 3.11 and 3.12 PCRF-associated mycosis in chameleon (*Chamaeleo calyptratus*). Formation of granulomas and severe scabby crusts, parakeratosis, edema, and subepidermal infiltration

Fig. 3.13 PCRF-associated mycosis in monitor lizard (*Varanus exanthematicus*). Depigmentation and desquamation of the epidermis and severe peripheral edema along with muscle infiltration

iguana) and bearded dragons (*Pogona vitticeps*). *Chrysosporium*-related fungal elements were cytologically detected in the following:

Blue-tongued skink (*Tiliqua scincoides*) – periorbital infection together with finger infection.

Fig. 3.14 Huge scabby crusts resembling the structure of the squama in green iguana affected by PCRF

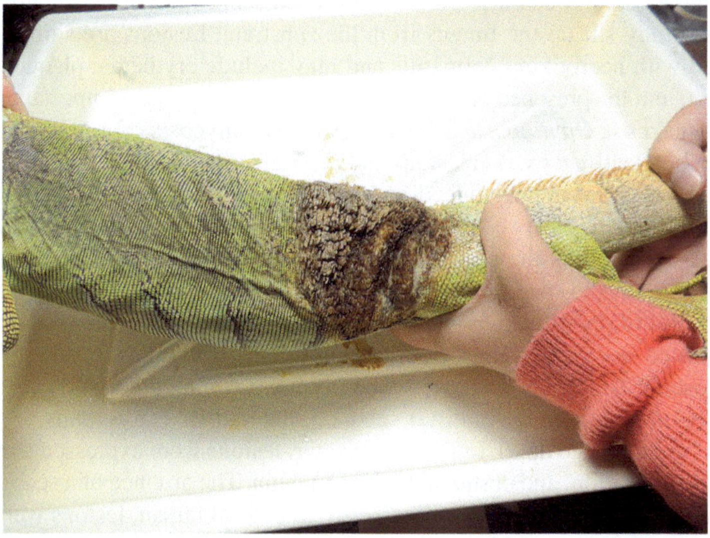

Fig. 3.15 The thickness of the crusts correlates with the aggressiveness of the PCRF infection in green iguana

Ocellated lizard (*Lacerta lepida*) – local lesions of dark color on lateral and abdominal scales (3–5 mm in size), with the formation of rugged scabs and local infiltration of subcutaneous tissue.

Armenian rock lizard (*Lacerta rudis*) – similar lesions as found in *Lacerta lepida*. Probably the same strain of pathogenic fungus, because animals were kept in closely spaced terrariums and common tools for cleaning were used.

Western fence lizard (*Sceloporus occidentalis*) – merging parakeratosis foci on the skin of the neck and sides, with the formation of scabs and local infiltration of muscles.

Parrot-beaked tortoise (*Homopus areolatus*) – vesicular merging foci on the skin of the thighs and tail, differing in color, but without the formation of crusts. Perhaps due to an early stage of the infection.

Chinese softshell turtle (*Pelodiscus sinensis*) – local ulcerative-necrotic lesions on the skin of the zygomatic area of the head, covered with a strong scab of yellow color.

The determined cases of PCRF infections in these animal species are not described in the literature. Although mycological culturing of these samples was not carried out, characteristic arthrospores and aleurioconidia were observed in direct microscopical examination. It makes possible to presume a preliminary diagnosis of PCRF by cytological examination of skin lesions. Along with the lacertian, both captive and wild snakes can be affected by PCRF. At least 9 species of snakes have been diagnosed with CANV till 2011 (Table 3.2). Although the distribution of CANV lesions in snakes seems to be primarily associated with the head [3, 25, 27, 29], in some cases, it can also be found on ventrum [26]. The active burrowing by the snakes may stir up the fungus from the substrate. Lesions are similar to those described with necrotizing dermatitis and may include erythema, plaque and crust formation, and the presence of vesicles. Among PCRF, the most important pathogen for wild snakes is *Ophidiomyces ophiodiicola*. The mycosis caused by *O. ophiodiicola* can be identified by skin lesions and thick blisters that can distort the face of a snake and even prevent it from feeding which often leads to starvation. The outcome of the disease varies between species, but the mortality rate is especially high in rattlesnakes, including the eastern massasauga rattlesnake. According to the latest data, about 30 snake species are infected in the United States (15 states), and 3000 snake species of rest of the world are vulnerable for the disease. The disease has also been reported in captive snakes from England, Germany, and Australia [25].

In Moscow Zoo, CANV was observed in snakes common boa (*Boa constrictor*), woma python (*Aspidites ramsayi*), and coastal taipan (*Oxyuranus scutellatus*) (authors' unpublished data). In *Boa constrictor*, lesions were expressed as deformation, depigmentation, and desquamation of squama. The absence of a scab indicated a subacute course of the disease (Fig. 3.16). In coastal taipan, lesions were located on the ventral and lateral sides of the body and were manifested in the form of local and confluent foci of necrosis penetrating to the intercostal muscles. Peripheral local edema and the formation of scabby crusts were also seen (Fig. 3.17). There is a single report of PCRF (CANV) infection in captive saltwater crocodiles (*Crocodylus porosus*) [7]. The lesions were associated with the skin and mortalities ($n = 548$) were recorded. Infections were recorded twice on the same farm in a span of 3 years. High-density production facilities, such as those used for the crocodilian leather industry, probably made crocodiles highly susceptible to PCRF infections. Moreover, a case of PCRF mycosis was also diagnosed in *Caiman crocodilus* in Moscow Zoo by the authors.

In general, there are few reports on the epidemiology and prevalence of PCRF infections among reptiles in the literature. Most publications represent the descriptions of individual clinical cases or outbreaks. To clarify the situation, we conducted

Fig. 3.16 PCRF-associated mycosis in common boa (*Boa constrictor*). Deformation, depigmentation, and desquamation of squama

Fig. 3.17 PCRF-associated mycosis in coastal taipan (*Oxyuranus scutellatus*). Lesions on ventral and lateral side of the body are manifested as necrotic foci penetrating to the intercostal muscles. Edema and scabby crusts are also seen

an etiological study of skin lesions of reptiles kept in captivity in the Moscow region [15]. The mycological examination of clinical samples from 109 reptiles having skin lesions was performed. Seventeen reptile species were presented including green iguana (*Iguana iguana*) (35 animals), central bearded dragon (*Pogona vitticeps*) (9 animals), red-eared slider (*Trachemys scripta elegans*) (14 animals), Chinese softshell turtle (*Pelodiscus sinensis*) (10 animals), monitor lizard (*Varanus* spp.) (5 animals), frill-necked lizard (*Chlamydosaurus kingii*) (4 animals), spiny-tailed lizards (*Uromastyx* spp.) (4 animals), geckos (*Gekko* spp.) (9 animals), chameleons (*Chameleo* spp.) (8 animals), skinks (*Scincidae* spp.) (7 animals), python (*Python regius*) (3 animals), and caiman (*Caiman crocodilus*) (1 animal). Fungal infections were diagnosed in 86 reptiles, which accounted for 79% of the total number of animals examined. Eighteen fungal species were isolated, among which the *Chrysosporium* anamorph *of Nannizziopsis vriesii* (CANV) dominated (37%). The most susceptible reptile species was found to be green iguana (89% of CANV cases).

Identification of fungal species in the aforementioned study was performed on the basis of conventional mycological techniques (morphological characteristics, etc.). Three most typical CANV cultures isolated from captive green iguanas were identified by sequencing of ITS region. In the NCBI GeneBank database, the best match was found with the sequence KX755439 belonging to the species *Nannizziopsis guarroi* (unpublished data 2017).

Although reptiles are the main targets of PCRF, several cases have also been described in warm-blooded animal species. *Chrysosporium pannicola* (formerly *C. evolceanui*) was isolated from the skin of a dog [31] and from a horse [32]. Similarly, the probable cases of mycosis caused by *C. tropicum* were reported in two breeds of chickens [33] and in a dog [34]. Recently, Cook et al. (2015) reported the disseminated infection in a German shepherd dog caused by *Chrysosporium* spp. The diagnosis was based on a positive fungal culture and cytological investigations of intralesional fungi associated with granulomatous splenitis and neutrophilic lymphadenitis. The patient showed rapid clinical improvement on oral posaconazole. Based on colonial and microscopic features, the fungus was identified as *Chrysosporium* spp. Unfortunately, further speciation of the isolate and antifungal susceptibility testing could not be performed [35].

Several cases of mycoses caused by genus *Nannizziopsis* have been described in humans. Diagnosed fungal species apparently are specific for humans [8]. Human cases of *Nannizziopsis* mycoses are summarized in Table 3.3.

Most of these human cases occurred as opportunistic infections in immunocompromised patients. Suchonwanit et al. (2015) reported a case of primary cutaneous *Chrysosporium* infection following ear piercing in an immunocompetent patient [40]. A 25-year-old healthy woman presented with a 2-year history of an itchy erythematous plaque on the right ear pinna. PCR was performed on the colony sample using the gene fragment, and BLASTN search against the GeneBank database revealed a 98% nucleotide sequence identity to *Chrysosporium* spp.

Apparently, the probability of transmission of PCRF infection from domestic reptiles in humans is very low. *N. guarroi*, the main causative agent of mycoses in bearded dragons and iguanas, has not yet been isolated from humans [8]. The environment and wild animals seem to be a more likely source of infection for humans [4]. However, precautions when dealing with reptiles are still advisable.

3.5 Ecology of PCRF and Predisposing Factors

The source of the etiologic agents of contagious PCRF-associated mycoses is not well defined as CANV is not a member of the resident or transient microbiota of reptile skin. Pare et al. [22] evaluated the mycobiota of skin in different reptiles and found the rare presence of CANV in comparison to *Aspergillus* spp., *Paecilomyces*

Table 3.3 Cases of human mycoses caused by the species of genus *Nannizziopsis* [37]

Species	Disease, dissemination	Immune status	References
N. infrequens	Localized, bronchial wash specimen, M, 40 years old, USA, IA, 2004	HIV positive	Sigler et al. (2013) [8] (elucidated further from Brandt et al. 2005) [36], who named isolate as *Nannizziopsis vriesii*)
N. hominis	Disseminated disease. Right thigh mass with lung lesion, M, USA, CA, 1994	HIV positive	[8]
N. hominis	Disseminated inguinal node, Nigerian, M, 32 years old, with disseminated adenopathy, USA, MA, 2000	Immunocompetent	[8]
N. obscura	Localized disease. Abscess right ankle, African, M, 24 years old (isolated twice), USA, NY, 1984	Immunocompetent	Sigler et al. (2013) (elucidated further from Stillwell et al. [1984] [38] who named isolate as *Chrysosporium* spp. [8]
N. vriesii	Disseminated disease. Lung infiltration and a brain abscess in a Nigerian, M, 38 years old, Germany, 2005	HIV positive	Steininger et al. (2005) [39]
N. obscura	Disseminated disease. Thoracic collection, lymphadenopathy, and skin rash in Gambian, M, 34 years old, UK. 2015	Immunosuppressed for renal transplant	[37]

spp., and *Penicillium* spp. In total, 127 reptile (36 lizards, 91 snakes) sheds were evaluated, and CANV was isolated only from an African rock python (*Python sebae*). Thus, the rarity of CANV suggested that it is not an opportunistic fungal but an obligate pathogen that infects reptiles after exposure. In order to confirm its status, veiled chameleons (*Chamaeleo calyptratus*) experimentally infected with the fungus *Chrysosporium* anamorph of *Nannizziopsis vriesii* (CANV). Chameleons were inoculated by direct application of a conidial suspension on intact and abraded skin [24]. The CANV induced lesions in all experimental groups and was recovered from infected animals, thus fulfilling Koch's postulates. A breach in cutaneous integrity, as simulated by mild scarification, increased the risk of infection. CANV dermatomycosis was shown to be contagious and could readily spread within a reptile collection, either directly through contact with infective arthroconidia or indirectly via fomites. Dense tufts of arthroconidiating hyphae were demonstrated histologically on the skin surface of many animals that developed dermatomycosis, and these arthroconidia may act as infective propagules. The infection was similar to that described for clinical cases.

Recently, the pathogenicity of *Ophidiomyces ophiodiicola* (a member of CANV complex) was experimentally studied by Lorch et al. (2015) [41]. They experimentally infected captive-bred corn snakes (*Pantherophis guttatus*) in the laboratory with pure cultures of *O. ophiodiicola*. All snakes in the infected group ($n = 8$) developed gross and microscopic lesions identical to those observed in wild snakes with snake fungal disease (SFD). Furthermore, the same strain of *O. ophiodiicola* was recovered from lesions of all animals in the infected group. The host response to the infection included marked recruitment of granulocytes to sites of fungal invasion; increased frequency of molting and abnormal behaviors, such as anorexia; and resting in conspicuous areas of enclosures. While these responses may help snakes to fight infection, they could also impact host fitness and may contribute to mortality in wild snakes with chronic *O. ophiodiicola* infection. This experiment demonstrates that *O. ophiodiicola* is the causative agent of SFD and can elicit pathological changes and affects the fitness of wild snakes.

Since reptile owners usually have many animals in their collection, hence awareness regarding good hygiene practices is essential to prevent the dissemination of the disease. Although inadequate diet and husbandry, environmental stresses, trauma, and existing dermatitis are likely contributors, however, the circumstances under which mycotic diseases in reptile species occur are still unknown [24].

In Moscow Zoo, fungal infections caused by PCRF are very rare. In general, the disease is prevalent in captive animals of private owners or in the animals having been recently imported from dealers. PCRF mycoses occur more often during group maintenance of reptiles. It is important to determine more specific risk factors associated with this fungal pathogen. Substrate also plays an important role in disease dissemination; therefore, it is important to examine substrates under different conditions (e.g., temperature and humidity) for the presence of this organism. Pet owners should be advised to maintain the cleanliness of pet's habitat. Moreover, PCRF being keratophilic, it is possible that the food sources (e.g., crickets) can also be the source of infection for reptiles. Therefore, its association with foodstuffs must be evaluated and discussed [2]. Since little is known about dissemination of PCRF, thus further research is needed on the epidemiology of this infection.

3.6 Diagnosis of PCRF-Associated Mycoses

In case of superficial lesions (dermatitis) in reptiles, veterinary practitioners are inclined to predict the bacterial etiology of the disease. Although bacteria can be isolated from most skin lesions, they do not always possess the clinical significance. The primary etiological role may belong to pathogenic fungi. The clinical signs usually do not differentiate the bacterial infection from the fungal one. Thus, mycological diagnostic testing considering PCRF in cases presenting with dermatitis or necrosis is recommended [2].

Wearing medical clothes during examination and treatment of infected reptiles can help in reducing transmission to other uninfected reptiles, as transmission occurs by direct contact or indirectly via fomites. Primary mycological diagnosis can be carried out directly in the doctor's office. It consists of direct microscopy (cytology) of the samples from the affected areas. The samples (crusts, scabs, scales, etc.) are taken from the periphery of the lesion focus by sterilized forceps and (or) scalpel.

Direct microscopy is done in 10–15% solution of potassium hydroxide with subsequent moderate heating of the slide. In samples positive for PCRF, the fungal elements can be detected occasionally segmenting apart the arthrospores (Figs. 3.18 and 3.19). If numerous fungal fragments are found in affected tissues, the mycotic etiology of the infection can be confirmed. Full-thickness biopsy samples can be submitted for culture, histopathology, and/or polymerase chain reaction (PCR) testing. Fungal elements can be detected in histological sections (Fig. 3.20), but due to presence of inflammatory cells (macrophages and heterophils), identification of PCRF seemed difficult.

Granuloma formation is a common defense reaction in reptiles which consists of central fibrin, cell detritus, and fungal elements surrounded by heterophils, macrophages, and connective tissue. The infected epidermis is often ulcerated, with fibrin deposition, fungal hyphae, and conidia [20]. However, mycological examination should not be limited to microscopy of the samples. Identification of fungal species and determination of antifungal susceptibility can be obtained only by cultural study (inoculation of the samples on mycological nutrient media). PCRF cultures (tested in our laboratory) do not cause reddening of DTM (dermatophyte test media) in contrast to dermatophytes (*Trichophyton* spp., *Microsporum* spp.) (Fig. 3.8). Incubation of the inoculated media is carried out at 28–30 °C. The beginning of growth of PCRF colonies can be seen on the 4th to 6th days. The colonies on SDA or Malt Extract Agar (MEA) media are usually white, velvety, or powdery, with diameter up to 1 cm on the 7th day (Fig. 3.7). From the 7th to 10th days, fungal cultures begin to sporulate which is necessary for proper identification of fungus. In case of slow growth, the cultures are incubated for 14–21 days. A fragment of mycelium can be taken for microscopy from mature colonies. PCRF usually forms numerous unicellular aleurioconidia, located directly on unspecialized hyphae

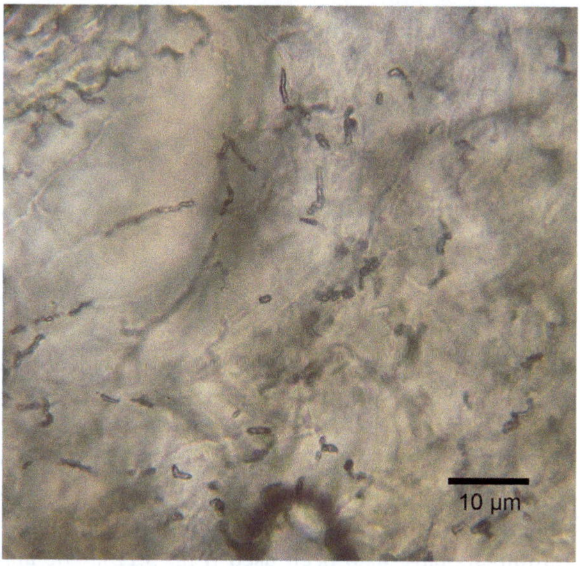

Fig. 3.18 Microscopic detection of branching hyaline hyphae in skin scales affected by PCRF. (Smear in 15% solution of potassium hydroxide, 400×)

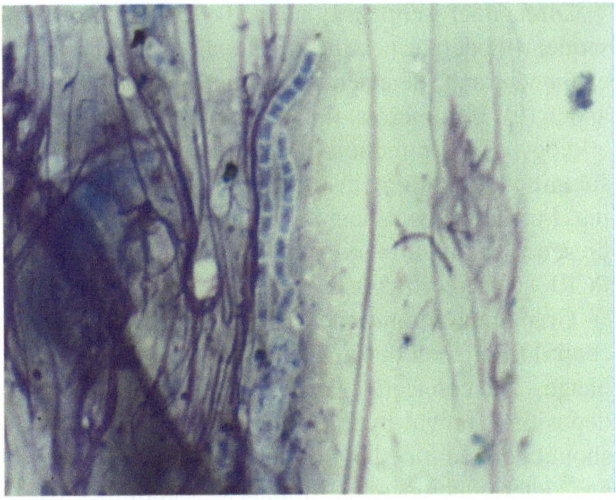

Fig. 3.19 Hyphae forming arthrospores in skin scales affected by PCRF. (Diff-Quik-stained cytological smear, 1000×)

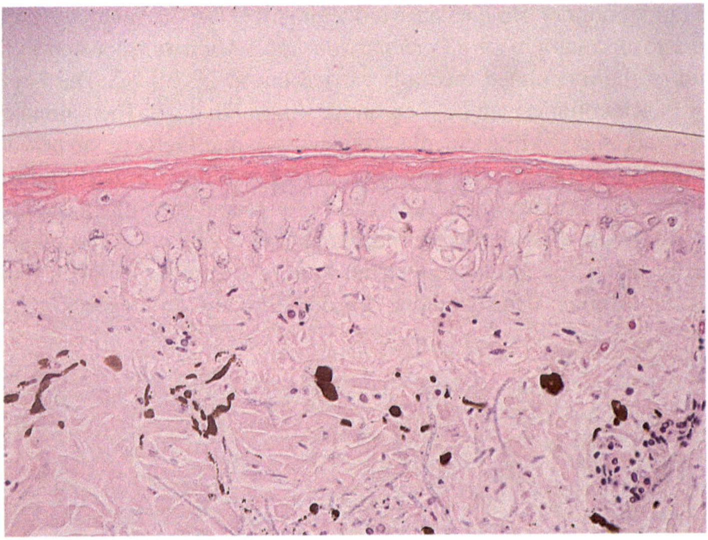

Fig. 3.20 Fungal elements in histological section of affected skin. (Hematoxylin and eosin 1000×)

(Fig. 3.5). Morphological features intrinsic for different species of PCRF have been described in Table 3.1.

When the culture of PCRF is isolated and identified, the antifungal susceptibility testing can be performed [20]. In our laboratory, we use the discs with clotrimazole (10 µg, HiMedia), nystatin (100 U, HiMedia), miconazole (10 µg, HiMedia), fluconazole (25 µg, HiMedia), ketoconazole (10 µg, HiMedia), itraconazole (10 µg, HiMedia), and voriconazole (1 µg, HiMedia) (Fig. 3.21). Moreover, E-test strips which quantitatively determine MIC (minimal inhibitory concentrations) for some antifungals have also been used (Fig. 3.22). The appropriateness of this test in a particular case should be discussed with the attending veterinarian. Considering

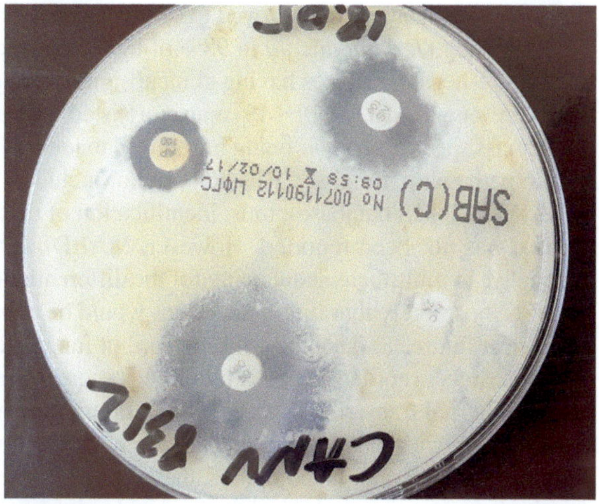

Fig. 3.21 Disc diffusion method for antifungal susceptibility testing of PCRF

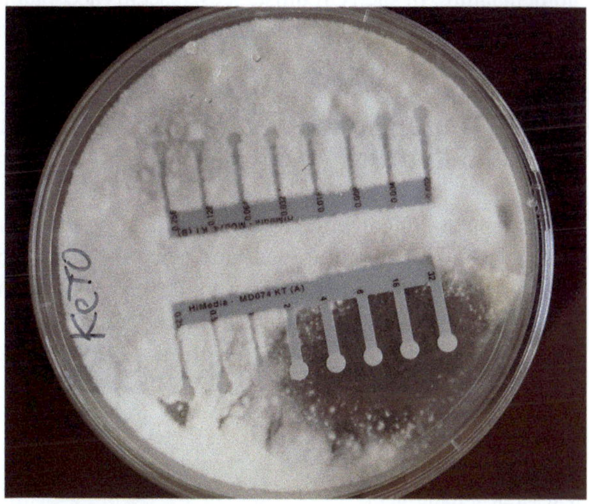

Fig. 3.22 Quantitative susceptibility testing of PCRF to ketoconazole using E-test strips. Minimal inhibitory concentration (MIC) value is 4.0 μg/ml for tested culture

osteomyelitis which is common in affected reptiles, radiographs can be used to assess the integrity of the bone while planning medical and surgical options. Advanced imaging such as computed tomography can also be used for better characterization of extent of lesions (especially bony involvement) [2].

Few publications indicate the applicability of PCR (polymerase chain reaction) for the diagnosis of PCRF mycoses [9]. For instance, Schmidt-Ukaj et al. (2016) applied PCR for identification of *Nannizziopsis chlamydospora* in 3 central bearded dragons (*Pogona vitticeps*) [42]. PCR kits that are useful for routine diagnosis of these mycoses are currently being developed. TaqMan real-time PCR has also been used to detect *Ophidiomyces ophiodiicola* (a member of PCRF group) in clinical samples [43]. One assay targets the internal transcribed spacer region (ITS) of the fungal genome while the other targets the more variable intergenic spacer region (IGS). Both assays performed equivalently and proved to be more sensitive than traditional culture methods, detecting *O. ophiodiicola* in 98% of the culture-positive samples and in 40% of the culture-negative snakes having clinical signs of the disease.

At present, MALDI-TOF MS (matrix-assisted laser desorption-ionization-time-of-flight mass spectrometry) is being introduced to identify pathogenic fungi. As obligate reptile pathogenic fungi (PCRF, etc.) are not included in the commercially available MALDI-TOF MS databases, thus, identification of reptile-associated fungi using this method has not been reported. However, MALDI-TOF MS is a rapid and reliable alternative to multilocus sequencing for the differentiation of reptile pathogenic fungi, and it is most likely that future databases would be expanded to cover reptile isolates in the near future. To date, the gold standard of fungal differentiation for fungal organisms that infect reptile species is multilocus sequencing of the large or small subunit and the internal transcribed spacer (ITS) region of the nuclear ribosomal gene [20].

As the above mentioned molecular methods are unavailable for identification of fungi in most veterinary laboratories, therefore, in routine practice, the identification of fungi is based on morphological features and cannot exactly correspond to the modern nomenclature. In such circumstances, it seems acceptable to use the traditional name of the pathogen (e.g., CANV *complex* or PCRF group) in the clinical context.

3.7 Therapy and Prevention of PCRF-Associated Mycoses in Reptiles

Treatment of fungal infections in reptiles includes the administration of effective antifungal agents (both topical and systemic) for a long term, along with maintenance of optimal environmental conditions. Debridement and surgical removal of crusts is the first step for successful treatment of deep fungal dermatitis (Fig. 3.23). Disinfection of skin lesions with a 0.125% chlorhexidine solution is also

Fig. 3.23 Debridement and surgical removal of crusts in green iguana affected by PCRF

Table 3.4 Antifungals employed for therapy of PCRF (CANV) mycoses in reptiles

Animals and clinical manifestation	Treatment	Outcome	References
Two green iguanas (*Iguana iguana*) – cutaneous hyalohyphomycosis	Oral ketoconazole and topical 2% chlorhexidine solution and terbinafine	Clinical cure	[9]
Bearded dragon (*Pogona vitticeps*) – dermatomycosis	Oral ketoconazole (20 mg/kg 24 h PO) and topical chlorhexidine and terbinafine	Lesions regressed; lost for follow-up	[10]
Bearded dragon (*Pogona vitticeps*) – focal maxillary swelling involving the skin and gingiva	Itraconazole and topical miconazole therapy	Failure (fatal)	[23]
Bearded dragon (*Pogona vitticeps*) – focally extensive discoloration and thickening of the skin	Itraconazole	Failure (euthanized after 10 weeks of therapy)	[23]
Bearded dragon (*Pogona vitticeps*) – hyperkeratotic exudative dermatitis on a swollen forelimb	Amputation and itraconazole	Clinical cure	[23]
Fourteen naturally infected bearded dragons (*Pogona vitticeps*)	Itraconazole (5 mg/kg q24h) or voriconazole (10 mg/kg q24h) until complete clearance of the fungus	2 out of 7 survived after itraconazole treatment. Only a single animal died in the voriconazole-treated group	[45]
Girdled lizard (*Cordylus giganteus*) – cutaneous hyalohyphomycosis	Voriconazole 10 mg/kg of body weight once daily for 10 weeks	Clinical cure	[11]

recommended [20]. There are relatively few reports that discuss effective dosages and dosage intervals of antifungal agents. Most of them are summarized in Table 3.4.

Most treatment regimens use systemic azole antifungals such as ketoconazole, itraconazole, or voriconazole. Treatment typically consists of both topical and systemic application of antifungals. In the past, ketoconazole was used as the drug of choice for treating fungal diseases in vertebrates; however, newer drugs with fewer side effects have been developed (e.g., itraconazole and voriconazole) to replace it. Itraconazole is often used for therapy of CANV infections, but treatment is not always successful [23]. Voriconazole is a second-generation triazole that is being used more frequently in human and veterinary medicine because it seems to have a lower incidence of side effects compared with other antifungals [44].

The oral treatment of CANV dermatomycosis in bearded dragons (*Pogona vitticeps*) with itraconazole (5 mg/kg, once a day) versus voriconazole (10 mg/kg orally, once a day) at an ambient temperature of 28–30 °C was compared [45]. Both drugs were found to treat the animals successfully (in 27 and 47 days for itraconazole and voriconazole, respectively); however, voriconazole appeared to be safer for the animals with 6 of 7 survivors as compared to 2 of 7 in the itraconazole group. Notably, hepatocellular injury may have occurred in approximately 50% of the animals in both groups as there were significant elevations of aspartate transaminase levels, while plasma concentrations of voriconazole in the bearded dragons showed more interindividual variation than itraconazole plasma concentrations. To minimize the risk of side effects with itraconazole, lower doses at less frequent intervals are recommended. Pulse therapy is another potential consideration for reducing toxic side effects associated with itraconazole, as it was demonstrated in mammalian hosts [46].

Side effects and toxicity of triazole antifungals in reptiles represent an important issue in veterinary medicine. The most common clinical signs associated with triazole toxicity are anorexia and depression. Affected animals tend to develop mild hepatitis as a result of the toxicity; therefore, liver functioning tests should be performed or are recommended before, during, and after treatment regimens. Behavioral observation is the best method for monitoring animals. Further, serial measurement of clinical chemistries and determination of level of drug (by performing liver biopsies) should be undertaken. Reptiles with hepatic disease secondary to drug toxicity are often depressed and anorexic. The clinical chemistries commonly used to asses sliver disease include aspartate aminotransferase (AST), gamma glutyltransaminase (GGT), and bile acids. Although AST and GGT are not liver specific, they may be helpful in combination with other parameters. Bile acid testing is the most useful for evaluating liver function and should be performed if secondary liver disease is suspected. Results of bile acid testing should be interpreted carefully taking into account the species of the animal. Liver biopsies can be used to assess liver disease and check for the accumulation of drugs beyond safe levels [2].

An increase in the level of AST (120–420 mmol/l) in 57.1% of cases ($n = 7$) was observed when ketoconazole therapy course in green iguanas lasted over 2 weeks. However, supportive therapy with heptral, B complex, and LRS (lactated Ringer's solution) usually allowed withstanding course duration up to 4–5 weeks with the dosages of ketoconazole 20 mg/kg daily. The toxicity of voriconazole for snakes was reported by Allender (2017) [47]. In his practice, 4 of 7 snakes died in 12 h after the commencement of voriconazole. However, voriconazole was successfully used for therapy of CANV in 9 green iguanas by us (Fig. 3.23, 3.24 and 3.25). The choice of the drug was determined by the susceptibility of the CANV cultures in vitro. Voriconazole was administered orally at doses of 5 mg/kg/day (4 cases) and 10 mg/kg/day (5 cases). The duration of the course was 3–7 weeks, usually until removal of the scabs and epithelialization on the periphery of the dermatitis foci. Therapy for two additional weeks is also recommended. Topical antifungals (ointment or emulsion) were applied along with the systemic therapy. The complete cure was observed in 3 cases. In other cases, the persistent remission was achieved.

In three cases, despite complete epithelialization, granulomas remained in the subepidermal layers, which were surgically removed (Fig. 3.25). Relapses of the mycosis occurred not only in animals having granulomas but also in completely cured reptiles. New lesions appeared both in the zone of primary infection and in

Fig. 3.24 Subepidermal darkly pigmented granuloma remained after antifungal therapy

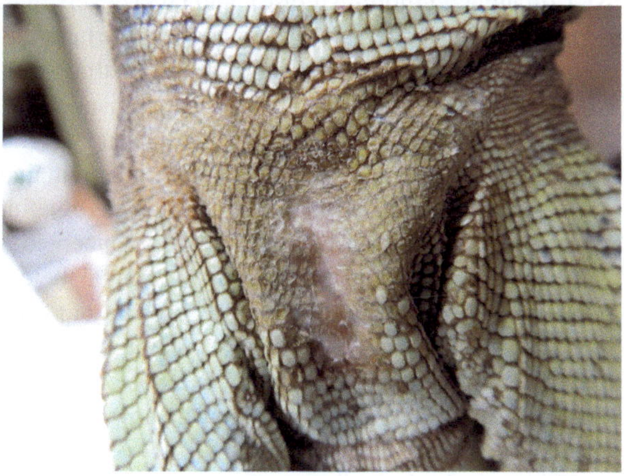

Fig. 3.25 Epithelialization of lesions after 3 weeks of combined antifungal therapy in green iguana

completely different areas of the body. In some cases, the antifungal susceptibility of CANV isolates also differed from the initial one.

Among systemic azoles, posaconazole could be a promising drug for the treatment of mycoses in reptiles. It was already used for successful treatment of fusariomycosis in marine turtles *Caretta caretta* [48]. Posaconazole has also been successfully used in wild and exotic animals for treatment of resistant mycoses, e.g., coccidioidomycosis in dolphins [49]. Recently, a successful treatment of *Nannizziopsis obscura* infection by posaconazole in immunocompromised human patient was reported [37]. Moreover, oral posaconazole therapy demonstrated rapid clinical improvement in disseminated *Chrysosporium* spp. infection in a German shepherd dog [35]. But as far as PCRF in reptiles is concerned, there is no report on posaconazole therapy. In some cases of PCRF mycoses in reptiles, voriconazole-resistant fungal isolates showed susceptibility against ketoconazole therapy. This indicates that the therapy should be based on laboratory data on the antifungal susceptibility of a particular fungal isolate.

In addition to azoles, terbinafine can also be effective in the treatment of mycoses caused by PCRF. The pharmacokinetics of terbinafine with the use of a subcutaneous implant and for nebulization therapy in the case of CANV in snakes was investigated by Allender [47]. The topical application of terbinafine in green iguanas in the Moscow Zoo with confirmed CANV mycoses showed efficiency in 44.4% of cases ($n = 9$). However, in comparison to azoles, terbinafine showed more potent dermato-toxicity. Usually pathological changes resembling a chemical burn appear on the skin of green iguanas as early as 1–2 weeks of therapy.

Terbinafine was successfully used by us for the systemic treatment of disseminated mycoses caused by *Paecilomyces lilacinus* in the group of green iguanas and the *Fusarium moniliforme* mycosis in the group of frilled-neck lizard (*Chlamydosaurus kingii*). Terbinafine was administered at a dose of 5 mg/kg by pulse therapy: daily, for 5 days, followed by a break for 7 days, etc. [50]. However, in all cases of oral administration of terbinafine, lizards developed severe depression, temporary paresis, vomiting attempts, and polyuria. But, no increase in AST, ALT (alanine aminotransferase), and GGT levels in the blood was observed.

In two cases of CANV treatment with terbinafine in green iguanas, by the 5th day of therapy, there was an increase in the level of uric acid up to 867 and 617 μmol/l, respectively. After the acute episode of renal failure, the animals were relieved with allopurinol (20 mg/kg/daily) and polyionic crystalloid solutions intravenously drip (30 ml/kg/daily). The animals remained stabilized, but therapy was not found effective. Therefore, the use of terbinafine for the systemic therapy in lizards should be prescribed only in accordance with laboratory indications.

Caspofungin, the most modern antifungal belonging to echinocandins, is of interest for the therapy of PCRF mycoses. In our practice, it was used for the treatment of CANV dermatomycosis in green iguana, but the clinical improvement was not significant. The drug was used at a dose of 1 mg/kg every 48 h by slow intravenous bolus administration for 14 days, and no distinct side effects were found associated with the use of caspofungin.

Chitin synthesis inhibitors (lufenuron and nikkomycin) have been used for the therapy of human and animal mycoses [51]. In our practice, we used the 5% emulsion of lufenuron (manufactured by Syngenta) to treat CANV dermatomycosis in two green iguanas. Despite the declared low toxicity of the drug for humans and bees, the topical application of emulsion diluted at 1:10 caused a severe toxic reaction (salivation, apnea, stupor) in lizards. Apparently, the reactions were associated with inhalation of drug vapors. Then the drug was used topically at a dilution of 1:1000, daily. Within 2 weeks, significant clinical improvements such as spontaneous removal of the scrotal crusts and partial epithelialization at the periphery of the dermatitis foci were noticed. However, further use of lufenuron as monotherapy failed to prevent the recurrence of the disease. Oral form of the drug (Program tablets for dogs, 67.8 mg) at doses of 10 mg/kg once a week for 4 weeks was not found effective.

As tools of adjunctive topical treatment of PCRF mycoses, nystatin, terbinafine, clotrimazole, and enilconazole were also used (authors' unpublished data). Unfortunately, we have no sufficient data to objectively compare the clinical efficacy of these drugs. However, based on limited observations, clinical improvement with nystatin was observed in 41% of reptiles ($n = 7$), terbinafine 44% ($n = 9$), clotrimazole 58% ($n = 12$), and enilconazole 66% ($n = 6$) (Fig. 3.25). These data need to be verified in the future, taking into account the background systemic therapy.

As mentioned earlier, adequate light and heat are essential for reptilian health and largely influence clinical recovery because metabolism of drugs, use of fluid therapy, and immune system of reptiles are heat dependent [52].

Taking into account the contagious nature of the disease, during the treatment of PCRF infections in reptiles, it is necessary to employ hygienic measures. If affected reptile is detected, the cage mate needs to be observed closely for signs of disease. It is desirable to isolate a sick animal to reduce the risk of cross-infection. Other exposed reptiles should have thorough veterinary examinations, and any skin lesions should be assessed to rule out fungal infection as the underlying cause. Reptile that comes with a skin lesion or that develops a skin lesion while in quarantine should not be released until mycosis is eliminated.

In the prevention of fungal infections, the quarantine of animals destined for sale, especially those imported, plays a crucial role. The standard veterinary quarantine at 30 days may not be sufficient to detect infected animals. The quarantine in the Moscow Zoo lasts up to 60 days, during which the latent PCRF infection usually turns into a clinically manifested form. Reptiles imported illegally can possess a dangerous source of fungal pathogens.

Proper sanitation and hygiene are key risk reducers. Disinfectants possessing antifungal activity should be used for decontamination of reptilian cages, animal care equipment, and house environment. Future studies on fungicidal activity of available disinfectants against *Chrysosporium*-related fungi are highly needed. In-depth studies on the ecology of the fungal pathogens, pathways of infection, and its transmission are also required.

Acknowledgments We sincerely thank Marina M. Manoyan (Laboratory of Mycological Expertise, FSFI VGNKI, Moscow, Russia) for assistance in mycological diagnostic examination of clinical samples and general assistance.

Conflicts of Interests There are none.

References

1. Casadevall A (2005) Fungal virulence, vertebrate endothermy, and dinosaur extinction: is there a connection? Fung Gen Biol 42:98–106
2. Mitchell M, Walden M (2013) *Chrysosporium* anamorph of *Nannizziopsis vriesii*. An emerging fungal pathogen of captive and wild reptiles.Vet Clin North Am Exot Anim Pract 16(3):659–668
3. Allender M, Dreslik M, Wylie S et al (2011) *Chrysosporium* sp. infection in eastern massasauga rattlesnakes. Emerg Infect Dis 17(21):2383–2384
4. Cabanes F, Sutton D, Guarro J (2014) *Chrysosporium*-related fungi and reptiles: a fatal attraction. PLoS Pathog 10(10):e1004367

5. Schildger B, Frank H, Gobel T et al (1991) Mycotic infections of the integument and inner organs in reptiles. Herpetopathologia 2:81–97
6. Paré J, Sigler L, Hunter D et al (1997) Cutaneous mycoses in chameleons caused by the *Chrysosporium* anamorph of *Nannizziopsis vriesii* (Apinis) Currah. J Zoo Wildl Med 28:443–453
7. Thomas A, Sigler L, Peucker S et al (2002) *Chrysosporium* anamorph of *Nannizziopsis vriesii* associated with fatal cutaneous mycoses in the salt-water crocodile (*Crocodylus porosus*). Med Mycol 40:143–151
8. Sigler L, Hambleton S, Pare J (2013) Molecular characterization of reptile pathogens currently known as members of the *Chrysosporium anamorph* of *Nannizziopsis vriesii* complex and relationship with some human-associated isolates. J Clin Microbiol 51:3338–3357
9. Abarca M, Martorell J, Castellá G et al (2008) Cutaneous hyalohyphomycosis caused by a *Chrysosporium* species related to *Nannizziopsis vriesii* in two green iguanas (*Iguana iguana*). Med Mycol 46:349–354
10. Abarca M, Martorell J, Castellá G et al (2009) Dermatomycosis in a pet inland bearded dragon (*Pogona vitticeps*) caused by a *Chrysosporium* species related to *Nannizziopsis vriesii*. Vet Dermatol 20:295–299
11. Hellebuyck T, Baert K, Pasmans F et al (2010) Cutaneous hyalohyphomycosis in a girdled lizard (*Cordylus giganteus*) caused by *Chrysosporium* anamorph of *Nannizziopsis vriesii* and successful treatment with voriconazole. Vet Dermatol 21:429–433
12. Johnson R, Sangster C, Sigler L et al (2011) Deep fungal dermatitis caused by the *Chrysosporium* anamorph of *Nannizziopsis vriesii* in captive coastal bearded dragons (*Pogona barbata*). Aust Vet J 89:515–519
13. Ovchinnikov R, Manoyan M, Gaynullina A (2007) Etiological role of non-dermatophytic fungi in superficial animal mycoses (*article in Russian*). Proc VGNKI 68:148–153
14. Ovchinnikov R, Manoyan M, Gaynullina A et al (2008) Mycoses in exotic reptiles. Vet Dermatol 19:59
15. Ovchinnikov R, Manoyan M, Vasyliev D et al (2014) *Chrysosporium* anamorph of *Nannizziopsis vriesii* as a dominating fungal pathogen in pet reptiles (*article in Russian*) Adv Med Mycol 13:371–373
16. Seifert K, Morgan-Jones G, Gams W et al (2011) The genera of *Hyphomycetes*. Centraalbureau voor Schimmelcultures, Utrecht
17. Pare J, Sigler L, Rosenthal K et al (2006) Microbiology: fungal and bacterial diseases of reptiles. In: Mader D (ed) Reptile medicine and surgery. Saunders Elsevier, St. Louis, pp 217–238
18. Stchigel A, Sutton D, Cano-Lira J et al (2013) Phylogeny of chrysosporia infecting reptiles: proposal of the new family *Nannizziopsiaceae* and five new species. Persoonia 31:86–100
19. Hedley J, Eatwell K, Hume L (2010) Necrotizing fungal dermatitis in a group of bearded dragons (*Pogona vitticeps*). Vet Rec 166:464–465
20. Schmidt V (2015) Fungal infections in reptiles—an emerging problem. J Exotic Pet Med 24(3):267–275
21. Martel A, Fonteyne P, Chiers K et al (2006) Nasal *Nannizziopsis vriesii* granuloma in an Amevia lizard (*Amevia chaitzarni*) Vlaams Diergeneeskd Tijdschr. 75:306–307
22. Pare J, Sigler L, Rypien K et al (2003) Cutaneous mycobiota of captive squamate reptiles with notes on the scarcity of *Chrysosporium* anamorph *Nannizziopsis vriesii*. J Herp Med Surg 13:10–15
23. Bowman M, Pare J, Sigler L et al (2007) Deep fungal dermatitis in three inland bearded dragons (*Pogona vitticeps*) caused by the *Chrysosporium* anamorph of *Nannizziopsis vriesii*. Med Mycol 45:371–376
24. Pare J, Coyle K, Sigler L et al (2006) Pathogenicity of the *Chrysosporium* anamorph *Nannizziopsis vriesii* for veiled chameleons (*Chamaeleo calyptratus*). Med Mycol 44:25–31
25. Lang H (2017) Mysterious disease attacks rattlesnakes. Natl Geogr. Available online. http://news.nationalgeographic.com/2017/06/snake-fungal-disease-treatmentconservation/. Accessed 28 March 2018

26. Nichols D, Weyant R, Lamirande B et al (1999) Fatal mycotic dermatitis in captive brown tree snakes (*Boiga irregularis*). J Zoo Wildl Med 30:111–118
27. Bertelsen M, Crawshaw G, Sigler L et al (2005) Fatal cutaneous mycosis in tentacle snakes (*Erpeton tentaculatum*) caused by *Chrysosporium* anamorph of *Nannizziopsis vriesii*. J Zoo Wildl Med 36:82–87
28. Toplon D, Terrell S, Sigler L et al (2013) Dermatitis and cellulitis in leopard geckos (*Eublepharis macularius*) caused by the *Chrysosporium* anamorph of *Nannizziopsis vriesii*. Vet Pathol 50(4):585–589
29. Eatwell K (2010) Suspected *Chrysosporium* anamorph of *Nannizziopsis vriesii* (CANV) dermatitis in an albino Boa constrictor (*Constrictor constrictor*). J Small Anim Pract 51:290
30. Pare J, Jacobson E (2007) Mycotic diseases of reptiles. In: Jacobson E (ed) Infectious diseases and pathology of reptiles: color atlas and text. CRC Press, Boca Raton, pp 527–570
31. Hajsig M, de Vries G, Sertic V et al (1974) *Chrysosporium evolceanui* from pathologically changed dog skin. Veter Arhiv 44:209–211
32. Grahn B, Wolfer J, Keller C et al (1993) Equine keratomycosis: clinical and laboratory findings in 23 cases. Prog Vet Comp Ophthalmol 3:2–7
33. Saidi S, Bhatt S, Richard J et al (1994) *Chrysosporium tropicum* as a probable cause of mycosis of poultry in India. Mycopathologia 125:143–147
34. Watt P, Robins G, Galloway A et al (1995) Disseminated opportunistic fungal disease in dogs: 10 cases (1982–1990). J Am Vet Med Assoc 207:67–70
35. Cook E, Meler E, Garrett K et al (2015) Disseminated *Chrysosporium* infection in a German shepherd dog. Med Myc Case Rep 10:29–33
36. Brandt M, Gaunt D, Iqbal N et al (2005) False-positive *Histoplasma capsulatum* gen-probe chemiluminescent test result caused by a *Chrysosporium* species. J Clin Microbiol 43:1456–1458
37. Baggott A, McGann H, Barton R et al (2017) Disseminated *Nannizziopsis obscura* infection in a renal transplant patient- the first reported case. Med Myc Case Rep 17:20–24
38. Stillwell W, Rubin B, Axelrod J (1984) *Chrysosporium*, a new causative agent in osteomyelitis. Clin Orthop 184:190–192
39. Steininger C, van Lunzen J, Sobottka I et al (2005) Mycotic brain abscess caused by opportunistic reptile pathogen. Emerg Infect Dis 11:349–350
40. Suchonwanit P, Chaiyabutr C, Vachiramon V (2015) Primary cutaneous *Chrysosporium* infection following ear piercing: a case report. Case Rep Dermatol 7:136–140
41. Lorch J, Lankton J, Werner K et al (2015) Experimental infection of snakes with *Ophidiomyces ophiodiicola* causes pathological changes that typify snake fungal disease. MBio 6:e01534–e01515
42. Schmidt-Ukaj S, Loncaric I, Spergser J et al (2016) Dermatomycosis in three central bearded dragons (*Pogona vitticeps*) associated with *Nannizziopsis chlamydospora*. J Vet Diagn Investig 28:319–322
43. Bohuski E, Lorch J, Griffn K et al (2015) TaqMan real-time polymerase chain reaction for detection of *Ophidiomyces ophiodiicola*, the fungus associated with snake fungal disease. BMC Vet Res 11:95
44. Sim R (2016) Voriconazole. J Exotic Pet Med 25:242–347
45. Van Waeyenberghe L, Baert K, Pasmans F et al (2010) Voriconazole, a safe alternative for treating infections caused by the *Chrysosporium* anamorph of *Nannizziopsis vriesii* in bearded dragons (*Pogona vitticeps*). Med Mycol 48:880–885
46. Bensignor E (2006) Oral itraconazole as a pulse therapy for the treatment of canine Malassezia dermatitis: a randomized, blinded, comparative trial. Prat Med Chir Anim Comp 41(2):69–72
47. Allender M (2017) Snake fungal disease: implications for pet snakes. In: Proc. NAVC conference Orlando February 4–8 2017, pp 1357–1358
48. Garcia-Hartmann M, Hennequin C, Catteau S et al (2017) Clusters of *Fusarium solani* infection in juvenile captive born *Caretta caretta* sea turtles. J Mycol Med 27:113–118

49. Walters C, Townsend J, Staggs L et al (2009) Posaconazole for the treatment of zygomycosis in cetaceans. IAAAM Conference Proceedings 2009. Available online: https://www.vin.com/apputil/content/defaultadv1.aspx?meta=Generic&pId=11285. Accessed 28 Mar 2018
50. Vasyliev D, Manoyan M, Dyagilets E (2005) Cases of disseminated dermatomycosis in reptiles: challenges in verifcation and therapy (article in Russian). Sci Res Zoo 18:78–87
51. Ben-Ziony Y, Boaz A (2000) Use of lufenuron for treating fungal infections of dogs and cats: 297 cases (1997–1999). J Am Vet Med Assoc 217:1510–1513
52. Latney L, Wellehan J (2013) Selected emerging infectious diseases of squamata. Vet Clin Exot Anim 16:319–338

Aspergillosis in Humans and Animals

4

Seyedmojtaba Seyedmousavi

Abstract

Aspergillosis is a one-health infectious disease caused by fungi belonging to the genus *Aspergillus*, a group of ubiquitous saprophytes recovered from a variety of substrates in the environment. *Aspergillus* species cause a wide range of diseases in human and various animal species, including acute, chronic, noninvasive localized infections, fatal disseminated diseases, as well as allergic reactions. There are more than 250 validly described species in the genus *Aspergillus* based on polyphasic taxonomy, and they are subdivided into 22 distinct sections. Species identification of *Aspergillus* species can be challenging. A two-step approach has been suggested for molecular identification of *Aspergillus* species in the clinical setting. The first step is to sequence ITS, the barcoding marker for the identification to the intersection level, followed by sequencing of partial β-tubulin for individual species identification within the sections. The matrix-assisted laser desorption/ionization-time-of-flight (MALDI-TOF) mass spectrometry (MS)-based strategy has also shown promising results to discriminate the intra-section level of clinically relevant species of *Aspergillus*, including the non-*fumigatus* "cryptic" species.

The antifungal triazoles are preferred agents for treatment and prevention of infections caused by *Aspergillus* species. However, with wide application of azoles in medical practice and agriculture, emergence of triazole resistance in *A. fumigatus* caused by mutations in the *Cyp51A* gene has become a global public health concern.

S. Seyedmousavi (✉)
National Institutes of Health (NIH), Bethesda, MD, USA
e-mail: Seyedmousavi@nih.gov

K. Singh, N. Srivastava (eds.), *Recent Trends in Human and Animal Mycology*,
https://doi.org/10.1007/978-981-13-9435-5_4

Keywords
Aspergillus fumigatus · Antifungal resistance · Emerging non-*fumigatus* species · Human aspergillosis · Mammalian aspergillosis

4.1 Introduction

Aspergillus species are ubiquitous saprophytes recovered from a wide variety of environments and substrates on the earth throughout the year [1]. The great majority of aspergilli are prevalent in soil-decaying vegetation, seeds, grains, and foodstuffs. Only a few species are well known as important opportunistic zoopathogens [2–4]. There are more than 250 validly described species in the genus *Aspergillus* based on polyphasic taxonomy, and they are subdivided into 22 distinct sections [5]. Of these, 13 sections, including *Candidi*, *Circumdati*, *Flavipides*, *Fumigati*, *Nidulantes*, *Nigri*, *Ornati*, *Restricti*, *Tanneri*, *Terrei*, *Usti*, *Versicolores* and *Warcupi*, contain clinically relevant species [6, 7]. Of note, identification at the species level is critical for proper disease management [8].

In humans, inhalation of *Aspergillus* conidia into the lungs may cause multiple diseases, which depends on the immunological status of the host, including invasive and chronic pulmonary aspergillosis, aspergilloma, and different forms of hypersensitivity diseases such as allergic asthma, hypersensitivity pneumonitis, and allergic bronchopulmonary aspergillosis [9, 10].

Aspergillosis in animals covers a wide range of diseases from localized conditions to fatal disseminated infections, as well as allergic reactions caused by fungi belonging to the genus *Aspergillus* [11]. Similar to infections in humans, animals exhibiting inability to produce a normal immune response are at higher risk of infection. Aspergillosis may also occur in healthy animals under environmental stress and other immunocompromising conditions [12, 13].

This chapter represents a summary of comparative knowledge on *Aspergillus* infections in humans and various types of animals.

4.2 Aspergillosis in Humans

In humans, invasive aspergillosis occurs in patients with iatrogenic immunosuppression (secondary immunodeficiencies) during the treatment of neoplastic or autoimmune diseases and in recipients of allogeneic hematopoietic stem cell transplantation (HSCT) and solid organ transplantation in the setting of neutropenia induced by myeloablative chemotherapy [14, 15], and/or corticosteroid treatment, which qualitatively and quantitatively affects neutrophils and monocytes/macrophages [16, 17]. The recruitment and activation of neutrophils and Ly6Chi inflammatory monocytes at the site of *Aspergillus*-infected tissue are critical for effective inhibition of *Aspergillus* conidia and invasive filamentous elements via distinct mechanisms [8].

In addition, the number of primary immunodeficiencies (PIDs) recognized as underlying conditions predisposing to *Aspergillus* infections are also increasing, most often in children and young adults [18]. PIDs are congenital genetic disorders, mostly due to single-gene abnormalities that cause enhancement in susceptibility to autoimmunity and/or infectious disease [19–21]. As a result, patients may suffer from recurrent, protracted, or severe infections caused by opportunistic pathogens including molds and yeasts [22, 23]. Among the known congenital immunodeficiencies, CGD (chronic granulomatous disease); AD (autosomal-dominant) hyper-IgE syndrome (HIES); AD deficiency in GATA2 (also known as MonoMAC, syndrome of monocytopenia, B cell, and NK cell lymphopenias); AD or AR (autosomal-recessive) severe congenital neutropenia (SCN) syndromes; LAD (AR type I leukocyte adhesion deficiency), also called CD18 deficiency; and CARD9 (caspase recruitment domain-containing protein 9) deficiency are discussed here. Patients with these conditions are at risk for developing infection by *Aspergillus* species [18, 24].

Individuals with chronic respiratory disease are also susceptible to airborne fungal infections including allergic bronchopulmonary aspergillosis (ABPA) [25], severe asthma with fungal sensitization (SAFS) [26], and chronic pulmonary aspergillosis [27]. CPA, a chronic progressive infection that destroys lung tissue in non-immunocompromised patients, is thought to affect about three million people worldwide [27, 28]. CPA complicates individuals with preexisting lung disease such as pulmonary tuberculosis (TB), nontuberculous mycobacterial infection, asthma, allergic bronchopulmonary aspergillosis (ABPA), chronic obstructive pulmonary disease (COPD), sarcoidosis, and pneumothorax [29]. If untreated, 50–85% of patients with CPA will die within 5 years [30, 31].

4.3 Aspergillosis in Animals

Similar to infections in humans, animals exhibiting inability to produce a normal immune response are at higher risk of infection [11]. Aspergillosis may also occur in healthy animals under immunocompromising conditions. In addition to environmental stressors, tuberculosis is a well-known underlying disease in chronic necrotizing pulmonary aspergillosis and aspergilloma [12, 13], with the clinical symptoms of aspergillosis being characterized by limited invasiveness that occurs in mildly immunocompromised animals [13, 32, 33].

4.3.1 Invertebrates

In invertebrates, *A. sydowii* caused a large epizootic affecting sea fan corals (*Gorgonia* spp.) [34], first documented in 1995 near Saba the Bahamas and subsequently spreading throughout the Caribbean basin, including in the Florida Keys [35, 36]. Studies also suggest that sponges may be potential new reservoir of a marine pathogen, *A. sydowii* [37]. It is believed that virulence of *A. sydowii* increases with temperature, probably because the rate of pathogen development continued to increase in a temperature range where coral defenses became less potent [38]. However, it remains difficult to distinguish between the role of the environment in allowing opportunistic pathogens to increase and creation of a niche for new pathogenic microorganism that causes coral disease [39].

Aspergillus species are also known to infect honeybee (*Apis mellifera*) brood, causing stonebrood disease over all larval stages [40, 41]. *Aspergillus* species with the ability to produce mycotoxins such as *A. flavus*, *A. fumigatus*, and *A. niger* have been suggested to be the primary cause of this disease [42]. In some countries, stonebrood is a notifiable disease that has to be reported to the authorities if it occurs.

Despite above described diseases of *Aspergillus* species in honeybees and sea fan corals, aspergillosis has not been reported from any other invertebrate animal in natural conditions. However, a variety of different insect species such as *Drosophila melanogaster* and *Galleria mellonella* have also been employed to study fungal pathogen-host interactions [43, 44].

4.3.2 Reptiles

In reptiles, *Aspergillus* disease may be promoted by immunocompromising conditions, such as husbandry deficiencies or inappropriate temperatures, humidity, or poor enclosure hygiene [45]. *Aspergillus* species such as *A. fumigatus*, *A. niger*, and *A. terreus* have been isolated from both cutaneous and disseminated infections [46] in turtles (*Sternotherus odoratus*) [47], crocodiles [48], San Esteban chuckwallas (*Sauromalus varius*) [49], free-ranging gopher tortoise (*Gopherus polyphemus*) [50], and captive snakes (*Eunectes murinus*) [51].

4.3.3 Birds

Avian aspergillosis is predominantly a disease of the respiratory tract, but all organs can be involved, leading to a variety of acute or chronic manifestations [52, 53]. All avian species should be considered as susceptible. *Aspergillus fumigatus* is the major respiratory pathogens and has been involved in significant common-source sapronotic die-offs of domestic and free-ranging wild birds [54]. Other *Aspergillus* species like *A. flavus*, *A. niger*, *A. nidulans*, and *A. terreus* may also be isolated from cases of aspergillosis (sometimes in mixed infections) in commercial poultry, but much less frequently than *A. fumigatus* [55, 56]. In tropical countries, *A. flavus* is probably more prevalent than *A. fumigatus* [57]. Clinical manifestations of aspergillosis in birds depend on the infective dose, spore distribution, preexisting diseases, and immune response of the host [52, 53, 58–60]. It is believed that impaired immunity and the inhalation of a large inoculum of conidia are important causative factors [61]. Active fungal proliferation and sporulation of *A. fumigatus* on organic material produce large amounts of small-sized conidia that are easily dispersed in air, then potentially inhaled, and deposited deep in the respiratory tract. Discriminatory molecular genotyping based on multilocus microsatellite panels has demonstrated that the environment of diseased animals may be a source for *A. fumigatus* infection and that either multiple [62–64] or single genotype-linked infections could occur in confirmed cases [65]. Susceptible hosts will develop polymorphic clinical forms in relation to either localized or disseminated lesions [53, 66]. Economic significance of aspergillosis is most readily apparent in poultry production, where disease occurs late in the growing cycle [67]. In spontaneous outbreaks, the mortality ranged between 4.5% and 90%, with the age of diseased birds varying from 3 days to 20 weeks [56–68].

Of note, both host and fungus characteristics explain the particular susceptibility of birds to *Aspergillus* infection [69, 70]. Environmental stressors may also play a role, e.g., in poultry farms, where many environmental stressors may be present, including excessive ammonia and moisture, inappropriate temperature, and degraded litter. Furthermore, feed contamination with mycotoxins and/or competing pathogens may affect avian immunocompetence. In wild birds, it has been shown that there is a significant link between resource allocation and the costs of immunity, especially in defense against pathogens in environments where multiple factors change in time and space [71].

4.3.4 Dogs

Sinonasal, bronchopulmonary, and disseminated infections are major forms of aspergillosis in dogs [72–74], and a breed or gender predisposition can be recognized [75]. Factors that may predispose dogs to infection include injury to any of the mucous membranes, the use of catheters, administration of antibiotics, and immunosuppressive drugs or the presence of other diseases. Sinonasal aspergillosis is usually seen in dolichocephalic and mesocephalic dogs and is very rare in brachycephalic dogs. German Shepherds and Rottweilers are the commonly affected breeds. Dogs of any age may be affected, but approximately 40% are 3 years or younger and 80% are 7 years or younger [75]. In several studies of dogs with chronic nasal disease, sinonasal aspergillosis occurred with a frequency of 7–34% [76]. It is the second most common cause of nasal discharge in dogs after nasal neoplasia [77]. *Aspergillus fumigatus* is most frequently isolated, although various other species including *A. niger*, *A. nidulans*, *A. versicolor*, and *A. flavus* have been reported. *Penicillium* spp. and other fungi are much less frequently detected [78, 79]. Bronchopulmonary aspergillosis is a rare disease in dogs [80–82]. The clinical signs are nonspecific, including depression, fever, and cough [80]. Disseminated aspergillosis in dogs is relatively infrequent, but it is a potentially fatal disease, which most often is seen in German Shepherds and is usually due to *A. terreus* and *A. deflectus*, followed in order of decreasing frequency by *A. fumigatus*, *A. niger*, and *A. flavipes* [83]. Otomycosis due to *Aspergillus* species has occasionally been described in dogs previously been treated with various topical and oral antibiotics, which may have predisposed them to develop a secondary *Aspergillus* infection [84].

4.3.5 Cats

Sinonasal and sinoorbital infections are two forms of *Aspergillus* disease in cats [73, 74]. Orbital aspergillosis is characterized by progression of sinonasal aspergillosis to the preorbital area, which is challenging to treat and the prognosis for resolution of infection is generally poor [85–87]. Cats stressed by underlying disease (such as viral infection) or immunosuppression are more susceptible to infection [85–87]. Viral diseases (due to feline immunodeficiency virus and feline leukemia virus) [88, 89] may cause a severe immunodeficiency and short-term reduction of the number of neutrophils and of lymphocyte responsiveness [90]. Moreover, an inherited susceptibility are presumed to influence the incidence of aspergillosis in purebred cats of brachycephalic conformation [91]. Ulcerative keratomycosis is common in cats and is frequently associated with feline herpes viral infection [92]. *Aspergillus felis* has been the most frequently reported etiologic agent of sinoorbital aspergillosis in cats, followed by cryptic species of the section *Fumigati*, including *A. udagawae* and *A. viridinutans* [87, 93].

4.3.6 Ruminants

In ruminants, *Aspergillus* species, particularly *A. fumigatus*, are known worldwide to cause pneumonia, gastroenteritis, mastitis, placentitis, and abortions [94]. Dairy cows in early lactation show increased susceptibility to *Aspergillus* infections [94]. Other factors that seem to predispose to aspergillosis include the presence of other diseases such as tuberculosis, cholangiocarcinoma, and intense antimicrobial therapy [32, 95].

Bronchopulmonary aspergillosis is a fatal disease in ruminants that may progress rapidly [95]. Clinical signs of disease include pyrexia; rapid, shallow, and stertorous respiration; nasal discharge; and a moist cough. The lungs are firm, heavy, and mottled and do not collapse. In subacute to chronic mycotic pneumonia, the lungs contain multiple discrete granulomas, and the disease grossly resembles tuberculosis [96–98]. Overall, histologic examination of the lungs in pulmonary aspergillosis indicates abundant hyphae and high numbers of associated oxalate crystals [96]. In cows, the gastrointestinal tract, and almost exclusively the omasum, is the primary site of mycotic lesions caused by *A. fumigatus* [95]. The reported incidence of mycotic mastitis in cows caused by *A. fumigatus* has increased subsequent to the antibiotic treatment of animals, although the number of reported cases is lower than in small ruminants [99–102]. Fungal placentitis due to *Aspergillus* species is an important cause of abortion in cattle, which generally occurs as an uncomplicated infection in the third trimester of pregnancy [103].

Secretory products of *Aspergillus* such as gliotoxin and tremorgens are toxic to cattle. *A. fumigatus*-contaminated silage was found to contain fumigaclavine A and C and several fumitremorgins [104]. Cattle-consuming silages containing these mycotoxins demonstrated signs of generalized deterioration, protein deficiency, malnutrition, diarrhea, irritability, abnormal behavior, and occasionally death. A neurological syndrome also has been observed in dairy cattle associated with consumption of patulin- and clavatol-contaminated foodstuffs produced by *A. clavatus* [105, 106].

4.3.7 Horses

Aspergillus species cause guttural pouch infections, keratomycosis, and pneumonia in horses [107–111]. The inflammation of the intestines, prolonged administration of antibiotics, immunosuppressive state of the host, and the presence of endocrinopathies and/or neoplasia often are predisposing factors thought to weaken the immune system of the horse, favoring penetration and growth of opportunistic fungi such as *Aspergillus* [112–115]. In horses, aspergillosis can be rapidly fatal when the infection invades the lungs. Pulmonary aspergillosis in horses may present with mild respiratory signs, tachypnea associated with adventitious lung or pleural sounds, and fever [116–118]. Nasal aspergillosis is another uncommon presentation of disease in horses with a wide range of clinical signs, characterized by dyspnea and nasal discharge [119]. In horses, keratomycosis is a relatively common disease, particularly in warm climates, usually following a corneal injury by plant material. In guttural pouch infections, inflammation of the cranial nerves leads to the development of dysphagia (with nasal discharge), laryngeal hemiplegia, facial paresis, or Horner's syndrome. Typical lesions are characterized by clearly demarcated, yellow-brown, necrotic tissue firmly adherent to the surface of the medial compartment of one guttural pouch [120, 121]. As long as the underlying structures (vessels and nerves) are not affected, the infection remains asymptomatic. However, the erosion of the internal carotid or maxillary artery leads to the sudden development of profuse epistaxis in a horse at rest [122, 123].

4.3.8 Marine Mammals

In marine mammals, stranded northern bottlenose whale (*Hyperoodon ampullatus*), killer whale (*Orcinus orca*), and stranded harbor porpoise (*Phocoena phocoena*) aspergillosis can be primary or secondary to any chronic infection, physiologic stress, or immunosuppression [124–126]. Pulmonary infections due to *A. fumigatus* or less frequently *A. niger* or *A. terreus* occur in cetaceans [127, 128]. Other organs may also be affected [124, 126, 129].

4.3.9 Nonhuman Primates

Aspergillosis also may occur in various nonhuman primate species, particularly in immunocompromised hosts as a posttransplant infectious complication [130, 131] or following metabolic disorders [132]. In addition, the underlying infection with Simian Immunodeficiency Virus (SIV) could be considered a risk factor [13, 133].

Environmental exposure to *Aspergillus* species is an important source of infection in nonhuman primates such as rhesus macaques (*Macaca mulatta*), cynomolgus macaques (*Macaca fascicularis*), and baboons (*Papio* spp.), which are extensively used in research models of solid organ transplantation [13, 130, 133, 134]. Similar to ruminants, when diagnosing pulmonary aspergillosis in primates, the possibility of concurrent involvement of tuberculosis should be considered [13].

4.4 Diagnosis and Species Identification

Since *Aspergillus* species are ubiquitous environmental airborne contaminants, a positive culture from a nonsterile clinical specimen does not indicate aspergillosis. However, isolation of *Aspergillus* species in cultures even from non-sterile specimens can be of diagnostic importance. The non-*fumigatus Aspergillus* species in the section *Fumigati* are very difficult to differentiate from *A. fumigatus* with certainty based solely on morphological and physiological characteristics since many characteristics are variable between strains. However, most of them grow slower than *A. fumigatus* at 37 °C, and thermotolerance can be used as a presumptive diagnostic criterion to differentiate them from *A. fumigatus*, the only *Aspergillus* species of clinical importance that grows readily at temperatures higher than 45 °C [135].

The current species recognition and identification of *Aspergillus* spp. are generally polyphasic based on morphology, physiology, and molecular data, and this requires skills and equipment that may not be universally available [136]. A two-step approach has been suggested in the clinical setting for molecular species identification of *Aspergillus.* The first step is to sequence ITS, the barcoding marker for the identification to the intersection level, followed by sequencing of partial β-tubulin for individual species identification within the sections [137].

The matrix-assisted laser desorption/ionization-time-of-flight (MALDI-TOF) mass spectrometry (MS)-based strategy has shown promising results to discriminate the intra-section level of clinically relevant species of *Aspergillus*, including the non-*fumigatus* "cryptic" species [138, 139]. Currently, four different MALDI-TOF MS benchtop platforms are commercially available for the routine identification of microbial pathogens in diagnostic laboratories; they include the Andromas (Andromas SAS, Paris, France), the Axima@Saramis (Shimadzu/Anagnos Tec, Duisburg, Germany), the Bruker Biotyper (Bruker Daltonics, Bremen, Germany), and the Vitek MS (bioMérieux, Marcy l'Etoile, France) systems. Of these, the latter two systems, Bruker Biotyper and Vitex MS, are the only systems that have been approved by the US Food and Drug Administration for identification of bacteria and yeasts [140]. The main concerns regarding the MALDI-TOF MS-based mold species identification is the molecular components similarities among sister species which may hinder the species differentiation.

In addition, *Aspergillus* galactomannan ELISA has shown promising result in the diagnosis of *Aspergillus* infection in small animals. Immunofluorescent procedures can also be used to identify hyphae in tissue sections [141].

4.5 Treatment

Many of the antifungal agents that are used in humans are also used in animals for the treatment of *Aspergillus* infections. These are the polyenes (e.g., amphotericin B) and azoles, including both the imidazoles and triazoles [142]. Table 4.1 summarizes the uses of various antifungals that have proved to be successful in various animals against *Aspergillus* infections.

4.6 Risk of Antifungal Resistance

Recent changes in the taxonomy of *Aspergillus* have major implications for our understanding of drug susceptibility profiles [143]. New sibling species of *A. fumigatus* exhibit in vitro susceptibility profiles that differ significantly from that of *A. fumigatus*. While acquired resistance is an emerging problem in *A. fumigatus* [144, 145], other *Aspergillus* species may be intrinsically resistant to amphotericin B and azoles [143]. MICs of *A. flavus* clinical isolates to amphotericin B are consistently twofold dilution steps higher than those of *A. fumigatus* [146]. Using CLSI methodology [147], *A. nidulans* was shown to have MIC values of 1–2 mg/l for amphotericin B, which is higher than commonly observed with *A. fumigatus* [148]. Itraconazole and voriconazole cross-resistance and variable susceptibilities for caspofungin were observed in vitro against *A. felis*, another possibly intrinsically resistant sibling of the *A. fumigatus* species complex [93, 149, 150]. In the section *Usti*, azoles are not active against *A. calidoustus* with MICs of ≥8 mg/l, while other classes of antifungal drugs also appear less active [151]. Resistance of *A. terreus* to amphotericin B is well known [152]. Based on susceptibility to azoles, three different susceptibility patterns were distinguished in the black aspergilli (section *Nigri*). Some isolates showed low-azole MICs and others high MICs, and a third group showed an uncommon paradoxical effect. However, these groups did not coincide with species boundaries, making it difficult to interpret as an intrinsic or acquired property [153].

Several studies have analyzed *Aspergillus* isolates from different animals for resistance to commercially available antimycotic agents, and many of them reported surprisingly high levels of azole resistance in *Aspergillus* species. There is no evidence of emerging azole resistance among *A. fumigatus* isolates from dogs and cats, and topical azole therapy should be effective against most isolates [154]. However, acquired resistance to itraconazole and voriconazole has been reported for avian *A. fumigatus* strains obtained from domestic and wild birds in Belgium and the Netherlands [155], where azole resistance is widespread both in clinical and environmental isolates. The source of these resistant isolates is unclear. However, two of the four resistant strains were isolated from birds that received itraconazole. This is important, and a fungicide-driven route of resistance selection in *A. fumigatus* may have implications for the management of aspergillosis in animals. Another possibility in these birds can also be considered as an indication of the presence of acquired resistance in the surrounding environment [156]. Of note, resistance to medical

Table 4.1 Recommended indications of antifungals against *Aspergillus* infections in animals

Antifungal drug	Animal species	Recommended dosages
Amphotericin B	Birds	Conventional AmB: IV 1.5 mg/kg q8h 3–5 days
		Nebulization: 15 min 1 mg/kg q24 h 10–14 days
	Dogs	Conventional AmB: 0.5 mg/kg IV q48 (slow infusion) to a cumulative dose of 4–8 mg/kg
		Liposomal AmB: 3 mg/kg/day IV, at a rate of more than 90–120 mg/kg, 3 times a week, up until 12 treatments
	Cats	Conventional AmB: 0.25 mg/kg IV q48 (slow infusion) to a cumulative dose of 4–8 mg/kg
		Liposomal AmB: 1 mg/kg/day IV, at a rate of more than 90–120 mg/kg, 3 times a week, up until 12 treatments
	Horses	Conventional AmB: 0.3 mg/kg IV for 3 consecutive days, and repeat after 24–48 h drug-free interval; long-term treatment needed
Ketoconazole	Birds	30 mg/kg q 12 h 14–30 days
Fluconazole	Dogs	Nasal infection: 1.25–10 mg/kg/day q12h, 50 mg/cat q12h, for 3–6 months
	Cats	CNS infection: 1.25–10 mg/kg/day q12h, 50 mg/cat q12h, for 3–6 months
Itraconazole	Birds	Treatment: 5–15 mg/kg q12h with food for 7–21 days or 10 mg/kg q24 h for 3 weeks
		Prevention: 10 mg/kg q24 h for 10 days, or 20 mg/kg q24 h, or 15–25 mg/kg/day, for 1 week
	Dogs	2.5 mg/kg q12h or 5 mg/kg q24h PO (give with food) for 15–30 days
	Cats	2.5 mg/kg q12h, or 5 mg/kg q24h, or 50–100 mg/cat PO (give with food), for 15–30 days
	Horses	2.5 mg/kg q12 or 5 mg/kg q24h PO
Voriconazole	Birds	10–18 mg/kg q12h
	Dogs	4–5 mg/kg q12h PO
	Cats	4–5 mg/kg q12h PO
	Horses	Systemic infection: 2–4 mg/kg q24h, or 3 mg/kg q24h PO, topical solution for keratitis and intracorneal administration
Posaconazole	Dogs	5–10 mg/kg q12–24 h
	Cats	5 mg/kg q24h
Clotrimazole	Birds (raptors)	Nebulization: 45 min q24h
	Dogs	Single or multiple intranasal local instillation, topical treatment
	Cats	Single or multiple intranasal local instillation, topical treatment
Miconazole	Birds	45 min/day in raptors
Enilconazole	Birds	Nebulization: 0.1 ml/kg for 30 min q24h (5 days on/2 days off)
		Disinfection of environment: flush with solutions as recommended for use in poultry houses
	Dogs	10 mg/kg q12h instilled into nasal sinus for 14 days (10% solution diluted 50/50 with water)
	Cats	10 mg/kg q12h instilled into nasal sinus for 14 days (10% solution diluted 50/50 with water)

triazoles may be associated with resistance selection to azole fungicides in the environment [157]. In humans, azole-resistant *Aspergillus* disease can be observed in patients without previous azole therapy, indicating that hosts inhale both azole-susceptible and azole-resistant *A. fumigatus* conidia [158].

In another study, Ziółkowska et al. (2014) investigated the in vitro susceptibility of *A. fumigatus* strains isolates obtained from oral cavity, lungs, and air sacs of healthy domestic geese, birds with aspergillosis, and from their environment. All of the strains were susceptible to enilconazole, itraconazole, and voriconazole but, irrespective of source, showed various degree of resistance to miconazole, clotrimazole (MIC_{90} = 16 μg/ml), and amphotericin B (MIC_{90} = 16 μg/ml) [159]. To assess the potential risk of azole-resistance emergence in avian farms where azole compounds were used for the control of avian mycoses, a drug susceptibility study including *A. fumigatus* isolates from birds and avian farms was conducted in France and Southern China [160]. A total number of 175 *A. fumigatus* isolates were analyzed. No resistant isolate was detected, and the distribution of MICs was similar for isolates collected in farms with or without azole chemoprophylaxis. For 61 randomly selected isolates, the full coding sequence of *Cyp51A* gene was determined to detect mutations. Nine amino acid alterations were found in the target enzyme, three of which were new mutations.

Of note, invasive infections caused by azole-resistant *A. fumigatus* are challenging to treat due to the lack of therapeutic options. In humans, combination of an azole with echinocandins or lipid formulations of amphotericin B can be used, and 5-flucytosine has also been recommended to be added to other therapies in patients with central nervous system infections caused by resistant isolates [161]. However, both antifungals have limitations, including toxicities, which may prohibit their long-term use in both humans and animals. Depending on the mechanism of resistance, higher doses of certain triazoles may be attempted, and there is a recent report of the successful treatment of invasive aspergillosis caused by an *A. fumigatus* isolate harboring a TR_{46}/Y121F/T289A mutation in a bottlenose dolphin with high-dose posaconazole [162]. Here, the oral solution of posaconazole was incorporated into gelatin capsules and administered with a goal of achieving trough concentrations of >3 mg/l, which was achieved after prolonged administration and resulted in clinical improvement.

4.7 Concluding Remarks

Overall, *Aspergillus* species are capable of causing different clinical diseases in a wide range of living organisms. *A. fumigatus* remains the most frequently causative fungal agent; however, role of the newly identified *Aspergillus* species in causing disease in animals remains unclear, considering the fact that clinical observations and evolution of infections caused by non-*fumigatus Aspergillus* species may differ significantly from those by *A. fumigatus*. To prevent invasive *Aspergillus* infection in animals, treatment should be based on eliminating predisposing factors such as improper husbandry, and appropriate samples should be taken for culture and susceptibility testing and selection of antifungal agents. To reduce risk of mycotoxins, preventive strategies to inhibit growth of *Aspergillus* species and to reduce mycotoxin loads in animal feed are required.

Potential Conflict of Interest This research was supported by the Intramural Research Program of the National Institutes of Health, Clinical Center, Department of Laboratory Medicine.

References

1. Kwon-Chung KJ, Sugui JA (2013) *Aspergillus fumigatus*-what makes the species a ubiquitous human fungal pathogen? PLoS Pathog 9:e1003743
2. Pitt JI (1994) The current role of *Aspergillus* and *Penicillium* in human and animal health. J Med Vet Mycol 32:17–32
3. Heitman J (2011) Microbial pathogens in the fungal kingdom. Fungal Biol Rev 25:48–60
4. Seyedmousavi S, Guillot J, Arne P et al (2015) *Aspergillus* and aspergilloses in wild and domestic animals: a global health concern with parallels to human disease. Med Mycol 53:765–797
5. Lamoth F (2016) *Aspergillus fumigatus*-related species in clinical practice. Front Microbiol 7:683
6. Peterson SW, Varga J, Frisvad JC et al (2008) Phylogeny and subgeneric taxonomy of *Aspergillus*. Wageningen Academic Publishers, Wageningen
7. Sugui JA, Kwon-Chung KJ, Juvvadi PR et al (2014) *Aspergillus fumigatus* and related species. Cold Spring Harb Perspect Med 5:a019786
8. Seyedmousavi S, Lionakis MS, Parta M et al (2018) Emerging *Aspergillus* species almost exclusively associated with primary immunodeficiencies. Open Forum Infect Dis 5:ofy213
9. Greub G, Bille J (1998) *Aspergillus* species isolated from clinical specimens: suggested clinical and microbiological criteria to determine significance. Clin Microbiol Infect 4:710–716
10. Latge JP (1999) *Aspergillus fumigatus* and aspergillosis. Clin Microbiol Rev 12:310–350
11. Seyedmousavi S, Bosco SDG, de Hoog S et al (2018) Fungal infections in animals: a patchwork of different situations. Med Mycol 56:165–187
12. Zmeili OS, Soubani AO (2007) Pulmonary aspergillosis: a clinical update. QJM 100:317–334
13. Yamauchi H, Takai Y, Yamasaki H et al (2011) Thoracic mass in a cynomolgus macaque (*Macaca fascicularis*). Vet Pathol 48:E1–E5
14. Denning DW (1998) Invasive aspergillosis. Clin Infect Dis 26:781–803. quiz 804-805
15. Marr KA, Carter RA, Boeckh M et al (2002) Invasive aspergillosis in allogeneic stem cell transplant recipients: changes in epidemiology and risk factors. Blood 100:4358–4366
16. Lionakis MS, Kontoyiannis DP (2003) Glucocorticoids and invasive fungal infections. Lancet 362:1828–1838
17. Fukuda T, Boeckh M, Carter RA et al (2003) Risks and outcomes of invasive fungal infections in recipients of allogeneic hematopoietic stem cell transplants after nonmyeloablative conditioning. Blood 102:827–833
18. Beauté J, Obenga G, Le Mignot L et al (2011) Epidemiology and outcome of invasive fungal diseases in patients with chronic granulomatous disease: a multicenter study in France. Pediatr Infect Dis J 30:57–62
19. Romani L (2011) Immunity to fungal infections. Nat Rev Immunol 11:275–288
20. Kumar A, Teuber SS, Gershwin ME (2006) Current perspectives on primary immunodeficiency diseases. Clin Dev Immunol 13:223–259
21. Antachopoulos C (2010) Invasive fungal infections in congenital immunodeficiencies. Clin Microbiol Infect 16:1335–1342

22. Almyroudis NG, Holland SM, Segal BH (2005) Invasive aspergillosis in primary immunodeficiencies. Med Mycol 43:S247–S259
23. Segal BH, Holland SM (2000) Primary phagocytic disorders of childhood. Pediatr Clin N Am 47:1311–1338
24. Pana ZD, Farmaki E, Roilides E (2014) Host genetics and opportunistic fungal infections. Clin Microbiol Infect 20:1254–1264
25. Denning DW, Pleuvry A, Cole DC (2013) Global burden of allergic bronchopulmonary aspergillosis with asthma and its complication chronic pulmonary aspergillosis in adults. Med Mycol 51:361–370
26. Agarwal R (2011) Severe asthma with fungal sensitization. Curr Allergy Asthma Rep 11:403–413
27. Denning DW, Pleuvry A, Cole DC (2011) Global burden of chronic pulmonary aspergillosis as a sequel to pulmonary tuberculosis. Bull World Health Organ 89:864–872
28. Bongomin F, Gago S, Oladele RO et al (2017) Global and multi-national prevalence of fungal diseases-estimate precision. J Fungi (Basel) 3:E57
29. Smith NL, Denning DW (2011) Underlying conditions in chronic pulmonary aspergillosis including simple aspergilloma. Eur Respir J 37:865–872
30. Kosmidis C, Denning DW (2015) The clinical spectrum of pulmonary aspergillosis. Thorax 70:270–277
31. Lowes D, Al-Shair K, Newton PJ et al (2017) Predictors of mortality in chronic pulmonary aspergillosis. Eur Respir J 49:1601062
32. Dominguez MC, Chavez G, Trigo FJ et al (2001) Concurrent cholangiocarcinoma, peritonitis, paratuberculosis, and aspergillosis in a goat. Can Vet J 42:884–885
33. Jensen HE, Schønheyder H, Jorgensen JB (1990) Intestinal and pulmonary mycotic lymphadenitis in cattle. J Comp Pathol 102:345–355
34. Kim K, Harvell CD (2004) The rise and fall of a six-year coral-fungal epizootic. Am Nat 164:S52–S63
35. Paddack MJ, Reynolds JD, Aguilar C et al (2009) Recent region-wide declines in Caribbean reef fish abundance. Curr Biol 19:590–595
36. Nagelkerken I, Grol MG, Mumby PJ (2012) Effects of marine reserves versus nursery habitat availability on structure of reef fish communities. PLoS One 7:e36906
37. Ein-Gil N, Ilan M, Carmeli S et al (2009) Presence of *Aspergillus sydowii*, a pathogen of gorgonian sea fans in the marine sponge *Spongia obscura*. ISME J 3:752–755
38. Altizer S, Ostfeld RS, Johnson PT et al (2013) Climate change and infectious diseases: from evidence to a predictive framework. Science 341:514–519
39. Rypien KL, Andras JP, Harvell CD (2008) Globally panmictic population structure in the opportunistic fungal pathogen *Aspergillus sydowii*. Mol Ecol 17:4068–4078
40. Bailey L (1963) Infectious Diseases of the Honeybee. Land Books Ltd, London, p 176
41. Gilliam M, Vandenberg JD (1997) Fungi. In: Morse RA, Flottum K (eds) Honey bee pests, predators & diseases, 3rd edn. A. I. Root Company, Medina, pp 81–110
42. Burnside CE (1930) Fungous diseases of the honey bee. US Department of Agriculture Technical Bulletin, p 149
43. Lionakis MS, Kontoyiannis DP (2012) *Drosophila melanogaster* as a model organism for invasive aspergillosis. Methods Mol Biol 845:455–468
44. Arvanitis M, Glavis-Bloom J, Mylonakis E (2013) Invertebrate models of fungal infection. Biochim Biophys Acta 1832:1378–1383
45. Jacobson ER, Cheatwood JL, Maxwell LK (2000) Mycotic diseases of reptiles. Semin Avian Exot Pet Med 9:94–101
46. Girling SJ, Fraser MA (2009) Treatment of *Aspergillus* species infection in reptiles with itraconazole at metabolically scaled doses. Vet Rec 165:52–54
47. Frye FL, Dutra FR (1974) Mycotic granulomata involving the forefeet of a turtle. Vet Med Small Anim Clin 69:1554–1556
48. Buenviaje GN, Ladds PW, Melville L et al (1994) Disease-husbandry associations in farmed crocodiles in Queensland and the Northern Territory. Aust Vet J 71:165–173

49. Tappe JP, Chandler FW, Liu SK et al (1984) Aspergillosis in two San Esteban chuckwallas. J Am Vet Med Assoc 185:1425–1428
50. Myers DA, Isaza R, Ben-Shlomo G et al (2009) Fungal keratitis in a gopher tortoise (*Gopherus polyphemus*). J Zoo Wildl Med 40:579–582
51. Miller DL, Radi ZA, Stiver SL et al (2004) Cutaneous and pulmonary mycosis in green anacondas (*Euncectes murinus*). J Zoo Wildl Med 35:557–561
52. Arne P, Thierry S, Wang D et al (2011) *Aspergillus fumigatus* in Poultry. Int J Microbiol 2011:746356
53. Beernaert LA, Pasmans F, Van Waeyenberghe L et al (2010) *Aspergillus* infections in birds: a review. Avian Pathol 39:325–331
54. Mc Dougle HC, Vaught RW (1968) An epizootic of aspergillosis in Canada geese. J Wildl Manag 32:577–578
55. Martin MP, Bouck KP, Helm J et al (2007) Disseminated *Aspergillus flavus* infection in broiler breeder pullets. Avian Dis 51:626–631
56. Akan M, Haziroglu R, Ilhan Z et al (2002) A case of aspergillosis in a broiler breeder flock. Avian Dis 46:497–501
57. Wang DY, Hadj-Henni L, Thierry S et al (2012) Simple and highly discriminatory VNTR-based multiplex PCR for tracing sources of *Aspergillus flavus* isolates. PLoS One 7:e44204
58. Sato Y, Itagaki T (2010) Fungal airsacculitis associated with multiple helminth infestations in a black-eared kite (*Milvus migrans*). Avian Dis 54:965–968
59. Tarello W (2011) Etiologic agents and diseases found associated with clinical aspergillosis in falcons. Int J Microbiol 2011:176963
60. Verstappen FALM, Dorrestein GM (2005) Aspergillosis in Amazon parrots after corticosteroid therapy for smoke-inhalation injury. J Avian Med Surg 19:138–141
61. Hoerr FJ (2010) Clinical aspects of immunosuppression in poultry. Avian Dis 5:2–15
62. Alvarez-Perez S, Mateos A, Dominguez L et al (2010) Polyclonal *Aspergillus fumigatus* infection in captive penguins. Vet Microbiol 144:444–449
63. Olias P, Gruber AD, Hafez HM et al (2011) Molecular epidemiology and virulence assessment of *Aspergillus fumigatus* isolates from white stork chicks and their environment. Vet Microbiol 148:348–355
64. Burco JD, Etienne KA, Massey JG et al (2012) Molecular sub-typing suggests that the environment of rehabilitation centers may be a potential source of *Aspergillus fumigatus* infecting rehabilitating seabirds. Med Mycol 50:91–98
65. Lair-Fulleringer S, Guillot J, Desterke C (2003) Differentiation between isolates of *Aspergillus fumigatus* from breeding turkeys and their environment by genotyping with microsatellite markers. J Clin Microbiol 41:1798–1800
66. Beytut E, Ozcan K, Erginsoy S (2004) Immunohistochemical detection of fungal elements in the tissues of goslings with pulmonary and systemic aspergillosis. Acta Vet Hung 52:71–84
67. Kunkle RA, Sacco RE (1998) Susceptibility of convalescent turkeys to pulmonary aspergillosis. Avian Dis 42:787–790
68. Throne Steinlage SJ, Sander JE, Brown TP et al (2003) Disseminated mycosis in layer cockerels and pullets. Avian Dis 47:229–233
69. Fedde MR (1998) Relationship of structure and function of the avian respiratory system to disease susceptibility. Poult Sci 77:1130–1138
70. Brown RE, Brain JD, Wang N (1997) The avian respiratory system: a unique model for studies of respiratory toxicosis and for monitoring air quality. Environ Health Perspect 105:188–200
71. Buehler DM, Tieleman BI, Piersma T (2010) How do migratory species stay healthy over the annual cycle? A conceptual model for immune function and for resistance to disease. Integr Comp Biol 50:346–357
72. Sharman MJ, Mansfield CS (2012) Sinonasal aspergillosis in dogs: a review. J Small Anim Pract 53:434–444
73. Barrs VR, Halliday C, Martin P et al (2012) Sinonasal and sino-orbital aspergillosis in 23 cats: aetiology, clinicopathological features and treatment outcomes. Vet J 191:58–64

74. Barrs VR, Talbot JJ (2014) Feline aspergillosis. Vet Clin North Am Small Anim Pract 44:51–73
75. Sharp NJH, Harvey CE, Sullivan M (1991) Canine nasal aspergillosis/penicilliosis. Compend Contin Educ Pract Vet 13:41–49
76. Sharman M, Paul A, Davies D et al (2010) Multi-centre assessment of mycotic rhinosinusitis in dogs: a retrospective study of initial treatment success (1998 to 2008). J Small Anim Pract 51:423–427
77. Tasker S, Knottenbelt CM, Munro EA et al (1999) Aetiology and diagnosis of persistent nasal disease in the dog: a retrospective study of 42 cases. J Small Anim Pract 40:473–478
78. Zhang S, Corapi W, Quist E et al (2012) *Aspergillus versicolor*, a new causative agent of canine disseminated aspergillosis. J Clin Microbiol 50:187–191
79. Talbot JJ, Johnson LR, Martin P et al (2014) What causes canine sino-nasal aspergillosis? A molecular approach to species identification. Vet J 200:17–21
80. Adamama-Moraitou KK, Pardali D, Day MJ et al (2011) *Aspergillus fumigatus* broncho-pneumonia in a Hellenic Shepherd dog. J Am Anim Hosp Assoc 47:e13–e18
81. Kim SH, Yong HC, Yoon JH et al (2003) *Aspergillus niger* pulmonary infection in a dog. J Vet Med Sci 65:1139–1140
82. Southard C (1987) Bronchopulmonary aspergillosis in a dog. J Am Vet Med Assoc 190:875–877
83. Schultz RM, Johnson EG, Wisner ER et al (2008) Clinicopathologic and diagnostic imaging characteristics of systemic aspergillosis in 30 dogs. J Vet Intern Med 22:851–859
84. Coyner K (2010) Otomycosis due to *Aspergillus* spp. in a dog: case report and literature review. Vet Dermatol 21:613–618
85. Barachetti L, Mortellaro CM, Di Giancamillo M et al (2009) Bilateral orbital and nasal aspergillosis in a cat. Vet Ophthalmol 12:176–182
86. Hamilton HL, Whitley RD, McLaughlin SA (2000) Exophthalmos secondary to aspergillosis in a cat. J Am Anim Hosp Assoc 36:343–347
87. Kano R, Itamoto K, Okuda M et al (2008) Isolation of *Aspergillus udagawae* from a fatal case of feline orbital aspergillosis. Mycoses 51:360–361
88. Tell LA (2005) Aspergillosis in mammals and birds: impact on veterinary medicine. Med Mycol 43:S71–S73
89. Smith LN, Hoffman SB (2010) A case series of unilateral orbital aspergillosis in three cats and treatment with voriconazole. Vet Ophthalmol 13:190–203
90. Little S, Bienzle D, Carioto L et al (2011) Feline leukemia virus and feline immunodeficiency virus in Canada: recommendations for testing and management. Can Vet J 52:849–855
91. Barrs VR, Ujvari B, Dhand NK et al (2015) Detection of *Aspergillus*-specific antibodies by agar gel double immunodiffusion and IgG ELISA in feline upper respiratory tract aspergillosis. Vet J 203:285–289
92. Labelle AL, Hamor RE, Barger AM et al (2009) *Aspergillus flavus* keratomycosis in a cat treated with topical 1% voriconazole solution. Vet Ophthalmol 12:48–52
93. Barrs VR, van Doorn TM, Houbraken J et al (2013) *Aspergillus felis* sp. nov., an emerging agent of invasive aspergillosis in humans, cats, and dogs. PLoS One 8:e64871
94. Puntenney SB, Wang Y, Forsberg NE (2003) Mycotic infections in livestock: recent insights and studies on etiology, diagnostics and prevention of hemorrhagic bowel syndrome. Tuscon, University of Arizona, Department of Animal Science, Southwest Nutrition and Management Conference, p 4963
95. Jensen HE, Olsen SN, Aalbaek B (1994) Gastrointestinal aspergillosis and zygomycosis of cattle. Vet Pathol 31:28–36
96. Muntz FH (1999) Oxalate-producing pulmonary aspergillosis in an alpaca. Vet Pathol 36:631–632
97. Pickett JP, Moore CP, Beehler BA et al (1985) Bilateral chorioretinitis secondary to disseminated aspergillosis in an alpaca. J Am Vet Med Assoc 187:1241–1243
98. Severo LC, Bohrer JC, Geyer GR et al (1989) Invasive aspergillosis in an alpaca (*Lama pacos*). J Med Vet Mycol 27:193–195

99. Pérez V, Corpa JM, Garcia Marin JF et al (1999) Generalized aspergillosis in dairy sheep. Zentralbl Veterinarmed B 46:613–621
100. Pérez V, Corpa JM, Garcia Marin JF (1998) Mammary and systemic aspergillosis in dairy sheep. Vet Pathol 35:235–240
101. Las Heras A, Dominguez L, Lopez I et al (2000) Intramammary *Aspergillus fumigatus* infection in dairy ewes associated with antibiotic dry therapy. Vet Rec 147:578–580
102. Garcia ME, Duran C, Cruzado M et al (2004) Evaluation of molecular and immunological techniques for the diagnosis of mammary aspergillosis in ewes. Vet Microbiol 98:17–21
103. Knudtson WU, Kirkbride CA (1992) Fungi associated with bovine abortion in the northern plains states (USA). J Vet Diagn Investig 4:181–185
104. Cole RJ, Kirksey JW, Dorner JW et al (1977) Mycotoxins produced by *Aspergillus fumigatus* species isolated from molded silage. J Agric Food Chem 25:826–830
105. Finnie JW, Windsor PA, Kessell AE (2011) Neurological diseases of ruminant livestock in Australia. II: toxic disorders and nutritional deficiencies. Aust Vet J 89:247–253
106. Loretti AP, Colodel EM, Driemeier D et al (2003) Neurological disorder in dairy cattle associated with consumption of beer residues contaminated with *Aspergillus clavatus*. J Vet Diagn Investig 15:123–132
107. Dobesova O, Schwarz B, Velde K (2012) Guttural pouch mycosis in horses: a retrospective study of 28 cases. Vet Rec 171:561
108. Blomme E, Del Piero F, La Perle KMD et al (1998) Aspergillosis in horses: a review. Equine Vet Educ 10:86–93
109. Sansom J, Featherstone H, Barnett KC (2005) Keratomycosis in six horses in the United Kingdom. Vet Rec 156:13–17
110. Wada S, Hobo S, Ode H et al (2013) Equine keratomycosis in Japan. Vet Ophthalmol 16:1–9
111. Scotty N (2005) Equine keratomycosis. Clin Tech Equine Pract 4:29–46
112. Guillot J, Collobert C, Gueho E et al (1997) *Emericella nidulans* as an agent of guttural pouch mycosis in a horse. J Med Vet Mycol 35:433–435
113. Carrasco L, Mendez A, Jensen HE (1996) Chronic bronchopulmonary aspergillosis in a horse with Cushing's syndrome. Mycoses 39:443–447
114. Hattel AL, Drake TR, Anderholm BJ et al (1991) Pulmonary aspergillosis associated with acute enteritis in a horse. J Am Vet Med Assoc 199:589–590
115. Lepage OM, Perron MF, Cadore JL (2004) The mystery of fungal infection in the guttural pouches. Vet J 168:60–64
116. Sweeney CR, Habecker PL (1999) Pulmonary aspergillosis in horses: 29 cases (1974–1997). J Am Vet Med Assoc 214:808–811
117. Pace LW, Wirth NR, Foss RR et al (1994) Endocarditis and pulmonary aspergillosis in a horse. J Vet Diagn Investig 6:504–506
118. Johnson PJ, Moore LA, Mrad DR et al (1999) Sudden death of two horses associated with pulmonary aspergillosis. Vet Rec 145:16–20
119. Kendall A, Brojer J, Karlstam E et al (2008) Enilconazole treatment of horses with superficial *Aspergillus* spp. rhinitis. J Vet Intern Med 22:1239–1242
120. Cook WR (1966) Observations on the aetiology of epistaxis and cranial nerve paralysis in the horse. Vet Rec 78:396–406
121. Markus R, Deegen E, Drommer W et al (2005) Guttural Pouch Mycosis. J Equine Vet Sci 25:150–156
122. Lane JG (1989) The management of guttural pouch mycosis. Equine Vet J 21:321–324
123. Ludwig A, Gatineau S, Reynaud MC et al (2005) Fungal isolation and identification in 21 cases of guttural pouch mycosis in horses (1998–2002). Vet J 169:457–461
124. Dagleish MP, Foster G, Howie FE et al (2008) Fatal mycotic encephalitis caused by *Aspergillus fumigatus* in a northern bottlenose whale (*Hyperoodon ampullatus*). Vet Rec 163:602–604
125. Barley J, Foster G, Reid B et al (2007) Encephalitis in a northern bottlenose whale. Vet Rec 160:452

126. Abdo W, Kawachi T, Sakai H et al (2012) Disseminated mycosis in a killer whale (*Orcinus orca*). J Vet Diagn Investig 24:211–218
127. Reidarson TH, Mcbain JF, Dalton LM (1999) Diagnosis and treatment of fungal infections in marine mammals. In: Fowler ME, Miller RE (eds) Zoo and wild animal medicine current therapy, 4th edn. WB Saunders, London, pp 478–485
128. Migaki G, Jones SR (1983) Mycotic diseases in marine mammals. In: Howard EB (ed) Pathobiology of marine mammal diseases, CRC Press, vol 2. Boca Raton, FL, pp 1–27
129. Dagleish MP, Patterson IA, Foster G et al (2006) Intracranial granuloma caused by asporogenic *Aspergillus fumigatus* in a harbour porpoise (*Phocoena phocoena*). Vet Rec 159:458–460
130. Haustein SV, Kolterman AJ, Sundblad JJ et al (2008) Nonhuman primate infections after organ transplantation. ILAR J 49:209–219
131. Starzl TE, Fung J, Tzakis A et al (1993) Baboon-to-human liver transplantation. Lancet 341:65–71
132. Jurczynski K, Gruber-Dujardin E, Widmer D et al (2012) Invasive aspergillosis in a putty-nosed monkey (*Cercopithecus nictitans*) with adrenocortical Cushing's syndrome. J Med Primatol 41:172–175
133. Wilk J, Lewis A, Lukas V (2008) Dermatitis in a rhesus macaque (*Macaca mulatta*) experimentally infected with simian immunodeficiency virus. J Med Primatol 37:25–28
134. Rubin RH (2002) Overview: pathogenesis of fungal infections in the organ transplant recipient. Transpl Infect Dis 4:12–17
135. Chang YC, Tsai HF, Karos M et al (2004) *THTA*, a thermotolerance gene of *Aspergillus fumigatus*. Fungal Genet Biol 41:888–896
136. Geiser DM, Klich MA, Frisvad JC et al (2007) The current status of species recognition and identification in *Aspergillus*. Stud Mycol 59:1–10
137. Balajee SA, Houbraken J, Verweij PE et al (2007) *Aspergillus* species identification in the clinical setting. Stud Mycol 59:39–46
138. Alanio A, Beretti JL, Dauphin B et al (2011) Matrix-assisted laser desorption ionization time-of-flight mass spectrometry for fast and accurate identification of clinically relevant *Aspergillus* species. Clin Microbiol Infect 17:750–755
139. Sleiman S, Halliday CL, Chapman B et al (2016) Performance of matrix-assisted laser desorption ionization-time of flight mass spectrometry for identification of *Aspergillus*, *Scedosporium*, and *Fusarium* spp. in the australian clinical setting. J Clin Microbiol 54:2182–2186
140. Sanguinetti M, Posteraro B (2017) Identification of molds by matrix-assisted laser desorption ionization-time of flight mass spectrometry. J Clin Microbiol 55:369–379
141. Thirion-Delalande C, Guillot J, Jensen HE et al (2005) Disseminated acute concomitant aspergillosis and mucormycosis in a pony. J Vet Med A Physiol Pathol Clin Med 52:121–124
142. Seyedmousavi S, Wiederhold NP, Ebel F et al (2018) Antifungal use in veterinary practice and emergence of resistance. In: Seyedmousavi S, de Hoog GS, Guillot J, Verweij PE (eds) Emerging and epizootic fungal infections in animals. Springer, Cham, pp 359–402
143. Van Der Linden JW, Warris A, Verweij PE (2011) *Aspergillus* species intrinsically resistant to antifungal agents. Med Mycol 49:S82–S89
144. Verweij PE, Howard SJ, Melchers WJ et al (2009) Azole-resistance in *Aspergillus*: proposed nomenclature and breakpoints. Drug Resist Updat 12:141–147
145. Verweij PE, Mellado E, Melchers WJ (2007) Multiple-triazole-resistant aspergillosis. N Engl J Med 356:1481–1483
146. Gomez-Lopez A, Garcia-Effron G, Mellado E et al (2003) *In vitro* activities of three licensed antifungal agents against spanish clinical isolates of *Aspergillus* spp. Antimicrob Agents Chemother 47:3085–3088
147. CLSI (2008) Reference method for broth dilution antifungal susceptibility testing of filamentous fungi; Approved standard-Second Edition. CLSI Document. M38-A2., vol 28 no.16, Clinical and Laboratory Standards Institute, Wane, PA

148. Kontoyiannis DP, Lewis RE, May GS et al (2001) *Aspergillus nidulans* is frequently resistant to amphotericin B. Mycoses 45:406–407
149. Pelaez T, Alvarez-Perez S, Mellado E et al (2013) Invasive aspergillosis caused by cryptic *Aspergillus* species: a report of two consecutive episodes in a patient with leukaemia. J Med Microbiol 62:474–478
150. Coelho D, Silva S, Vale-Silva L et al (2011) *Aspergillus viridinutans*: an agent of adult chronic invasive aspergillosis. Med Mycol 49:755–759
151. Varga J, Houbraken J, Van Der Lee HA et al (2008) *Aspergillus calidoustus* sp. nov., causative agent of human infections previously assigned to *Aspergillus ustus*. Eukaryot Cell 7:630–638
152. Lass-Florl C, Alastruey-Izquierdo A, Cuenca-Estrella M et al (2009) *In vitro* activities of various antifungal drugs against *Aspergillus terreus*: Global assessment using the methodology of the European committee on antimicrobial susceptibility testing. Antimicrob Agents Chemother 53:794–795
153. Alcazar-Fuoli L, Mellado E, Alastruey-Izquierdo A et al (2009) Species identification and antifungal susceptibility patterns of species belonging to *Aspergillus* section *Nigri*. Antimicrob Agents Chemother 53:4514–4517
154. Talbot JJ, Kidd SE, Martin P et al (2015) Azole resistance in canine and feline isolates of *Aspergillus fumigatus*. Comp Immunol Microbiol Infect Dis 42:37–41
155. Beernaert LA, Pasmans F, Van Waeyenberghe L et al (2009) Avian *Aspergillus fumigatus* strains resistant to both itraconazole and voriconazole. Antimicrob Agents Chemother 53:2199–2201
156. Bromley MJ, van Muijlwijk G, Fraczek MG et al (2014) Occurrence of azole-resistant species of *Aspergillus* in the UK environment. J Glob Antimicrob Resist 2:276–279
157. Verweij PE, Snelders E, Kema GH (2009) Azole resistance in *Aspergillus fumigatus*: a side-effect of environmental fungicide use? Lancet Infect Dis 9:789–795
158. ECDC (2013) Risk assessment on the impact of environmental usage of triazoles on the development and spread of resistance to medical triazoles in *Aspergillus* species. Stockholm
159. Ziolkowska G, Tokarzewski S, Nowakiewicz A (2014) Drug resistance of *Aspergillus fumigatus* strains isolated from flocks of domestic geese in Poland. Poult Sci 93:1106–1112
160. Wang DY, Gricourt M, Thierry S et al (2014) Mutations in the cyp51A gene and susceptibility to itraconazole in *Aspergillus fumigatus* isolated from avian farms in France and China. Poult Sci 93:12–15
161. Verweij PE, Ananda-Rajah M, Andes D et al (2015) International expert opinion on the management of infection caused by azole-resistant *Aspergillus fumigatus*. Drug Resist Updat 21-22:30–40
162. Bunskoek PE, Seyedmousavi S, Gans SJ et al (2017) Successful treatment of azole-resistant invasive aspergillosis in a bottlenose dolphin with high-dose posaconazole. Med Mycol Case Rep 16:16–19

Some Clinically Significant Genera of Dematiaceous Hyphomycetes: An Update

5

Shanker Mohan Singh and Richa Gumasta

Abstract

Dematiaceous (melanized or phaeoid) fungi are common in the environment. However, clinical diseases caused by them are uncommon. Despite their rarity, they are being increasingly recognized as causal agents of disease in man and animal. There is growing awareness among medical fraternity about the clinical significance of these melanized fungi in medical practice. In this article, we have reviewed some clinically significant publications reporting some emerging phaeoid genera up to 2017. The genera reviewed are *Alternaria*, *Aureobasidium*, *Bipolaris*, *Exserohilum*, *Curvularia*, *Ochroconis*, *Exophiala*, *Phialophora*, *Chaetomium*, *Neoscytalidium*, *Leptosphaeria*, *Microascus*, *Lecythophora*, *Phaeoacremonium*, *Scedosporium*, *Veronaea*, *Fonsecaea*, *Wallemia sebi*, *Verruconis*, etc. In addition, laboratory diagnosis and future areas of research have also been dealt along with conclusion.

Keywords

Clinical significance · Phaeoid · Genera · Review

5.1 Introduction

In the recent years, there has been spectacular decrease in incidence and mortality due to microbial diseases largely because of effective preventive measures based on sound knowledge of pathogenesis and immunology of these diseases particularly in the west. It is, however, ironical that in such an era when so many advances have

---

S. M. Singh (✉) · R. Gumasta
Medical Mycology Laboratory, Department of Biological Science, RD University, Jabalpur, India

Fungal Disease Diagnostic and Research Centre, Jabalpur, India

K. Singh, N. Srivastava (eds.), *Recent Trends in Human and Animal Mycology*,
https://doi.org/10.1007/978-981-13-9435-5_5

been made in the rapid cure of microbial as well as nonmicrobial diseases, certain opportunistic mycotic infections have increased on the global basis. This notable growth in the prevalence and incidence of opportunistic infections has been attributed to direct consequences of these medical advances which directly or indirectly affect immune system of man. Some of these are broad-spectrum antibiotics, adrenal corticosteroids, immunosuppressive drugs for organ transplantation, chemotherapy for cancer, prolonged use of indwelling catheters, prosthetic heart valve surgery, open heart surgery, renal transplantation, pacemaker wires, etc. Apart from these, a number of conditions may promote the development of fungal infections in the compromised host. These include malignant diseases, blood dyscrasias, diabetes mellitus, endocrinopathies, autoimmune disorder, AIDS, etc.

Out of numerous fungi in nature, infections caused by dematiaceous genera have been and continue to be unusual clinical phenomenon, particularly with increasing number of immunocompromised patients resulting in a resurgence of interest in these pathogens. Periodically, some reviews [1–9] have been written on the infections caused by dematiaceous fungi. Since many new findings have accumulated, a review of the current status of available information on some emerging medically important genera of the clinical-type phaeohyphomycosis was considered worth undertaking. Such a review would serve to bring together widely scattered data for critical analysis and would point out the unexplored areas of research.

5.1.1 *Alternaria*

Alternaria infection has been mostly described in Europe. In France, 23 cases have been published followed by 6 cases from Japan and 5 from the USA [4, 10–12]. In India, only three cases have been reported [13, 14]. This fungal infection affects both men (22 cases) and women (17 cases) aged 8 months to more than 80 years. They can be encountered in rural populations, especially in people having occupations that make them an easy prey to traumas. In few cases, no predisposition or deficiency likely to have caused the disease was encountered [4]. Inversely, in other patients, there is a clue to different types of predispositions like kidney transplantation, Cushing's syndrome [4], lymphoma [15], and Hodgkin's disease plus other chronic diseases like lupus, diabetes, etc. [4]. Table 5.1 shows cases described in different parts of the world. In most of the cases, the infection was confined to the dermal region with the fungal element spreading to the dermal layer accompanied by inflammatory reaction followed by an epidermal infection in which the hyphae were confined to stratum corneum [4]. Cases of sinusitis and nail infection are found to be rare [4, 13]. It is interesting to note that no systemic infection has been reported by *Alternaria* spp. so far. The clinical pictures described are variable. Typically, the initial sign is a macula gradually converting to usually squamous, sometimes crust like, indurated, verrucous, purplish erythematous nodule. The nodules can be either single or scattered all over the body or clustered at a site [4]. In maximum number of cases, elbow, forearm, wrist, and back of the hands were involved [4, 16]. Knees and pretibial region were the next in the order of frequency [15]. Interestingly, in a

Table 5.1 Clinical, epidemiological, and therapeutic data of some patients infected with species of different genera of dematiaceous hyphomycetes reported up to 2017

Fungal pathogen	Total cases	Age (years)	Sex M:F	Disease entity	Underlying disease	Therapy	References
Alternaria alternata	18	7–78	10:8	Erythematous nodular, ulcerated lesion on the extremities	Hodgkin's disease, Cushing's syndrome, lymphoma	Tropical pimaricin. Amp-B, miconazole, ketoconazole, Surgical excision	[4, 16]
A. chlamydospora	02	25, 50	1:1	Erythematous papules on legs, breast, back, and abdomen	–	–	[13]
A. chartarum	01	63	0:1	01 cm nodule on nasal septum	Kidney transplant	Oral ketoconazole for 3 months	[266]
A. dianthicola	01	56	1:0	–	Immunodeficient	Excision, cured	[4]
A. humicola	01	53	0:1	Onychomycosis	Rheumatism and diabetes mellitus	Antifungal are ineffective	[14]
A. infectoria	01	74	1:0	Multiple skin lesion and pulmonary infiltrate	Heart transplant	Posaconazole	[20]
A. rosae	01	66	1:0	Nodule on thumb	Stem cell transplant	Posaconazole	[22]
A. malorum	01	27	1:0	Subcutaneous necrotic lesions, disseminated over time	Immunocompetent	Amp-B and itraconazole	[21]
A. pleuriseptata	01	32	1:0	Onychomycosis	–	–	[14]
A. stemphylioides	01	87	1:0	Granulomatous dermatitidis	Cushing disease trauma	–	[15]
A. tenuissima	18	7–80	8:9	Nodular, sometimes ulcerative in the extremities	Lymphocytic lymphoma, leukemia, lupus, diabetes, asthma	Ketoconazole curve and failure, curve with amp-B	[10, 265]

(continued)

Table 5.1 (continued)

Fungal pathogen	Total cases	Age (years)	Sex M:F	Disease entity	Underlying disease	Therapy	References
Aureobasidium pullulans		23		Infection of the lymphatic system	Erythema nodosum leprosum	Cefepime and vancomycin	[268]
A. melanogenum	01			Superficial phaeohyphomycosis			[249]
Bipolaris spicifera	10	5–76	1: 0	Necrotizing pneumonia, fungemia, vasculitis, meningoencephalitis, sinusitis, peritonitis, brain abscess	Leukemia, breast carcinoma, cardiac transplant, allergic rhinitis	Miconazole, amp-B, and ketoconazole, excision, mixed results death and cure	[23, 269]
B. spicifera	01	56	1:0	Plaque papillomatosis on left foot	–	Itraconazole and terbinafine	[271]
B. hawaiiensis	05	15–46	1:4	Meningoencephalitis, nasal phaeohyphomycosis, cutaneous lesion	Lymphocarcinoma, diabetes	Cure with surgical excision +0.03% nystatin solution	[23]
B. australiensis	01	65	0:1	Skin lesion	Viral vascular dermatitidis	Acyclovir 200 mg daily for 7 days	[269]
Botryosphaeria dothidea	01	82	1:0	Black-pigmented area on the right thumbnail, possibility of malignant melanoma	Chronic heart failure and hemiparesis	10% EFCZ solution	[273]
Chaetomium globosum	01	14	0:1	Phaeohyphomycotic painful erythema, necrosis of face	–	–	[274]
Chaetomiaceae family Fungi	07	8–22 equine	Animal	Encephalitis neurologic signs	–	–	[275]

Cladosporium bantianum	03	33–59	2:1	Brain abscess	Present	Neurosurgical excision, amp-B + 5FC: death	[272, 276–278]
C. carrionii	01	36	1:0	Chromoblastomycosis, onychomycosis	Not present	Not done	[278]
C. cladosporioides	1	18	1:0	Skin lesions		Ketoconazole + itraconazole	[279]
C. oxysporum	01	57	0:1	Bronchopulmonary disorder, histologically not confirmed	Present	None	[281]
C. sphaerospermum	07	24–57	4:1	Onychomycosis, cutaneous lesion	Trauma, diabetes	None	[280]
C. trichoides	23	6–63	18:3	Brain abscess	None: suspected in some	Amp-B and 5FC: mostly death, some survived	[281, 282, 283]
C. deVriesii	01	26	0:1	Lump in the left breast	None	Surgical excision, 5FC and thiabendazole, recovered	[40]
Colletotrichum coccodes	1			Keratitis	Immunosuppressed heart transplant recipient	Surgical excision and antifungal agents	[285]
C. gloeosporioides	1	68	1:0	Lesion on the 5th finger of the right hand	Lesions removed surgically	Amp-B and voriconazole	[286]

(continued)

Table 5.1 (continued)

Fungal pathogen	Total cases	Age (years)	Sex M:F	Disease entity	Underlying disease	Therapy	References
C. truncatum	02	41–70	1:0	Keratitis	Immunocompetent person without any history of trauma or comorbidity	Amp-B and voriconazole	[287]
Corynespora cassiicola	1	37	0:1	Tissue necrosis, edematous papillary dermis		Oral terbinafine with topical povidone-iodine	[288]
Curvularia spicifera	01	Pediatric patient	–	Infection on the nasal septum	Acute myeloid leukemia	Surgical debridement plus antifungal therapy	[289]
Cyphellophora pluriseptata				Extensive infiltration of the left ear	Leprosy		[290]
Exophiala dermatitidis	1	8	0:1		None		[109]
E. dermatitidis	01	8	2:7	Skin lesion, brain lymph node abscess, hepatic involvement	Immunocompetent	Amp-B and voriconazole	[110]
E. dermatitidis	01	Teenage girl		Phaeohyphomycosis of breast	Fibroadenoma	–	[293]
E. equine	1	75	0:1	Subcutaneous abscesses on both forearms	Lesion was initiated by inoculation with a spine from a tree	Fluconazole	[291]
E. jeanselmei					Lung transplant patient	Cyst excised and treated with oral fluconazole and oral itraconazole	[94, 95, 294]
E. oligosperma	1	57	0:1	Granuloma in the superficial dermal layer	Pruritic, hyperkeratotic, brownish, erythematous lesion of the left cheek		[295]

E. lecanii-corni	1	56	0:1	Asymptomatic skin lesions on both cheeks	Phlebosclerotic colitis and gastroesophageal reflux disease	Treatment with itraconazole 200 mg/day	[296]
E. spinifera	1	26	0:1		None		[284, 291]
E. xenobiotica	1	70	1:0	Cystic nodular lesion on the dorsum of his right thumb	Rheumatoid arthritis, lung cancer with vertebral bone metastasis		[113, 226]
Exserohilum rostratum	04	10–63	3:1	Osteomyelitis, sinusitis	Heart transplantation, leukemia, rhinitis	Amp-B and surgery, cured	[23,–25]
E. rostratum	01	60	1:0	Sinusitis	Cord blood transplantation for myelodysplastic syndrome	Liposomal amp-B improved sinusitis	[292]
E. mcginnissi	01	27	1:0	Sinusitis	Allergic rhinitis	Surgery, cured	[25]
E. longirostratum	01	20	0:1	Endocarditis and osteomyelitis	Cardiac surgery	Amp-B and ketoconazole, cure	[264]
Fonsecaea monophora	1	71	0:1	Cerebral phaeohyphomycosis	Chronic diabetes mellitus and hypertension	Surgery, cured	[297]
Graphium basitruncatum	01	–	–	Subcutaneous infection	Heart transplant patient	–	[299]
Lasiodiplodia theobromae	1	30	0:1	Fungal sinusitis	Intermittent bleeding and nasal discharge, headache on the left side	Endoscopic surgery and antifungal treatment	[179, 298]
Microsphaeropsis arundinis	3			Soft tissue infection	Renal transplant	Triazole treatment	[300]

(continued)

Table 5.1 (continued)

Fungal pathogen	Total cases	Age (years)	Sex M:F	Disease entity	Underlying disease	Therapy	References
Nattrassia mangiferae	1	17	1:0	Cerebral phaeohyphomycosis	Systemic lupus erythematosus (SLE)	Amp-B	[302]
Neoscytalidium species	5			Cutaneous infections mimicking dermatophyte lesions	Renal transplant recipients	Antifungal therapy and surgical excision	[184, 301]
Ochroconis gallopavum	1			Systemic phaeohyphomycosis	Advanced HIV disease		[303]
Paraconiothyrium cyclothyrioides	1	49	1:0	Cutaneous lesions in his lower extremities	Type II diabetes, hypertension and atrial fibrillation	Vancomycin (1.25 g every 24 h) and silver sulfadiazine	[305]
Phaeoacremonium parasiticum	12	49	1:0	Impressive, large, inflammatory and draining cystic tumors on the left foot	Renal transplant	Itraconazole and amphotericin B	[304]
P. rubrigenum	1	76	0:1	Multiple, proliferating subcutaneous nodules on her right leg	Adult Still's disease	Oral itraconazole (400 mg)	[306]
Phialemoniopsis hongkongensis				Forearm nodule	Liver cirrhosis, ankylosing spondylosis, and tuberculosis	Itraconazole	[308]
Phialemoniopsis ocularis	1	67	1:0	Swelling of the right foot		Oral voriconazole	[309]
Phialophora richardsiae	13	15–70	9:4	Cyst	Trauma, diabetes, lymphoma, renal transplant	Excision + antifungal, cure sometimes death	[133, 134, 135]
P. parasitica	08	Mostly not reported	Not reported	Subcutaneous abscess mycetoma	Trauma, renal transplant	Excision and ketoconazole	[310]

P. mutabilis	02	Same	Same	Endocarditis	Transplantation	Death due to obstruction	[307]
P. bubakii	02	Same	Same	Subcutaneous abscess and onychomycosis	Trauma		[154]
P. verrucosa	01	Same	Same	Cyst, chromoblastomycosis			[143]
P. repens	01	Same	Same	Cyst	Lepromatous leprosy		[156]
Phoma herbarum	01		0:1	Hatchery-reared chinook salmon fingerlings	Infection in swim bladders spread to the kidneys, gastrointestinal tract, and surrounding musculature		[312]
P. insulana	01			Chromoblastomycosis verrucous nodules on skin			[313]
Pleurostoma ootheca	01	59	1:0	Suppurative tumefaction of left ankle	Kidney graft	Posaconazole	[314]
Pseudochaetosphaeronema martinelli	1	73, 72	1:1	A skin lesion located at the left knee	Rheumatoid arthritis		[315]
Pyrenophora phaeocomes and Drechslera nobleae	1	21	Appaloosa mare	Cutaneous mass at the base of the right side of the neck		Surgical excision and injectable detomidine hydrochloride, butorphanol tartrate, lidocaine	[316]
Ramichloridium mackenziei	1	66	0:1	Cerebral abscess		Intravenous amp-B lipid complex and voriconazole, died	[191, 317]

(continued)

Table 5.1 (continued)

Fungal pathogen	Total cases	Age (years)	Sex M:F	Disease entity	Underlying disease	Therapy	References
Rhytidhysteron rufulum	2			Subcutaneous lesions	Immunocompetent and gave no history of trauma		[318]
Scedosporium apiospermum	1	49		Cerebral mycosis	Renal transplant recipient	Liposomal amp-B and voriconazole	[320]
Thielavia subthermophila				Fatal brain infection		Surgical excision and amp-B, failed	[321]
Veronaea botryose	01	71	1:0	Cutaneous nodules	Heart transplant	Surgical excision, posaconazole	[230]
Verruconis gallopava				Disseminated	Renal transplant	Combination therapy	[322]
Wallemia sebi	1	34	0:1	A nonhealing ulcer on the dorsum of the left foot			[323]

number of cases, the lesions were mainly located on the face; unusual site reported was breast [4]. The infection at different sites of the body indicates that the pathogen may have entered the body through some trauma, but unfortunately such an inoculation was seldom mentioned in the case histories.

In most of the cases, the histopathological examination revealed a well-delimited dermal granuloma containing hyphae and single-celled structures [4]. Recently, in an experimental animal model, several giant cells containing fungal elements of *A. chlamydospora* along with spiral shape of fungal hyphae were observed in the granuloma on intraperitoneal inoculation. *A. chlamydospora* also exhibited neurotropism in animal model [17].

Alternaria strains reported as human pathogen belong to different species. However, majority of the cases were caused by either *A. tenuissima* or *A. alternata*. Unfortunately, several authors have identified them only up to the genus level. Experimental infection of small mammals frequently resulted in superficial lesion [2]. Ohashi (1960) reported an epidemic internal disease caused by *A. alternata* [18]. His experimental infections resulted in a high death rate among guinea pigs. Dubois et al. (2005) reported a case of cutaneous phaeohyphomycosis due to *A. infectoria* [19, 20]. *A. malorum* has also been reported to cause subcutaneous lesion [21]. Interestingly in 2017, Liu et al. (2017) reported a case of cutaneous infection in a stem cell transplant patient due to *A. rosae* [22].

Regarding treatment, it was observed that corticosteroids often produced ulceration. Injection of amphotericin B (amp-B) or miconazole into the lesions was effective occasionally [2, 4]. Ketoconazole exhibited diverse effects in the treatment. In some cases, it showed no improvement, while in others, definite improvement pertaining to healing was evident [2, 4, 15]. High doses of itraconazole only could cure a female patient who had several episodes of recurrence following excision and ketoconazole treatment [4].

5.1.2 *Bipolaris* and *Exserohilum*

Table 5.1 summarizes cases infected by *Bipolaris* and *Exserohilum* species. The total number of patients infected with *Bipolaris* species, namely, *B. hawaiiensis* and *B. australiensis*, were 16. Clinically significant species of *Bipolaris* causing human disease are *B. spicifera*, *B. hawaiiensis*, and *B. australiensis*.

In about 21 patients, an underlying local or systemic predisposing factor was present. Local factors included an injured cornea, allergic rhinitis, or nasal polyposis, and systemic factors included peritonitis following continuous ambulatory peritoneal dialysis (CAPD), cardiac surgery, chronic bronchitis with bronchiectasis, lymphoma, immunosuppressive therapy, leukemia, and myelodysplasia. Immunocompromised patients were more susceptible to systemic infection. *B. hawaiiensis* and *B. spicifera* demonstrated particular attraction for central nervous system [23]. Nevertheless, it is interesting to know that in about ten cases, no predisposing factors were identified. This indicates that isolation of *Bipolaris* and *Exserohilum* from clinical specimens, even of otherwise healthy people, may suggest a serious infection. We, therefore, suggest that critical clinical and laboratory evaluations of such patients should be done before arriving at a firm diagnosis.

Bipolaris and *Exserohilum* are responsible for a number of clinical syndromes [23–25]. Infection of the nose and paranasal sinuses was the most common form followed by infection of the central nervous system, skin, cornea, peritoneum, endocardium endocarditis, and bones. One immunocompromised patient also had positive blood cultures of *B. spicifera*. The prognosis was usually good since the infection resolved in about 24 cases including those with serious manifestations such as encephalitis, endocarditis, osteomyelitis, peritonitis, and paranasal involvement [3, 26]. Four patients, all with predisposing factors, died. One case developed corneal perforation and in another sputum examination disclosed masses of dematiaceous mycelia for over two years.

Regarding treatment, systemic amphotericin B is probably the most effective method of treating infections caused by *Bipolaris* or *Exserohilum* species. Four out of six patients who had not received systemic amphotericin B treatment died. Inversely, 16 patients received this drug with or without surgical excision and 14 of them recovered. Surgical excision alone of patients with nasal and paranasal infection cured four patients, and in other four cases, surgical excision along with systemic therapy was necessary. The role of imidazole derivatives in these infections is not clear.

In India, to our knowledge, the first case of bilateral nasal phaeohyphomycosis caused by *B. hawaiiensis* was diagnosed in a woman with chronic renal failure admitted in 1985 to the Christian Medical College Hospital, Vellore, and successfully treated with nystatin solution [27]. Lastly, we believe that infections caused by *Bipolaris* and *Exserohilum* species particularly in India are underdiagnosed. We recommend that it must be considered in different diagnoses, particularly in immunocompromised patients. Three species of *Exserohilum* causing human infections are *E. rostratum*, *E. longirostratum*, and *E. mcginnisii*. They have been reported as etiologic agents of sinusitis [25], CNS, keratitis [28], and disseminated case [29] along with cutaneous and subcutaneous diseases [30, 31]. In India, first report of *E. rostratum* cutaneous infection was reported by Agrawal and Singh [32].

5.1.3 *Cladosporium*

There are over 500 described species of *Cladosporium*. Most of these are ubiquitous saprophytes, occurring most abundantly in the air [2, 33–39]. Among these, *C. trichoides*, *C. carrionii*, *C. deVriesii*, *C. herbarum*, *C. cladosporioides*, *C sphaerospermum*, and *C. oxysporum* have been reported pathogenic to humans [40, 45].

Twenty-six cases of cerebral phaeohyphomycosis have been attributed to *Cladosporium trichoides* [41–45]. However, since 1960, two new names, *Cladophialophora bantianum* and *Xylohypha bantiana*, have been used for this fungus, and the controversy surrounding these names is still unsettled [2, 31, 46–49].

As the taxonomy of *C. trichoides* is controversial, we prefer to use its original name in this article on the basis of its priority [48].

In Table 5.1, cultural and histopathologically documented cases of brain abscess caused by *C. trichoides* have been included. Cases not validly reported or lacking cultural confirmation and incorrect fungal identification have not been included in this review. Out of thirty-three documented cases, 23 involved the central nervous system, 5 involved the skin, 2 involved the nails, and in 1 case bronchopulmonary disorder was suspected though it was not histologically confirmed. Majority of the patients had no apparent physiologic factors predisposing to infection. Headache was the most common presenting symptom of CNS infection followed by focal deficits, hemiparesis, temperature range of 36.9–39.5 °C, cranial nerve defects, papilledema, lethargy, meningismus, and ataxia. It is interesting to note that the initial diagnosis of CNS phaeohyphomycosis or other fungal infection was not described in any case, and diagnosis was not established until surgery. The initial diagnosis was bacterial brain abscess or meningitis, space-occupying mass, brain tumor, and viral encephalitis. Frontal and parietal lobes and cerebellum were mostly infected. Lesions on the brain were of two types: (1) discrete, well demarcated, and encapsulated, associated with good prognosis and (2) lesions not well demarcated and often with satellite lesions, not resectable, and associated with poor prognosis.

Review of CNS phaeohyphomycosis revealed that the survival rate of patients is 35–45%. At present, it is not possible to distinguish specific dematiaceous hyphomycetes in vivo [49]. Therefore, cultural identification is important in determining prognosis as well as therapy. Dixon and Polak (1987) reported that dematiaceous fungi demonstrate different responses to antifungal agents. This dependency of specific cultural identification of etiologic agent for therapeutic decisions may lessen in future with the availability of new broad-spectrum antifungals [50].

CNS phaeohyphomycosis generally occurred in males without predilection for racial or geographic location. As *C. trichoides* can be isolated from detritus [51, 52], traumatic or inhalational exposure to soil may be significant. Experimental infections have demonstrated that respiratory tract may be a portal of entry [53]. However, only mice treated with corticosteroids were found to have cerebral infection after respiratory challenge. These observations suggest that defective host immunity and exposure to a source where organism thrives in nature are important in establishing CNS infection in human [41].

Neurosurgical resection is an important determinant for cure because patients who did not undergo surgery died and were diagnosed postmortem. Unfortunately, antifungal therapy was not associated with improved survival of the patient. However, amphotericin B (10 mg/kg) reduced mortality by 20%. At present, a course of 5 flucytosine (FC) and amphotericin B (amp-B) seems best especially for poorly resectable lesions.

C. carrionii is an established causal agent of the clinical type chromoblastomycosis of man [2]. This fungus causes granulomatous skin lesions leading to warty, cauliflower-like tumors [2]. Interestingly, Barde and Singh [54] reported for the first time *C. carrionii* as a causal agent of phaeohyphomycosis. Like *Phialophora verrucosa* [1], *C. carrionii* could also play a dual role as pathogens of two clinical types, chromoblastomycosis and phaeohyphomycosis.

Cladosporium sphaerospermum is a ubiquitous saprophyte, occurring on all sorts of substrate. It was originally described from citrus leaves, but the neotype strain CBS 193.54 was isolated from human nails. A *C. sphaerospermum*-like strain described as *Hormodendrum cladosporioides* was involved in psoriasis-like, later eczematous skin lesions on humans [2]. The original strain of *H. langeronii* CBS 189.54, reidentified as *C. sphaerospermum* by other authors [55, 56], was isolated from human nodular lesions. Likewise, we have also diagnosed several cases of phaeohyphomycosis caused by *C. sphaerospermum* [unpublished data]. We are of the opinion that such cases are rather underdiagnosed.

C. herbarum has been implicated as pathogen under unusual circumstances [57]. *C. cladosporioides* was implicated as the cause of a pulmonary ball; however, pathogenicity of these strains have been doubted by later authors. In 1984, Gonzalez et al. (1984) reported subcutaneous phaeohyphomycosis caused by *C. deVriesii* in a 26-year-old woman from Grand Cayman Island. She was successfully treated with surgical excision along with 5 FC and thiabendazole [40].

In India, Chandramukhi and Gokul [58] reported that at National Institute of Mental Health and Neuroscience, Bangalore, four cases of CNS infection caused by *C. bantianum* were encountered in 10 years. All the infected patients were males. The clinical picture in two cases was abnormal behavior and stroke syndrome. In one case, it was brain abscess; and in another case, chronic meningitis along with chronic brain abscess was observed. Interestingly, none had any predisposing factors. However, none of these reported cases in the review were documented by the authors. In India, Tamsikar et al. (2005) reported a case of sebaceous cyst due to *C. cladosporioides* [59].

5.1.4 *Curvularia*

Some of the species of the genus *Curvularia* are opportunistic pathogens causing phaeohyphomycosis. They may cause variety of mycoses in man and animals including insects [60] like keratitis, sinusitis [61], endocarditis [62], cutaneous and subcutaneous infection [63], systemic infection [60], onychomycosis [64], and CNS infection [65]. Clinical isolates include *C. senegalensis*, *C. brachyspora*, *C. clavata*, *C. verruculosa*, and *C. inaequalis* [66]. Among these, *C. lunata* is the most common clinical species. Forester et al. (1975) reported *C. lunata* to cause mycotic keratitis [67]. Rohweeder et al. (1979) reported *C. lunata* as a causal agent of disseminated infection in a football player who was a 25-year-old immunocompetent male [68]. Deep soft tissue abscess, pulmonary suppuration, paravertebral abscess, and cerebral abscess all followed leg ulcers from neglected abrasion. From India, Barde and Singh [69] reported *C. lunata* from a case of onychomycosis involving all the finger and nails with a primary debilitating disease. This fungus was also involved in the case of mycotic keratitis reported by Agarwal et al. (1982) for the first time from India [70]. Singh et al. (1991) reported *C. lunata* causing disseminated phaeohyphomycosis among a group of *Nezara viridula* (Insecta: Heteroptera) parasitizing vegetable crop *Vigna unguiculata* [60]. Dark lesions were seen on pronotum and abdominal sterna. Histopathology revealed that almost all internal organs and tissues were extensively damaged. *C. lunata* exhibited predilection for chitinous tissues and elicited cellular immune response by granulocytes (phagocytosis). This was the first report of phaeohyphomycosis in an insect, extending the disease to invertebrates.

Nityananda et al. (1964) reported *C. geniculata* from mycotic keratitis in human. Kaufman reported *C. geniculata* infection on a homografted aortic valve [71, 72]. Forester et al. (1975) reported that *C. pallescens*, *C. senegalensis*, and *C. verruculosa* may also cause nonspecific corneal ulcers. *C. pallescens* was also described as the cause of a pulmonary mycetoma with cerebral metastasis in a human patient [67, 73]. Vishnoi et al. (2005) studied animal pathogenicity of a clinical isolate of *C. geniculata* from cancer patient in albino rats [74]. They reported that *C. geniculata* was capable to cause disease in the animal. The target organs were lungs, liver, and kidneys in both healthy and diseased rats.

We were referred a case of 29-year-old female suffering from sinusitis caused by *C. lunata*. She underwent an operation for right and left lateral rhinotomy and ethmoidectomy on 22-7-1988. The granuloma was surgically removed, and the patient was put on amphotericin B treatment for 7 months taking 1.425 g of the drug. After a trouble-free period of 20 months, she again developed similar symptoms and was readvised surgical excision and antifungal treatment. At this juncture, on January 1991, we were consulted for antifungal therapy. Our experience suggests that in vitro amphotericin B is not active against *C. lunata*, and this was the probable reason why amp-B could not kill the fragments of the fungus left after surgery leading to the recurrence of the lesion. Though we advised for the antifungal therapy, the patient could not be followed up (unpublished data).

5.1.5 *Dactylaria* and *Ochroconis*

Saccardo established *Dactylaria* in 1880 for *D. purpurella* which he had mistakenly described in 1877 as *Ascrothecium purpurellum*. Georg et al. (1964) described *Diplorhinotrichum gallopavum* Cook as the etiologic agent of fungal meningitis in Turkey. Bhatt and Kendric [75, 76] proposed the new combination *Dactylaria gallopava* after uniting *Diplorhinotrichun* under their amended definition of the genus *Dactylaria*. Later, deHoog (1983) proposed the genus *Ochroconis* for most species formerly classified under the name *Scolecobasidium* and classified three species under *Ochroconis* that were pathogenic to humans and animals, namely, *O. gallopavum*, *O. humicola*, and *O. tschawytschae* [2, 77]. Because of the morphological similarity of conidia produced by *O. gallopavum* and *Scolecobasidium constrictum*, these two species were reduced to varietal status under new combination *D. constricta* var. *gallopava* and *P. constricta* var. *constricta* [78, 79]. Cannon (1990) reported that the method of conidial secession in *Dactylaria* is schizolytic, while in *Ochroconis*, it is rhexolytic justifying the separation of the genera [80].

Dactylaria constricta is known as the causative agent of acute encephalitis in Turkey and Chicks [2]. Fukushiro et al. (1986) reported subcutaneous abscess caused by *O. gallopavum* (*D. constricta*) [81]. Similarly, Terreni et al. (1986) found *D. constricta* to cause disseminated infection in immunocompromised patients [82]. Dixon et al. (1987) evaluated the pathogenic potential of *D. constricta* and found that *D. constricta* formerly known as *D. gallopava* was capable of causing infections in the CNS, whereas the isolates of *D. constricta* formerly known as *Scolecobasidium constrictum* were nonpathogenic [83]. Sekhon et al. (1990) confirmed the above conclusion using antigenic relationship [84]. They found no antigenic relatedness between *D. gallopava* and *S. constrictum* and concluded that they should be retained as separate entities. Fukushiro et al. (1986) reported the first human infection caused by *O. gallopavum* in a patient with an acute myeloblastic leukemia [81]. The lesions were manifested as subcutaneous nodules and abscess. The second disseminated fatal infection caused by *O. gallopavum*, the first in the USA, was described by Terreni et al. (1990) in a 62-year-old man with chronic lymphocytic leukemia of T-cell type and also diabetes mellitus [85]. Sides et al. (1991) described a third human infection caused by this emerging pathogen in a patient from North Carolina, USA, with malignant lymphoma [86]. The patient, a 60-year-old man with a 9-year history of malignant lymphoma, developed an initial pulmonary infection with *Nocardia asteroides* which later disseminated to the central nervous system with multiple brain abscesses. He was treated successfully with intravenous trimethoprim sulfamethoxazole for 6 weeks. A follow-up CT scan showed complete resolution of the abscesses. However, 2 years later, he returned to the hospital with symptoms of loss of concentration; ataxia, leaning to the left; and confusion. Craniotomy revealed a right frontal lobe abscess. Histopathology of the tissue revealed dematiaceous fungus. The fungal isolate was identified as *O. gallopavum*. Later, despite treatment with amp-B, 5FC, and fluconazole, the patient gradually deteriorated and died.

Apart from these documented cases, Sides et al. (1991) have reported that A. A. Padhye of Centre of Disease Control, Atlanta, has identified increasing numbers of *O. gallopavum* isolates from clinical specimens [86]. In many instances, the fungus was not considered to be a pathogen but was treated as a saprophyte by the sender laboratory. This may, in part, be due to the failure of *O. gallopavum* to grow on media containing cycloheximide. Like all other neurotropic fungi, *O. gallopavum* exhibits its innate ability to grow at temperatures higher than 37°C. It is, therefore, suggested that when fungi with thermotolerant abilities are isolated from clinical specimens especially from immunocompromised patients, they should be fully investigated as opportunistic pathogens rather than being discarded as contaminants.

5.1.6 *Exophiala*

This genus was proposed in 1966 by Carmichael. It currently includes eight species, four of which are known to cause phaeohyphomycosis in humans. These are *E.*

jeanselmei (Langeron) McGinnis et Padhye, *E. moniliae* de Hoog, *E. spinifera* (Nelson et Conant) McGinnis, and *E. werneckii* (Horta) von Arx [87]. The most frequently documented human pathogen is *E. jeanselmei.* Its clinical picture varied from cutaneous to subcutaneous abscess. In most instances, the patient presented with solitary, discrete, asymptomatic, well-encapsulated subcutaneous nodules [88, 89]. The lesions may occur on the feet, legs, hands, arms, back, or other body sites. In many cases, the patient recalled localized trauma at the site of lesions. Such patients have been frequently misdiagnosed as ganglion cysts, Baker's cysts, etc. Dense collagenous connective tissue surrounds the abscess [88–90] in which macrophages, scattered giant cells, neutrophils, eosinophils, and lymphocytes may be seen. The fungus is prominent in the wall of the abscess [91]. Predisposing factors like diabetes, lupus, and leukemia have been found in some cases [9]. *E. jeanselmei* infection was described in a 53-year-old man in the USA who was undergoing treatment with steroid compounds against rheumatoid arthritis for 5 years. He developed subcutaneous abscess with sinus tracts at the site of the intramuscular injection. Another case was reported from Japan in a 61-year-old woman who had been receiving corticosteroids therapy for nephritic syndrome and developed a lesion that began as small, slightly erythematous papule on the planter aspect of the foot which gradually enlarged to flatly elevated plaque that discharged pus. In both cases, treatment consisted of complete excision of the lesion [9]. South et al. (1981) reported successful treatment of a patient having subcutaneous phaeohyphomycosis caused by *E. jeanselmei* with ketoconazole [92].

In India, Singh and coworkers (1992) reported a case of cutaneous phaeohyphomycosis due to *E. jeanselmei* in a 35-year-old male who had flat erythematous moist lesions on the interdigital area of the third digit of the left foot [93]. Earlier, Prabhakar et al. (1983) and Lal et al. (1984) had reported a case, each of phaeohyphomycosis and chromoblastomycosis, respectively [94, 95]. Similarly, Hemashettar and coworkers described a case of mycetoma pedis in a 35-year-old farmer due to *E. jeanselmei* [96, 97]. It is interesting to note that this species has the potentiality to cause phaeohyphomycosis, chromoblastomycosis, as well as mycetoma. However, parasitic form of *E. jeanselmei* has created certain controversy [98]. It is, therefore, important that while conceptually maintaining the term chromoblastomycosis and phaeohyphomycosis as separate entities, one must have in mind that the definition of clinical entity has to be determined by the host-parasite relationship. An interesting question also arises of how to treat such cases where fungus is the same but clinical types are different.

The only well-documented human infection caused by *E. moniliae* was reported by McGinnis et al. in 1981. They observed a primary subcutaneous lesion on the dorsum of the third digit of the left foot caused by this fungus in a 63 year-old male without predisposing factor and no history of a recent foot injury [4]. Matsumoto et al. (1984) described two cases of phaeohyphomycosis caused by *E. moniliae* from Japan [87]. The first case was of a 9-year-old girl with slow-growing submandibular lesion. The lesion was slightly elevated, flat, and erythematous. The patient had no predisposing factor. The second case was a 53-year-old male surgeon who had for the last 3 years slowly growing vegetative tumor on his left buttock. The tumor was dark brown, ellipsoidal, circumscribed, and a heaped-up granuloma-like plaque measuring 11 × 13 cm. The histological specimens of the lesion showed pseudocarcinomatous proliferation in the epidermis and multiple necrotizing granulomata in both upper and lower dermis. Histologically, dematiaceous mycelium and yeast-like cells were present predominantly; therefore, these cases were considered as cases of phaeohyphomycosis.

The first report of infection caused by *Phialophora spinifera* involved a fatal infection of a child from India [99]. de Hoog subsequently reidentified this fungus as *Exophiala spinifera* [2]. Later, *E. spinifera* was described causing a granulomatous mass in the nasal granuloma due to its ability to produce annellides on spine-like conidiophores [100]. In addition to these cases, Padhye and coworkers reported two cases in humans [101, 102]. Lacaz et al. (1984) reported a cutaneous infection in a 5-year-old girl, and Dai et al. (1987) reported a systemic infection in a 9-year-old boy [103, 104]. Recently, two cases of feline phaeohyphomycosis caused by *E. spinifera* have been described extending this disease to animals [105]. Despite the wide geographic distribution of *E. spinifera* in North, Central, and South America [2, 101, 102], along with Asia, [99, 104] and Australia [105], only eight human infections caused by this dematiaceous pathogen have been reported. With regard to the treatment of four human cases, first proved fatal. In the second one, nasal granuloma was diagnosed early and was successively excised, third was treated effectively with a combination of ketoconazole and flucytosine, and the last showed clinical improvement with itraconazole after treatment with amp-B, ketoconazole and flucytosine had proved ineffective. The reports of the other two human cases do not provide enough follow-up data for adequate assessment.

Tinea nigra palmaris caused by *Exophiala werneckii* almost exclusively occurs in tropical and subtropical countries and is particularly common in coastal areas. *E. werneckii* syn. *Cladosporium werneckii* is the main etiologic agent of the clinical type tinea nigra palmaris, a human disease characterized by blackish slightly scaly usually asymptomatic stains on the palms [2]. In India, two cases of tinea nigra from South have been reported [96, 106]. The strains occur superficially on one of the palms rarely on both or on the sole [2]. It remains doubtful whether all cases of similar lesions on other part of the body were caused by the same species [2]. It has been found that except for few cases, the cause of infections has been preceded by hand injuries [2]. Experimental infection is not always successful [2, 107]. Inoculation of feet usually gives no significant reaction [2]. But when the fungus is inoculated in the bleeding area of the palm, typical lesions do develop [2].

Exophiala dermatitidis, a normally saprophytic dematiaceous fungus, has also been reported as a cause of phaeohyphomycosis in humans. A review of the literature yielded nine validly documented cases of *E. dermatitidis* infections (Table 5.1). This dematiaceous fungus was first isolated by Kano in 1934 from a lesion on the cheek of a Japanese woman and was designated *Hormiscium dermatitidis* [108]. Most publications about human infections due to *E. dermatitidis* have come from Japan involving the skin which typically shows discoloration, granuloma, and ovoid cells in direct microscopy. In the case reported by Urabe et al. (1977), the brain was also found infected along with the skin, and intravenous treatment with amp-B was a failure [109]. Similarly, the lung, lymph nodes, and brain also showed involvement along with the skin in the patient reported by Harada et al. (1976) [110]. Autogenous vaccine showed rapid progress, but death occurred in 11 months due to *E. dermatitidis* in two sisters from Taiwan, one 19-year-old student in whom lymph nodes, bile duct, ilium, liver, and brain were involved and the other a 25-year-old nurse in whom *E. dermatitidis* showed involvement in the lymph nodes and brain [2]. Systemic therapy in the younger sister with nystatin, streptomycin, chloramphenicol, colistin, and intravenous amp-B proved unsuccessful. Similarly, intravenous treatment with amp-B and griseofulvin in case of the elder sister caused rapid progress with death in 5 months. Hohl et al. (1983) reported a subcutaneous cyst with occasional drainage on the knee due to *E. dermatitidis* in a diabetic man with impaired T-cell function and cutaneous allergy for the first time in North America [111]. Repeated cultures of the lesion were positive for *E. dermatitidis* despite therapy with amp-B. Polak experimentally inoculated *E. dermatitidis* in Swiss albino mice that produced acute, fatal cerebral infection closely resembling human phaeohyphomycosis [112]. No curative medical therapy is known for this infection, but surgical excision seems to be the treatment of choice for circumscribed *E. dermatitidis* infections. In view of lack of apparent efficacy of the currently known antifungal agents and the severity of the disease caused by *E. dermatitidis*, further experimental and clinical investigations seem to be warranted in the search for curative treatment. Today, *E. dermatitidis* should be considered invasive, capable of dissemination, and difficult to treat. *E. dermatitidis* was earlier known as *Wangiella dermatitidis*. A case of phaeohyphomycotic cyst caused by *E. xenobiotica* in patient with rheumatoid arthritis and lung cancer has been reported [113]. Likewise, a case of subcutaneous phaeohyphomycosis caused by *E. salmonis* has also been reported [114].

Exophiala asiatica is a newly described species causing a fatal, disseminated cerebral phaeohyphomycosis in China [115]. *E. spinifera* has also been reported in cases of subcutaneous infection [116, 117].

5.1.7 *Neoscytalidium (Syn Hendersonula toruloidea, Nattrassia mangiferae)*

Hendersonula toruloidea now known as *Nattrassia mangiferae* and also as *Neoscytalidium dimidiatum* is a well-recognized plant pathogen causing branch wilt, canker, and dieback disease of a wide range of trees [118]. Gentles and Evans were the first to suggest that this fungus could invade human skin and nails among immigrants from Pakistan, India, Fiji, and Kenya [119]. Since then, several reports have appeared regarding infections caused by this fungus [120, 121]. In India, Singh and Barde reported six cases and concluded that such infections may be common in India [122]. The species is classified in a coelomycete genus because of forming pycnidia. The hyphomycetous synanamorph *Scytalidium dimidiatum* is prevalent. A second species *S. hyalinum* also causes human infection [123].

The taxonomy of human pathogenic *Scytalidium* spp. is confusing because of its pleomorphic nature. Singler et al. (1997) reviewed the taxonomic status of this group and made four observations which make the taxonomy of this pathogen complicated [124]: first, the change of the pycnidial state *Hendersonula toruloidea* to *Nattrassia mangiferae*; second, the pathogen consisting of three cultural variants; third, similarities in morphology and clinical presentation between *N. mangiferae* and *Scytalidium hyalinum*; and fourth, the name *S. dimidiatum* that was given to the *Scytalidium* synanamorph of *N. mangiferae*. Dhindsa et al. (1998) reported a case of subcutaneous infection due to *Scytalidium* synanamorph of *N. mangiferae* in a DLE patient [125]. *N. dimidiatum* primarily produces infections mimicking those caused by dermatophytes [126, 127]. There are occasional reports of ocular infections [128]. The name *Nattrassia mangiferae* has now been placed in the new genus *Neofusicoccum* [129]. To merge the entirety of *Scytalidium* with *Fusicoccum* was found to be inappropriate by Crous et al. (2006) because they demonstrated *N. dimidiatum* to be phylogenetically distinct from *Neofusicoccum mangiferae*, thus suggesting *N. dimidiatum* to be the correct name for *Nattrassia mangiferae* [129].

5.1.8 *Phialophora*

Phialophora is a heterogenous assemblage of about 45 species [2]. Of these, nine species are known to be opportunistic pathogens of man and animals. These are *P. richardsiae*, *P. parasitica*, *P. mutabilis*, *P. bubakii*, *P. verrucosa*, *P. repens*, *P. hoffmannii*, *P. americana*, and *P. europaea* [130]. Some of human pathogens with phialidic conidiogenesis assigned to *Phialophora* have been moved to *Phaeoacremonium* [131] and *Pleurostomophora* [132].

Table 5.1 shows cases of mycosis due to *P. richardsiae*. Interestingly, eleven cases showed a marked resemblance in their clinical picture, i.e., the development of painless, well-defined, localized nodules often encapsulated by a fibrous layer. Central necrosis and inflammation occurred, and the nodule contained a viscous and yellowish fluid. This picture agrees well with the general description of phaeohyphomycotic cysts given by Zeifer and Connor in 1980 [89]. Unusual sites of infection were a lacrimal duct and prostate gland [2]. Two cases reported recently were remarkable because subcutaneous cysts or granuloma were not formed; instead, only the cutaneous layers of the skin were involved with only little inflammatory response [134]. The age of the patients varied from 15 to 80 years. Males outnumbered females (9:4). Maximum infections were seen on extremities like leg sole [135], finger [136], rarely knee [137], gluteal region [134], and lacrimal duct [2]. Similarly, the maximum number of patients had some predisposing factors responsible for infection. Two cases had diabetes, one had malignant lymphoma with diabetes and hepatosplenomegaly [135, 138, 139], one was a renal transplant patient [137], and the other developed a cystic mass on finger probably because he received Depo-Medrol injection as a treatment for extensor tendinitis. Trauma of the elbow probably helped in the development of phaeohyphomycotic cyst in a 79-year-old male [136]. Also, there were three cases reported [2, 134, 140] which were probably healthy as in their case histories, no predisposing factors were described. Regarding therapy, it is evident that the surgical excision of the localized cyst appears to be the best mode of therapy sometimes with or without antimycotic therapy [2, 137]. Relatively little information is available on the treatment of infections caused by dematiaceous fungi by antimycotics. Antifungal therapy of such infections may yield variable results within strains of a single species [141]. This necessitates in vitro testing of each clinical isolate [142]. Strain of *P. richardsiae* (CBS [143, 144]) isolated from a 15-year-old female [134] proved most susceptible to clotrimazole both in vitro and in vivo. On the contrary, the other strain of *P. richardsiae* isolated from a 50-year-old female, mother of the previous case, was most susceptible to amorolfine in vitro, but in both, the patients' cutaneous lesions were successfully treated with topical clotrimazole cream. Amorolfine was not attempted in the treatment of the second case [134]. *P. verrucosa* is primarily an agent of chromoblastomycosis [145], and other reported infections are endocarditis, keratitis, and osteomyelitis [146, 147]. A recently described species implicated in superficial infections is *P. europaea* [131].

Phialophora parasitica, for the first time, was described as the etiologic agent of subcutaneous infection in a kidney transplant patient undergoing therapy [148]. Since then, seven cases of *P. parasitica* infection have appeared in the literature (Table 5.1), and four of these were kidney transplant patients in whom *P. parasitica* was isolated from subcutaneous infection [149–152]. In two cases [153, 154], trauma appeared to be the predisposing factor, and in one case [155], the patient was healthy. Surgical excision with or without treatment with antimycotics like ketoconazole [152], ketoconazole-itraconazole [149], fluconazole, and amp-B, along with 5 FC [153], was the best treatment.

Two cases of *P. bubakii* have been reported in literature. Incidentally, in one of the cases, human nail was involved with trauma as the predisposing factor [155]. In another case, *P. bubakii* caused subcutaneous abscess in a renal transplant patient [154]. The cases of phaeohyphomycotic cyst caused by *P. verrucosa* [144], *P. repens* [156], and *P. gougerotii* [157] have been reported.

5.1.9 *Chaetomium*

Two ascomycetes genera, viz., *Chaetomium* and *Achaetomium*, are neurotropic and have been reported to cause human infection [158]. They form sexual-stage perithecia in culture. The human pathogen *C. globosum* is reported to cause invasive disease and is neurotropic [160]. It also causes phaeohyphomycotic cyst [159]. *C. perlucidum* and *C. atrobrunneum* are neurotropic and should be considered in the differential diagnosis of CNS fungal disease [161]. *Chaetomium strumarium* is also neurotropic and an agent of CNS phaeohyphomycosis [162, 163].

5.1.10 *Leptosphaeria*

Leptosphaeria senegalensis and *L. tompkinsii* are agents of black grain mycetoma restricted to Northern West Africa and India [161].

5.1.11 *Microascus*

It is a perithecial teleomorph with *Scopulariopsis* as anamorph. Several species of *Microascus* have been documented as agents of brain abscess [164], suppurative cutaneous granulomata in a patient with chronic granulomatous disease [165], endocarditis [166], and onychomycosis [167]. *M. cirrosus* was the etiologic agent of disseminated disease in a pediatric bone marrow recipient [168], and *M. trigonosporus* was reported in fatal pneumonia and in another bone marrow transplant recipient [169].

5.1.12 *Lecythophora*

Species of *Lecythophora*, viz., *L. mutabilis* and *L. hoffmannii*, are of clinical significance. They are reported to cause endophthalmitis [170], sinusitis [171], and prosthetic valve endocarditis [172].

5.1.13 *Phaeoacremonium*

The genus initially accommodated isolates with characters similar to both *Acremonium* and *Phialophora* [173, 174]. Human pathogens included in this genus are *P. parasiticum* [175], *P. alvesii* [176], *P. amstelodamense*, *P. griseorubrum*, *P. krajdenii*, *P. rubrigenum* [176], *P. tardicrescens*, and *P. venezuelense* [177]. Infections caused by *P. parasiticum* include subcutaneous abscesses [150], thorn-induced arthritis [153], and disseminated infection [178].

5.1.14 *Lasiodiplodia*

L. theobromae is a pycnidial fungus. It has been reported to cause subcutaneous disease [180], pneumonia in a liver transplant recipient [181], and mycotic keratitis [178,182]. This fungus was formerly known as *Botryodiplodia theobromae*.

5.1.15 *Macrophomina*

M. phaseolina is known to cause disseminated disease in a renal transplant recipient and cutaneous infection in a child with leukemia [183].

5.1.16 *Neoscytalidium*

N. dimidiatum was earlier known as *Scytalidium dimidiatum* [185, 186]. This species may also produce pycnidial synanmorph *Nattrassia mangiferae* and has now been placed in a new genus *Neofusicoccum* [129, 130]. *N. dimidiatum* cause skin infection which mimics the infection caused by dermatophytes. Other infections include mycotic keratitis and deep mycosis in immunocompromised patients [128,186,187].

5.1.17 *Rhinocladiella*

Four species of this genus are known agents of phaeohyphomycosis. *R. mackenziei* is a frequently fatal neurotropic fungus thought to be restricted to Middle Eastern countries and has now been reported from India causing brain abscesses in a man with no history of traveling abroad [189, 190]. *R. aquaspersa* is an occasional agent of chromoblastomycosis [190]. *R. basitona* was reported from Japan causing subcutaneous lesion in a man [192]. *R. similis* (*R. atrovirens*) is reported to cause mycetoma and cerebral phaeohyphomycosis in an AIDS patient [193, 194]. In 2013, Cai and coworkers for the first time isolated *R. basitona* from an immunocompetent child in China [195].

5.1.18 *Acrophialophora*

Acrophialophora fusispora is a rare species causing phaeohyphomycosis and is often confused with *Scedosporium prolificans*. It has been reported as an agent of cerebral phaeohyphomycosis in a leukemic child [195], as an agent of keratitis [196].

5.1.19 *Phialemonium*

The genus contains two species of clinical interest, *P. obovatum* and *P. curvatum*. Infection due to *P. curvatum* includes cutaneous and subcutaneous infections, disseminated disease, endophthalmitis, peritonitis, arthritis associated endovascular infections, and endocarditis [197, 202]. Rivero and coworkers have reviewed and published cases due to *Phialemonium* spp. [202].

5.1.20 *Scedosporium*

The genus *Scedosporium* and its teleomorph have been reviewed by Cortez and coworkers [203, 204]. *S. prolificans* appears to occupy a more restricted geographical range with infections occurring mainly in Australia, Spain, and the USA [205]. Infection with this organism is of major concern in all settings due to its resistance to antifungal therapy and high mortality associated [206–209]. The major risk factors are malignancy, organ transplant, pulmonary mycosis, bone and joint infection, cystic fibrosis, and hematological malignancies [210]. Molecular characterization suggests two or three distinct genotypes of *S. prolificans* which have been reported from France [211].

5.1.21 *Pseudoallescheria*

Teleomorphic genus *Pseudoallescheria boydii* (anamorph *Scedosporium boydii*), *P. apiosperma* (anamorph *Scedosporium apiospermum*), and *P. ellipsoidea* are also seldom reported to be pathogenic against humans. Other species of clinical interest in *P. boydii* species complex include *S. aurantiacum* [211] and *S. dehoogii* [212].

5.1.22 *Fonsecaea*

The genus *Fonsecaea* comprises of two species: *F. pedrosoi* and *F. monophora* [213, 214]. Former is known to cause chromoblastomycosis [215, 216], and the latter causes chromoblastomycosis, subcutaneous infection, and cerebral phaeohyphomycosis [217]. A genome sequence of strain type *F. multimorphosa* CBS 980.96 T was obtained. This species was first isolated from a cat with cerebral phaeohyphomycosis in Queensland, Australia [219].

5.1.23 *Phoma* and *Phoma*-like Pycnidial Fungi

All the above genera produce morphological similar pycnidia which are difficult to differentiate and identify. Their documentation and reporting as casual agents in disease are limited by a lack of adequate identification. Species in these similar genera are best differentiated by their sequencing. They are reported to cause subcutaneous disease [220], endophthalmitis [221, 222], and deep tissue infection [218, 221, 223–225].

5.1.24 *Veronaea*

V. botryosa infections were initially seen in China but now are a global concern. Literature reveals 12 cases reported since 1990 including three cases in transplant recipients [227–230]. Two cases are noteworthy as agents of subcutaneous disease in heart [230] and liver [231] transplant recipients. The present scenario suggests that *V. botryosa* is an emerging pathogen for immunocompromised patients. First report of *V. botryosa* as a causal agent of chromomycosis in frogs indicates animals are also susceptible to this pathogen [233]. The genus was examined at the molecular level by Arzanlose and coworkers [191]. *V. botryosa* has been reported as pathogen of white sturgeon [234]. The authors reported the diversity of *V. botryosa* from different hosts and evaluated its pathogenicity in *Acipenser transmontanus* [232].

5.2 Some Emerging Pathogenic Genera

Some other well-documented human pathogenic genera reported in literature are *Phaeoannellomyces elegans* [234], *Cephaliophora irregularis* Thaxter incorrectly identified as *Arthrobotrys oligospora* Fries [235, 236], *Scopulariopsis brumptii* [237], *Tetraploa aristata* [238, 239], *Thermomyces lanuginosus* Tsiklinsky [240], *Botryomyces caespitosus* de Hoog and Rubio [241], *Chaetomium globosum* [242, 243], *Aureobasidium pullulans* [244], *Ramichloridium cerophilum* [2], *Rhinocladiella compacta* [2], *Wallemia sebi* [2, 323], *Tritirachium oryzae* [245], *Phoma cava* [247], *P. cruris-hominis* [247], *P. eupyrena* [246, 247], *P. minutella* [248], *Aureobasidium melanogenum* [249], *Oidiodendron cerealis* [250], *Peyronellaea glomerata* [251], *Phaeosclera dermatioides* [252], *Phyllostictina citricarpa* [91], *Stenella araguata* [253], *Taeniolella stilbospora* [254], *Trichomaris invadens* [255], *Pseudomicrodochium suttonii* [256], *Pleurostomophora richardsiae* [319], *Cladorrhinum bulbillosum, Coniothyrium fuckelii, Dissitimurus exedrus, Oidiodendron cerealis, Phaeotrichonis crotolariae, Pleurophoma pleurospora, Sarcinomyces phaeomuriformis*, and *Scytalidium lignicola* [6]; all these fungi have one unifying character, i.e., they all form in vivo dematiaceous fungal hyphae or yeast-like cells. In some cases, however, the fungal element may not appear dematiaceous. For such cases, Wood and Bell [257] suggested the use of Fontana Masson Silver Stain. This stain is melanin specific and helps in confirming the dematiaceous nature of the fungi in tissue. The value of this stain was well documented in the case of dog infected by *Phialemonium obovatum* [258]. Some photomicrographs of medically important genera of phaeohyphomycoses have been provided in Figs. 5.1, 5.2, and 5.3. An updated list of phaeoid genera reported in human and animal diseases up to 2017 has been shown in Table 5.2.

5.3 Laboratory Diagnosis

Since the fungi discussed in this review are found in the environment and may also be encountered as contaminants of clinical specimens, their isolation may pose a diagnostic dilemma to determine if the isolate is a pathogen; a colonizer, i.e., not actually invading living tissues; or a contaminant. To resolve this problem, certain parameters were suggested [156, 260]. These were (1) the suspected organism should be repeatedly isolated from the clinical specimens, (2) fungal structures observed in direct microscopic examination should be morphologically compatible with the fungus isolated from the specimen, (3) histopathology of the biopsy specimen indicating tissue invasion, and (4) ability to grow at 37°C.

Usually, in vivo appearance of several dematiaceous opportunistic pathogens like *Phialophora and Exophiala* looks alike. This may suggest a presumptive

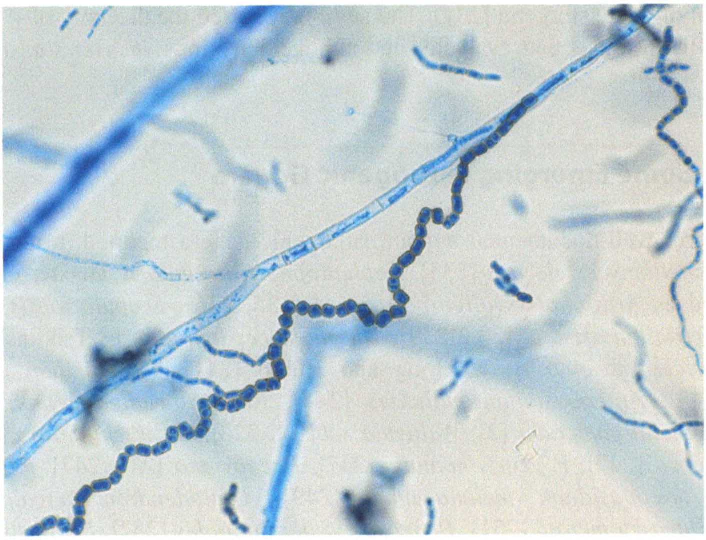

Fig. 5.1 *Neoscytalidium*. Broad septate hyphae and contiguous anthroconidia (lactophenol cotton blue, 400×). (Source: http://thunderhouse4-yuri.blogspot.in/2011/11/neoscytalidium-dimidiatum.html)

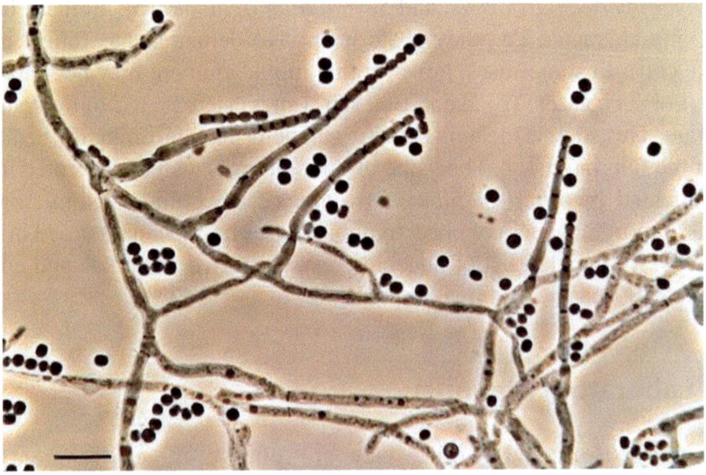

Fig. 5.2 Microscopic view of conidia-like arthrospores of *Wallemia sebi*. (Source: http://old.vscht.cz/obsah/fakulty/fpbt/ostatni/miniatlas/images/plisne/mikro/Wallemia%20sebi%20CCF%203005%20mikro.jpg)

identification of the pathogen and may help in the immediate initiation of appropriate therapy particularly when dealing with immunosuppressed patients where infection may rapidly progress to a fatal outcome. Nevertheless, precise identification of the pathogen is essential as even the strains of a single species respond differently

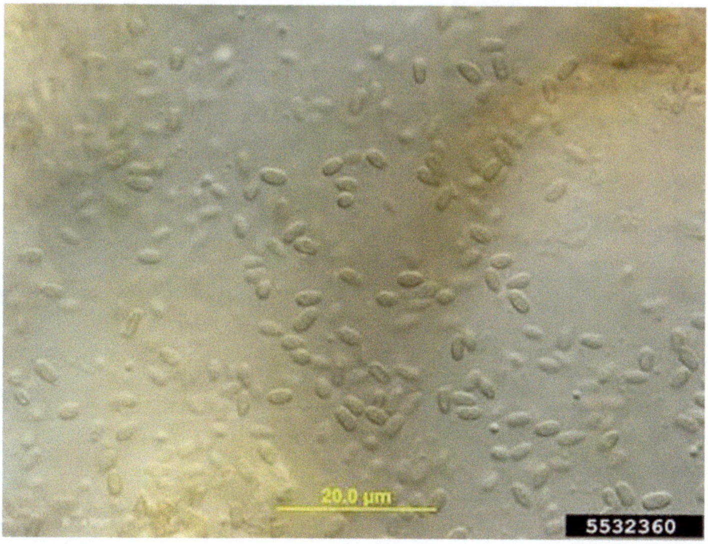

Fig. 5.3 Conidia of Coniothyrium fungi (*Coniothyrium* sp.) asexual spore. (Photo courtesy: Bruce Watt, University of Maine, Bugwood.org; source: https://www.invasive.org/browse/subthumb.cfm?sub=65488)

Table 5.2 Updated list of phaeoid genera reported in human and animal diseases up to 2017

S. No.	Genus	Species
1	*Achaetomium*	*A. strumarium*
2	*Acrophialophora*	*A. fusispora*
3	*Alternaria*	*A. alternata, A. chlamydospora, A. dianthicola, A. infectoria, A. longipes, A. tenuissima, A. rosae, A. dennisii, A. malorum*
4	*Anthopsis*	*A. deltoidea, A. malorum*
5	*Arnium*	*A. leporinum*
6	*Arthrinium*	*A. phaeospermum*
7	*Ascotricha*	*A. chartarum*
8	*Aureobasidium*	*A. pullulans, A. melanogenum*
9	*Bipolaris*	*B. australiensis, B. hawaiiensis, B. cynodontis, B. paperndorfii, B. spicifera*
10	*Biatriospora*	*B. mackinnonnii* [311]
11	*Botryomyces*	*B. caespitosus*
12	*Botryosphaeria*	*B. dothidea*
13	*Chaetomium*	*C. globosum, C. atrobrunneum, C. funicola, C. murorum, C. perlucidum, C. perpulchrum, C. funicolum, C. brasiliense*
14	*Cladophialophora*	*C. bantiana, C. boppi, C. carrionii, C. arxii, C. devriesii, C. emmonsii, C. modesta, C. mycetomatis, C. saturnica, C. samoensis*

(continued)

Table 5.2 (continued)

S. No.	Genus	Species
15	*Cladorrhinum*	*C. bulbillosum*
16	*Cladosporium*	*C. cladosporioides. C. herbarum, C. oxysporum, C. sphaerospermum*
17	*Colletotrichum*	*C. gloeosporioides, C. coccodes, C. graminicola, C. dematium, C. crassipes, C. truncatum*
18	*Coniothyrium*	*C. fuckelii*
19	*Corynespora*	*C. cassiicola*
20	*Curvularia*	*C. brachyspora, C. clavata, C. geniculata, C. lunata, C. pallescens, C. senegalensis, C. verruculosa, C. inaequalis*
21	*Cyphellophora*	*C. pleuriseptata, C. laciniata*
22	*Dichotomopthora*	*D. portulacae*
23	*Diachotomorphthoropsis*	*D. nymphaearum*
24	*Dissitimurus*	*D. exedrus*
25	*Drechslera*	*D. biseptata, D. nobleae, D. specifera* [267]
26	*Exophiala*	*E. asiatica, E. aenata, E. xenobiotica, E. moniliae* [259], *E. salmonis, E. bergeri, E. spinifera, E. dermatitidis, E. jeanselmei, E. werneckii, E. lecanii-corni, E. oligosperma, E. phaeomuriformis, E. pisciphila, E. angulospora, E. equine*
27	*Exserohilum*	*E. rostratum, E. longirostratum, E. mcginnisii*
28	*Fonsecaea*	*F. monophora, F. pedrosoi, F. multimorphosa*
29	*Graphium*	*G. basitruncatum*
30	*Honkongmyces*	*H. pedis*
31	*Hormonema*	*H. dermatioides*
32	*Hortaea*	*H. werneckii*
33	*Lasiodiplodia*	*L. theobromae*
34	*Lecythophora*	*L. hoffmannii, L. mutabilis*
35	*Leptosphaeria*	*L. senegalensis, L. tompkinsii*
36	*Macrophomina*	*M. phaseolina*
37	*Madurella*	*M. grisea, M. mycetomatis*
38	*Medicopsis*	*M. romeroi*
39	*Microascus*	*M. cinereus, M. cirrosus, M. trigonosporus*
40	*Microsphaeropsis*	*M. arundinis, M. olivacea*
41	*Myceliophthora*	*M. thermophila*
42	*Mycocentrospora*	*M. acerina*
43	*Mycoleptodiscus*	*M. indicus*
44	*Nattrassia*	*N. mangiferae*
45	*Neoscytalidium*	*N. dimidiatum*
46	*Neotestudina*	*N. rosatii*
47	*Nigrospora*	*N. sphaerica*
48	*Ochracladosporium*	*O. elatum*
49	*Ochroconis*	*O. gallopava, O. humicola, O. tshawytschae*
50	*Oidiodendron*	*O. cerealis*

(continued)

Table 5.2 (continued)

S. No.	Genus	Species
51	*Phaeoacremonium*	*P. alvesii, P. amstelodamense, P. griseorubrum, P. krajdenii, P. parasiticum, P. rubrigenum, P. sphinctrophorum, P. tardicrescens, P. venezuelense*
52	*Paraconiothyrium*	*P. cyclothyriodes*
53	*Phialemoniopsis*	*P. hongkongensis, P. ocularis*
54	*Phialemonium*	*P. curvatum, P. obovatum*
55	*Phaeosclera*	*P. dermatioides*
56	*Phaeotrichocanis*	*P. crotalariae*
57	*Phialophora*	*P. americana, P. bubakii, P. europaea, P. repens, P. verrucosa*
58	*Phoma*	*P. cruris-hominis, P. dennisiivar* var. *oculohominis, P. eupyrena, P. glomerata, P. herbarum, P. minutella, P. multispora, P. sorghina, P. insulana*
59.	*Piedraia*	*P. hortae*
60.	*Pleurophoma*	*P. cava*
61.	*Pleurophomopsis*	*P. lignicola*
62.	*Pleurostomophora*	*P. repens, P. richardsiae*
63.	*Pseudochaetosphaeronema*	*P. larense, P. martinelli*
64.	*Pseudomicrodochium*	*P. suttonii*
65.	*Pyrenochaeta*	*P. mackinnonii, P. romeroi, P. unguis-hominis*
66.	*Pyrenophora*	*P. phaeocomes*
67.	*Rhinocladiella*	*R. aquaspersa, R. basitona, R. mackenziei, R. similis*
68.	*Rhytidhysteron*	*R. rufulum*
69.	*Sarcinomyces*	*S. phaeomuriformis*
70.	*Scedosporium*	*S. prolificans*
71.	*Scopulariopsis*	*S. asperula, S. brumptii, S. fusca, S. brevicaulis,*
72.	*Scytalidium*	*S. dimidiatum*
73.	*Sphaeropsis*	*S. subglobosa*
74.	*Stenella*	*S. araguata*
75.	*Taeniolella*	*T. stillbospora*
76.	*Tetraploa*	*T. aristata*
77.	*Thermomyces*	*T. lanuginosus*
78.	*Thielavia*	*T. subthermophila*
79.	*Ulocladium*	*U. chartarum*
80.	*Veronaea*	*V. botryose*
81.	*Verruconis*	*V. gallopava*
82.	*Wallemia*	*W. sebi*

After Matsumoto et al. [6] and Revankar and Sutton [7]

to antimycotics [142, 261]. Molecular identification of most species relies on sequencing of ribosomal genes and comparison with published databases like GeneBank. However, over 16% of these deposits may be erroneous [262].

5.4 Future Areas of Research

For an organism to cause disease, it must enter the host, multiply in host tissues, resist host defense mechanism, and damage the host. Products of organisms which produce factors that assist in these processes are called virulence factors. Though numerous dematiaceous fungi have been reported as human pathogens, however, no mechanism of pathogenesis of any one of these fungi has been proposed. Similarly, there are other abilities of the organism which are not technically classified as virulence factors, but they deserve recognition and investigation for better understanding of the mechanism of pathogenesis. Some of these are critical mass phenomenon (mass effect of localized in vivo fungal cells on the vital organs and systems), induction of allergy (loss of immunologic response to mycoses), mycotoxins, space-occupying lesions, saprophytic to pathogenic conversion, genetics (surprisingly, little attention has been paid to genetics of pathogenesis), penetration of the host by the pathogen, dimorphism, morphological and biochemical aspects, etc. Immunological, serological, and epidemiological aspects of these infections are also little understood [2]. Therefore, there is tremendous scope for investigations on these lines in future.

The development of new agents for the treatment of mycotic infections has been slow in comparison to agents for the treatment of bacterial infections. There are two main reasons for this: (1) systemic fungal infections are much less common than bacterial infections, so less effort has been put in for developing new therapeutic agents, and (2) since fungi are eukaryotes, it has been difficult to develop antifungal agents that specifically affect fungal structures without affecting host cells. Although many antifungal drugs are available, the need of the hour is the development of new antifungal agents with a wide-spectrum activity, good absorption properties (from intestine), and rapid distribution in plasma and tissues after oral and intravenous administration along with no side effects.

Relatively little information is available on the treatment of infections by dematiaceous fungi. Antifungal therapy of such infections may yield variable results even within strains of a single species [141]. This necessitates in vitro testing of each clinical isolate. At times, however, the results of in vitro susceptibility have no predictive value particularly in systemic mycoses. Inhibitory capacities of the drugs seem to be more difficult to standardize in test with filamentous fungi. It is apparent that much more work is needed in this area particularly since the infections are becoming more prevalent and resistant organisms are being isolated with increasing frequency.

5.5 Conclusion

Interestingly, phaeohyphomycosis, the clinical type of great public health importance caused by diverse dematiaceous genera, is emerging into prominence on the global basis [9, 263, 264]. In India, no concerned effort has been made to determine its prevalence and distribution especially in immunosuppressed patients. Apart from our first reports on the emergence of this disease in India [13, 264], our experience suggests that this clinical type may be much more common in the population than expected.

Dematiaceous group of human pathogenic fungi in future will undoubtedly increase in both frequency and importance in contemporary medicine. A large number of cases with ulcers, abscesses, swellings, and tumors of mycotic etiology are misdiagnosed as other infections or sometimes as malignant lesions because adequate laboratory facilities for diagnosis of fungal infections and precise identification of etiologic agents are available at only few places in India.

We, therefore, suggests such facilities to be made available at more places and also request clinicians, pathologists, and mycologists to perform more critical mycological evaluation to define the role of fungi in diseases in a better way. It is only through the integration of each specialist's efforts that the victims of opportunistic fungi can be helped to overcome their infections. This step is very appropriate in view of the increased conditions prevailing, for opportunistic fungal invasion, the most important being AIDS, cancer, and other immunosuppressive diseases.

References

1. Ajello L (1986) Hyalohyphomycosis and phaeohyphomycosis: two global disease entities of public health importance. Eur J Epidemiol Microbiol 2:243–251
2. de Hoog GS (1983) On the potentially pathogenic dematiaceous hyphomycetes. In: Howard DH (ed) Fungi pathogenic for humans and animals, part A. Marcel Dekker, Inc, New York, pp 149–216
3. Fuste FJ, Ajello L, Threlkekt R et al (1973) *Drechslera hawaiiensis*: causative agent of a fatal fungal meningo-encephalitis. Sabouraudia 11:59–63
4. Bandillet G (1991) Les alternarioses cutanees. Revue de la literature. J Mycol Med 118:9–71
5. McGinnis MR, Hilger AE (1987) Infections caused by black fungi. Arch Dermatol 123:1300–1302
6. Matsumoto T, Aiello L, Matsuda T et al (1991) Developments in hyalohyphomycosis and phaeohyphomycosis. J Med Vet Mycol 32:329–349
7. Revankar SG, Sutton DA (2010) Melanized fungi in human diseases. Clin Microbiol Rev 23:884–928
8. Singh SM, Naidu J (1993) Some medically important genera of dematiaceous hyphomycetes. In: Rai B, Arora DK, Dubey NK, Sharma PD (eds) Fungal ecology and biotechnology. Rastogi Publications, Meerut, pp 153–171
9. Weitzman I (1986) Saprophytic molds as agents of cutaneous and subcutaneous infection in the immunocompromised host. Arch Dermatol 12:1161–1163
10. Bouteille B, Darde ML, Amici JM et al (1992) Alternariose cutanee. A proposed line observation. J Mycol Med 2:49–52

11. Contet-Audonneau N, Barbaud A, Guerin V et al (1991) Alternariose cutanee et syndrome de cushing: nouvelle observation. J Mycol Med 1:82–83
12. Drouhet E, Luciani J, Frantz P et al (1991) Cutaneous alternariosis and Kaposi's sarcoma in a renal transplant patient. J Mycol Med 118:84–87
13. Singh SM, Naidu J, Pouranik M (1990) Ungual and cutaneous phaeohyphomycosis caused by *Alternaria alternata* and *A. chlamydospora*. J Med Vet Mycol 28:275–278
14. Wadhwani K, Shrivastava AK (1985) Some cases of onychomycosis from north India in different working environments. Mycopathologia 92:149–155
15. Viviani MA, Tortoeano AM, Laria G et al (1988) Two main cases of cutaneous alternariosis with review of literature. Mycopathologia 96:3–12
16. Panagioutidous D, Kapetis E, Chryssomallis F et al (1984) Deux cos d'alternariose cutane in Greece. J Mycol Med 1:88–89
17. Naidu J, Singh SM, Dhindsa MK et al (2000) Pathogenicity of *Alternaria chlamydospora* for albino rats: study of a clinical isolate from human skin lesion. J Mycol Med 10:3–8
18. Ohashi Y (1960) On a rare disease due to *Alternaria tenuis* Nees (Alternariasis). Tohoku J Exp Med 72:78–82
19. Dubois D, Pihet M, Clec'h CL et al (2005) Cutaneous phaeohyphomycosis due to *Alternaria infectoria*. Mycopathologia 160:117–123
20. Lyskova P, Kubanek M, Hubka V et al (2017) Successful posaconazole therapy of disseminated alternariosis due to *Alternaria infectoria* in a heart transplant recipient. Mycopathologia 182:297–303
21. Mirhendi H, Fatemi MJ, Bateni H et al (2013) First case of disseminated phaeohyphomycosis in an immunocompetent individual due to *Alternaria malorum*. Med Mycol 51:196–202
22. Liu AW, Bateman AC, Greenbaum A et al (2017) Cutaneous phaeohyphomycosis in a hematopoietic stem cell transplant patient caused by *Alternaria rosae*: first case report. Transpl Infect Dis 19:e12698
23. Adam RD, Paquin ML, Petersen EA et al (1986) Phaeohyphomycosis caused by the fungi genera *Bipolaris* and *Exserohilum*. A report of 9 cases and review of the literature. Medicine 65:203–207
24. Douer D, Goldschrnied-Reouven A, Segev S et al (1987) Human *Exserohilum* and *Bipolaris* infections: report of *Exserohilum* nasal infection in a neutropenic patient with acute leukemia and review of the literature. J Med Vet Mycol 25:235–241
25. Padhye AA, Ajello L, Wieden MA et al (1986) Phaeohyphomycosis of the nasal sinuses caused by a new species of *Exserohilum*. J Clin Microbiol 24:245–249
26. Yoshimori RN, Moore RA, Itabashi HH et al (1982) Phaeohyphomycosis of brain: granulomatous encephalitis caused by *Drechslera spicifera*. Am J Clin Pathol 77:363–370
27. Koshi G, Anandi V, Kurien M et al (1987) Nasal phaeohyphomycosis caused by *Bipolaris hawaiiensis*. J Med Vet Mycol 25:397–402
28. Mathews MS, Maharajan SV (1990) *Exserohilum rostratum* causing keratitis in India. Med Mycol 37:279–283
29. Aquino VM, Norvell JM, Krisher K et al (1995) Fatal disseminated infection due to *Exserohilum rostratum* in a patient with aplastic anemia: case report and review. Clin Infect Dis 20:176–178
30. Hsu MM, Lee JY (1993) Cutaneous and subcutaneous phaeohyphomycosis caused by *Exserohilum rostratum*. J Am Acad Dermatol 28:340–344
31. McGinnis MR, Ranaldi MG, Winn RE (1986) Emerging agents of phaeohyphomycosis: pathogenic species of *Bipolaris* and *Exserohilum*. J Clin Microbiol 24:250–259
32. Agarwal A, Singh SM (1995) A case of cutaneous phaeohyphomycosis caused by *Exserohilum rostratum* its in vitro sensitivity and review of literature. Mycopathologia 131:9–12
33. Dereymaeker A, De Somer P (1955) Arachnidite fibro-purulente cerbello-cervical due a une moissiure (*Cladosporium)*. Acta Neurol Psychiatr Belg 55:629–632
34. Duque O (1961) Meningoencephalitis and brain abscess caused by *Cladosporium* and *Fonsecaea*. Am J Clin Pathol 36:505–517

35. Ellis MB (1971) Dematiaceous Hyphomycetes. Commonwealth Mycological Institute, Kew, Surrey, England, p 608
36. Ellis MB (1976) More dematiaceous hyphomycetes. Commonwealth Mycological Institute, Kew, Surrey, England, p 507
37. Meira GM, Neves ACA, Dias LB et al (1980) Cerebral cladosporiosis (dematiomycosis). A new case found in Brazil. Rev Inst Med Trop Sao Paulo 22:310–318
38. Salem FA, Kannangara DW, Nachum R (1983) Cerebral chromomycosis. Arch Neurol 40:173–174
39. Sandhyamani S, Bhatia R, Mohapatra LN et al (1981) Cerebral cladosporiosis. Surg Neurol 15:431–434
40. Gonzalez MS, Alfonso B, Seckinger D et al (1984) Subcutaneous phaeohyphomycosis caused by *Cladosporium devriesii*, sp. nov. J Med Vet Mycol 22:427–432
41. Brown JW, Nadell J, Sanders CV et al (1976) Brain abscess caused by *Cladosporium trichoides (bantianum)*: a case with paranasal sinus involvement. South Med J 69:1519–1523
42. Dixon DM, Walsh TJ, Merz WG et al (1989) Infection due to *Xylohypha bantiana* (*Cladosporium trichoides*). Rev Infect Dis 11:515–525
43. Fukomoto A, Mizuhara T, Matsuda Y et al (1980) Chromomycosis caused by *Cladosporium trichoides*. Jpn J Med Mycol 21:31–32
44. King AB, Collette TS (1952) Brain abscess due to *Cladosporium trichoides*. Report of the second case due to this organism. Bull Johns Hopkins Hosp 91:298–305
45. Middleton FG, Jurgenson PF, Utz JP et al (1976) Brain abscess caused by *C. trichoides*. Arch Intern Med 136:444–448
46. McGinnis MR, Borelli D (1981) *Cladosporium bantianum* and its synonym *Cladosporium trichoides*. Mycotaxon 13:127–136
47. McGinnis MR, Borelli D, Padhye AA et al (1986) Reclassification of *Cladosporium bantianum* in the genus *Xylohypha*. J Clin Microbiol 23:1148–1151
48. Kwon-Chung KJ, Wicker BL, Plaskowitz J (1989) Taxonomic clarification of *Cladosporium trichoides* Emmons and its subsequent synonyms. J Med Vet Mycol 27:413–426
49. Delfino DS, de Hoog GS, Polonelli L et al (2006) Survival of a neglected case of brain abscess caused by *Cladophialophora bantiana*. Med Mycol 44:651–654
50. Dixon DM, Polak A (1987) *In-vitro* and *in vivo* drug studies with three agents of central nervous system phaeohyphomycosis. Chemotherapy 33:129–140
51. Dixon DM, Shadomy HJ, Shadomy S (1977) Isolation of *Cladosporium trichoides* from nature. Mycopathologia 62:125–127
52. Dixon DM, Shadomy HJ, Shadomy S (1980) Dematiaceous fungal pathogens isolated from nature. Mycopathologia 70:153–161
53. Dixon DM, Merz WG, Elliot HL et al (1987) Experimental central nervous system phaeohyphomycosis following intranasal inoculation of *Xylohypha bantiana* in cortisone-treated mice. Mycopathologia 100:145–153
54. Barde AK, Singh SM (1984) *Cladosporium carrionii* Trejos 1954 infection of human nail. Mykosen 27:366–369
55. deHoog GS, Hermanides-Nijhof EJ (1977) The black yeast and allied hyphomycetes. Stud Mycol 15, Centraalbureau voor Schimmelcultures, Baarn
56. De Vries GA (1952) Contribution to the knowledge of the genus *Cladosporium* Link ex. Fr. thesis, CBS, Baarn, The Netherlands
57. Polak FM, Siliverio C, Bresky RH (1976) Corneal chromomycosis: double infection by *Phialophora verrucosa* (Medlar) and *Cladosporium cladosporioides* (Frescenius). Ann Opthalmol 8:139–144
58. Chandramukhi A, Gokul BN (1989) Progress in clinical neuroscience. Sinha KK (ed). Ranchi, pp 147–163
59. Tamsikar J, Naidu J, Singh SM (2005) Phaeohyphomycotic sebaceous cyst due to *Cladosporium cladosporioides*: case report and review of literature. J Mycol Med 16:55–57

60. Singh SM, Pathak SC, Kulkarni N et al (1991) First report of phaeohyphomycosis of *Nezara viridula* Linn. (Insecta: Heteroptera) caused by *Curvularia lunata*. Mycopathologia 116:37–43
61. Ebright JR, Chandrashekar PH, Marks S et al (1999) Invasive sinusitis and cerebritis due to *Curvularia clavata* in an immunocompetent adult. Clin Infect Dis 28:687–689
62. Bryan CS, Smith CW, Berg DE (1993) *Curvularia lunata* endocarditis treated with terbinafine: case report. Clin Infect Dis 16:30–32
63. Yau YC, de Nanassy J, Summerbell RC (1994) Fungal sternal wound infection due to *Curvularia lunata* in a neonate with congenital heart disease: case report and review. Clin Infect Dis 19:735–740
64. Janaki C, Sentamilselvi G, Janaki VR (1999) Case report. Eumycetoma due to *Curvularia lunata*. Mycoses 42:345–346
65. Carter E, Boudreaux C (2004) Fatal cerebral phaeohyphomycosis due to *Curvularia lunata* in an immunocompetent patient. J Clin Microbiol 42:5419–5423
66. Pimentel JD, Mahadevan K, Woodgyer A et al (2005) Peritonitis due to *Curvularia inaequalis* in an elderly patient undergoing peritoneal dialysis and a review of six cases of peritonitis associated with other spp. J Clin Microbial 43:4288–4292
67. Forester RK, Rebell G, Wilson LA (1975) Dematiaceous fungal keratitis. Clinical isolates and management. Br J Ophthalmol 59:372–376
68. Rohweeder JJ, Simmons JL, Colfer H et al (1979) Disseminated *Curvularia lunata* infection in a football player. Arch Intern Med 139:940–941
69. Barde AK, Singh SM (1983) A case of onychomycosis caused by *Curvularia lunata* (Wakker) Boedijn. Mykosen 26:311–316
70. Agarwal PK, Lal B, Shukla PK et al (1982) Clinical and experimental keratitis due to *Curvularia lunata* (Wakker) *Boedijn* var. *aeria* (Batista, Lima and Vasconcelos) Ellis. Sabouraudia 20:225–232
71. Nityananda K, Sivasubramaniam P, Ajello L (1964) A case of mycotic keratitis caused by *Curvularia geniculata*. Arch Ophthalmol 71:456–458
72. Kaufman SM (1971) *Curvularia endocarditis* following cardiac surgery. Am J Clin Pathol 56:466–470
73. Lampert RP, Hutto JH, Donnelly WH et al (1977) Pulmonary and cerebral mycetoma caused by *Curvularia pallescens*. J Pediatr 91:603–605
74. Vishnoi S, Naidu J, Singh SM et al (2005) Pathogenicity of *Curvularia geniculata* for albino rats. Study of clinical isolate from blood of a cancer patient. J Mycol Med 15:97–102
75. Georg LK, Bierer BW, Cooke WB (1964) Encephalitis in turkey poults due to a new fungus species. Sabouraudia 3:239–244
76. Bhatt GC, Kendrik WB (1968) The generic concept of *Diplorhinotrichum* and *Dactylaria*, and a new species of *Dactylaria* from soil. Can J Bot 46:1253–1257
77. Ajello L, McGinnis MR, Camper J (1997) An outbreak of phaeohyphomycosis in rainbow trout caused by *Scolecobasidium humicola*. Mycopathologia 62:15–22
78. Dixon DM, Salkin IF (1986) Morphologic and physiologic studies of three dematiaceous pathogens. J Clin Microbiol 24:12–15
79. Salkin IF, Dixon DM (1987) *Dactylaria constricta*: description of two varieties. Mycotaxon 29:377–338
80. Cannon PF (1990) Name changes in fungi of microbiological, industrial and medical importance. Mycopathologia 111:75–83
81. Fukushiro F, Udagawa Y, Kawashima Y et al (1986) Subcutaneous abscesses caused by *Ochroconis gallopavum*. J Med Vet Mycol 24:175–182
82. Terreni AA, Crymes WB, Morris PR et al (1986) Disseminated *Dactylaria gallopava* infection in an immunocompromised human being- Abstract annual meeting of the American Society for Microbiology, Washington, DC, pp 404–404
83. Dixon DM, Walsh TJ, Salkin IF et al (1987) *Dactylaria constricta*: another dematiaceous fungus with neurotropic potential in mammals. J Med Vet Mycol 25:55–58

84. Sekhon AS, Padhye AA, Standard PG et al (1990) Antigenic relationship of *Dactylaria gallopava* to *Scolecobasidium constrictum*. J Med Vet Mycol 28:59–66
85. Terreni AA, Disalvo A, Baker AS et al (1990) Disseminated *Dactylaria gallopava* infection in a diabetic patient with chronic lymphocytic leukemia of the T-cell type. Am J Clin Pathol 94:104–107
86. Sides EH III, Benson JD, Padhye AA (1991) Phaeohyphomycotic brain abscess due to *Ochroconis gallopavum* in a patient with malignant lymphoma of a large cell type. J Med Vet Mycol 29:317–322
87. Matsumoto T, Nishimoto K, Kimura K et al (1984) Phaeohyphomycosis caused by *Exophiala moniliae*. J Med Vet Mycol 22:17–26
88. Ravissi P, Vindas AJR (1981) Mycotic cysts. A histopathologic study. Bull Soc Pathol Exot 74:46–54
89. Zeifer A, Conner DH (1980) Phaeomycotic cyst. A clinicopathological study of twenty five patients. Am J Trop Med Hyg 29:901–911
90. Chandler FW, Kaplan W, Ajello L (1980) A colour atlas and text book of the histopathology of mycotic diseases. Wolfe Medical Publications Ltd, London
91. Calista D, Leardini M, Arcangeli F (2003) Subcutaneous *Exophiala jeanselmei* infection in heart transplant patient. Eur J Dermatol 13:489
92. South DA, Brass C, Stevens DA (1981) Chromohyphomycosis. Treatment with ketoconazole. Arch Dermatol 117:311–312
93. Singh SM, Pouranik M, Naidu J (1992) Cutaneous phaehyphomycosis caused by *Exophiala jeanselmei* var. lecanii-cornii (Benedek and Specht) de Hoog. Indian J Pathol Microbiol 35:269–273
94. Prabhakar Y, Rao R, Sharma S et al (1983) A rare case of phaeohyphomycosis caused by *Exophiala jeanselmei*. Indian J Dermatol Venerol Leprol 49:17–20
95. Lal S, Garg BR, Rao RS (1984) Chromomycosis caused by *Exophiala jeanselmei*. Indian J Dermatol Venereol Leprol 50:119–121
96. Hemashettar BM, Patil CS, Nagalotimath SJ et al (1986) Mycetoma due to *Exophiala jeanselmei* (a case report with description of the fungus). Indian J Pathol Microbiol 29:75–78
97. Hemashettar BM, Patil CS, Siddaramappa B et al (1985) A case of tinea nigra from South India. Indian J Dermatol Venerol Leprol 51:164–166
98. Naka W, Harada T, Nishikawa T et al (1986) A case of chromoblastomycosis with special reference to the mycology of the isolated *Exophiala jeanselmei*. Mykosen 29:445–452
99. Rajam RV, Kandhari DC, Thirumalachar MJ (1958) Chromoblastomycosis caused by a rare yeast like dematiaceous fungus. Mycopathol Mycol Appl 9:5–19
100. McGinnis MR, Padhye AA (1977) *Exophiala jeanselmei*, a new combination for *Phialophora jeanselmei*. Mycotaxon 5:341–352
101. Padhye AA, Ajello L, Chandler FW et al (1983) Phaeohyphomycosis in El Salvador caused by *Exophiala spinifera*. Am J Trop Med Hyg 32:799–803
102. Padhye AA, Kaplan W, Neuman MA et al (1984) Subcutaneous phaeohyphomycosis caused by *Exophiala spinifera*. J Med Vet Mycol 22:493–500
103. da Lacaz C, Porto E, de Andrade JG et al (1984) Faeohifomicose disseminada por *Exophiala spinifera*. Anais Brasil de Dermatol 59:238–243
104. Dai WL, Ren ZF, Wen JZ et al (1987) First case of systemic phaeohyphomycosis caused by *Exophiala spinifera* in China. Chin J Dermatol 20:13–16
105. Kettiewell P, McGinnis MR, Wilkinson GT (1989) Phaeohyphomycosis caused by *Exophiala spinifera* in two cats. J Med Vet Mycol 27:257–264
106. Dasgupta LR, Agrawal SC, Bedi BMS (1975) Tinea capitis in Pondicherry (South India). Sabouraudia 13:41–43
107. Smith JG, Sams WM, Roth FJ (1958) Tinea nigra palmaris; a disorder easily confused with junction nevus of the palm. JAMA 167:312–314
108. Kano KA (1934) Uber die chromoblastomykose durch einen noch nicht als pathogen beschriebenen pilz. Aichi Igakki Zasshi 41:1657–1673

109. Urabe H, Suenga Y, Yasuda M et al (1977) Recent advances in medical and veterinary mycology. Ikawa K (ed). University of Tokyo Press, Tokyo, pp 97–106
110. Harada S, Ueda T, Kusunoki T (1976) Systemic chromomycosis. J Dermatol 3:13–17
111. Hohl PE, Hollev HP Jr, Prevost E et al (1983) Infections due to *Wangiella dermatitidis* in humans: report of the first documented case from the United States and a review of the literature. Rev Infect Dis 5:854–864
112. Polak A (1984) Experimental infection of mice by *Fonsecaea pedrosoi* and *Wangiella dermatitidis*. J Med Vet Mycol 22:167–169
113. Morio F, Berre JY, Garcia-Hermoso D et al (2012) Phaeohyphomycosis due to *Exophiala xenobiotica* as a cause of fungal arthritis in an HIV-infected patient. Med Mycol 50:513–517
114. Yoon YA, Park KS, Lee JH et al (2012) Subcutaneous phaeohyphomycosis caused by *Exophiala salmonis*. Ann Lab Med 32:438–441
115. Li DM, Li RY, de Hoog GS et al (2009) *Exophiala asiatica*, a new species from a fatal case in China. Med Mycol 47:101–109
116. Harris JE, Sutton DA, Rubin A et al (2009) *Exophiala spinifera* as a cause of cutaneous phaeohyphomycosis: case study and review of the literature. Med Mycol 47:87–93
117. Singal A, Pandhi D, Bhattacharya SN et al (2008) Phaeohyphomycosis caused by *Exophiala spinifera*: a rare occurrence. Int J Dermatol 47:44–47
118. Punithalingam E, Waterston JM (1970) C.M.I. Descriptions of pathogenic fungi and bacteria no 274. Commonwealth Mycological Institute, Kew, Surrey, England, p 2
119. Gentles JC, Evans EGV (1970) Infections of feet and nails by *Hendersonula toruloidea*. Sabouraudia 8:72–75
120. Gugnani HC, Oyeka CA (1989) Foot infections due to *Hendersonula toruloidea* and *Scytalidium hyalinum* in coal miners. J Med Vet Mycol 27:169–179
121. Singh SM (1990) Nondermatophytic onychomycosis: a review. In: Hasija SK, Bilgrami HS (eds), (Prof GP Agrawal Festschrift), Perspectives in mycological research, vol 2.Today and Tomorrow Publisher, New Delhi, pp 293–310
122. Singh SM, Barde AK (1980) *Hendersonula toruloidea* infection of human skin and nails. Indian J Dermatol Venerol Leprol 46:350–355
123. Campbell CK, Mulder JL (1977) Skin and nail infection by *Scytalidium hyalinum*. Sabouraudia 15:161–166
124. Singler L, Summerbell RC, Poole L et al (1997) Invasive *Nattrassia mangiferae* infections. Case report, literature review and therapeutic and taxonomic appraisal. J Clin Microbiol 35:433–440
125. Dhindsa MK, Naidu J, Singh SM (1998) A case of subcutaneous infection n a patient with discoid lupus erythematosus caused by a *Scytalidium synanamorph* of *Nattrassia mangiferae* and its treatment. Med Mycol 36:425–427
126. Elewski BE (1996) Onychomycosis caused by *Scytalidium dimidiatum*. J Am Acad Dermatol 35:336–338
127. Little MG, Hammond ML (1995) *Scytalidium dimidiatum* in Australia. Australas J Dermatol 36:204–205
128. Al-Rajhi A, Awad AH, AL-Hedaithy SS et al (1993) *Scytalidium dimidiatum* fungal endophthalmitis. Br J Ophthalmol 77:388–390
129. Crous PW, Slippers B, Wingfield MJ et al (2006) Phylogenetic lineages in the Botryosphaeriaceae. Stud Mycol 55:235–253
130. deHoog GS, Mayser P, Hasse G et al (2000) a new species, *Phialophora europaea*, causing superficial infections in humans. Mycoses 43:409–416
131. Mostert L, Groenewald JZ, Summerbell RC et al (2005) Species of *Phaeoacremonium* associated with infections in humans and environmental reservoirs in infected woody plants. J Clin Microbiol 43:1752–1767
132. Vijay Krishna D, Mosterl L, Jeewon R et al (2005) *Pleurostomophora*, an anamorph of *Pleurostoma* (Calosphaeriales), a new anamorph genus morphologically similar to *Phialophora*. Stud Mycol 50:387–395

133. Singh SM, Agrawal A, Naidu J et al (1992) Cutaneous phaeohyphomycosis caused by *Phialophora richardsiae* and the effect of topical clotrimazole in its treatment. Ant Van Leeu 61:51–55
134. Corrado ML, Weitzman I, Stanek A et al (1980) Subcutaneous infection with *Phialophora richardsiae* and its susceptibility to 5-fluorocytosine, amphotericin B and miconazole. Sabouraudia 18:97–104
135. Torstrick RF, Harrison K, Heckmann JD et al (1979) Chronic bursitis caused by *Phialophora richardsiae.* A case report. J Bone Joint Surg 61:772–779
136. Gueho E, Bonnefoy A, Lubionski J et al (1989) Subcutaneous granuloma caused by *Phialophora richardsiae*: case report and review of the literature. Mycoses 32:219–223
137. Ikai K, Tomono H, Watanabe S (1988) Phaeohyphomycosis caused by *Phialophora richardsiae*. J Am Acad Dermatol 19:478–481
138. Yangco BG, Testrake D, Okafor J (1984) *Phialophora richardsiae* isolated from infected human bone: morphological, physiological and antifungal studies. Mycopathologia 86:103–111
139. Moskowitz LB, Cleary TJ, McGinnis MR et al (1983) *Phialophora richardsiae* in a lesion appearing as a giant cell tumor of the tendon sheath. Arch Path Lab Med 107:374–370
140. Corrado ML, Kramer M, Cummings M et al (1982) Susceptibility of fungi to amphotericin B, miconazole, ketoconazole and rifampin alone and in combination. Sabouraudia 20:109–113
141. Dixon DM, Polak A, Walshe TJ (1988) Phaeohyphomycosis. In: JM Torres-Rodriguez (ed) Proceedings of the Xth international congress ISHAM, JR Prous Science, Barcelona
142. McGinnis MR, Pasarell L (1998) *In-vitro* evaluation of terbinafine and itraconazole against dematiaceous fungi. Med Mycol 36:243–246
143. Kameda Y, Sasaki T, Takahashi Y et al (1984) A case of phaeomycotic cyst caused by *Phialophora verrucosa.* Jpn J Med Mycol 25:379–386
144. McGinnis MR (1983) Chromoblastomycosis and phaeohyphomycosis: new concepts, diagnosis, and mycology. J Am Acade Dermatol 8:1–16
145. Duggan JM, Wolf MD, Kauffman CA (1995) *Phialophora verrucosa* infection in an AIDS patient. Mycoses 38:215–218
146. Turiansky GW, Benson PM, Sperling LC et al (1995) *Phialophora verrucosa*: a new cause of mycetoma. J Am Acad Dermatol 32:311–315
147. Ajello L, George LK, Steigbigel RT et al (1974) A case of phaeohyphomycosis caused by a new species of *Phialophora.* Mycologia 66:490–498
148. Ferraro FA, Morgan MA (1989) A case of disseminated *Phialophora parasitica* infection. Arch Pathol Lab Med 113:1379–1381
149. Fincher RME, Fisher JF, Padhye AA et al (1988) Subcutaneous phaeohyphomycotic abscess caused by *Phialophora parasitica* in a renal allograft recipient. J Med Vet Mycol 26:311–314
150. Lavarde V, Bedrossian J, De Bievre C et al (1982) Un cas de phaeomycose a *Phialophora parasitica* chez un transplante. Deuxieme observation mondiale. Bull De la Soc Fr de Mycol 11:273–277
151. Ziza JM, Dupont E, Boissonnas A et al (1985) Osteoarthritis caused by dematiaceous fungi. Apropos of 3 cases. Ann Med Int 136:390–397
152. Kaell AT, Weitzman I (1983) Acute monoarticular arthritis due to *Phialophora parasitica.* Am J Med 74:519–522
153. Rowland MD, Farrar WE (1987) Case report: thorn induced *Phialophora parasitica* arthritis treated successfully with synovectomy and ketoconazole. Am J Med Sci 30:393–395
154. Porto E, Lacaz CS, Sabbaga E et al (1979) *Phialophora bubakii.* Isolation from a subcutaneous abscess in a kidney-transplant patient. Rev Inst Med Trop 21:106–109
155. Singh SM, Barde AK (1986) Opportunistic infections of skin and nails by non-dermatophytic fungi. Mykosen 29:272–277
156. Meyers WM, Dooley JR, Kwon-Chung KJ (1975) Mycotic granuloma caused by *Phialophora repens.* Am J Clin Pathol 64:549–555

157. Mariat F, Segretain G, Destombes P et al (1967) A subcutaneous mycotic cyst (simulating sporotrichosis) caused by *Phialophora gougerotii* (Matruchot 1910) Borelli 1955, observed in a Senegalese. Sabouraudia 5:209–219
158. Guppy KH, Thomas C, Thomas K et al (1998) Cerebral fungal infections in the immunocompromised host: a literature review and a new pathogen-*Chaetomium atrobrunneum*: case report. Neurosurgery 43:1463–1469
159. Lesire V, Hazouard E, Dequin PF et al (1999) Possible role of *Chaetomium globosum* in infection after autologous bone marrow transplantation. Intensive Care Med 25:124–125
160. Abbott SP, Sigler L, McAleer R et al (1995) Fatal cerebral mycoses caused by the ascomycete *Chaetomium strumarium*. J Clin Microbiol 33:2692–2698
161. deHoog GS, Guarro J, Gene J et al (2009) Atlas of clinical fungi, pilot version of 3rd edn., CD-ROM. Centraalbureau voor Schimmelcultures, Utrecht
162. Aribandi M, Bazan C III, Rinaldi MG (2005) Magnetic resonance imaging findings in fatal primary cerebral infection due to *Chaetomium strumarium*. Australas Radiol 49:166–169
163. Barron MA, Sutton DA, Veve R et al (2003) Invasive mycotic infections caused by *Chaetomium perlucidum*, a new agent of cerebral phaeohyphomycosis. J Clin Microbiol 41:5302–5307
164. Marques AR, Kwon-Chung KJ, Holland SM et al (1995) Suppurative cutaneous granulomata caused by *Microascus cinereus* in a patient with chronic granulomatous disease. Clin Infect Dis 2:110–114
165. Baddley JW, Moser SA, Sutton DA et al (2000) *Microascus cinereus* (Anamorph *Scopulariopsis*) brain abscess in a bone marrow transplant recipient. J Clin Microbiol 38:395–397
166. Celard M, Dannaoui E, Piens MA et al (1999) Early *Microascus cinereus* endocarditis of a prosthetic valve implanted after *Staphylococcus aureus* endocarditis of the native valve. Clin Infect Dis 29:691–692
167. Agarwal GP, Singh SM (1980) *Microascus cinereus* infection of human nail. Indian J Med Sci 34:263–265
168. Krisher KK, Holdridge NB, Mustafa MM et al (1995) Disseminated *Microascus cirrosus* infection in pediatric bone marrow transplant recipient. J Clin Microbiol 33:735–737
169. Mohammedi I, Piens MA, Audigier-Valette C et al (2004) Fatal *Microascus trigonosporus* (anamorph *Scopulariopsis*) pneumonia in a bone marrow transplant recipient. Eur J Clin Microbiol Infect Dis 23:215–217
170. Scott IU, Cruz-Villegas V, Flynn HW Jr et al (2004) Delayed-onset, bleb-associated endophthalmitis caused by *Lecythophora mutabilis*. Am J Ophthalmol 137:583–585
171. Marriott DJ, Wong KH, Aznar E et al (1997) *Scytalidium dimidiatum* and *Lecythophora hoffmannii*: unusual causes of fungal infections in a patient with AIDS. J Clin Microbiol 35:2949–2952
172. Drees M, Wickes BL, Gupta M et al (2007) *Lecythophora mutabilis* prosthetic valve endocarditis in a diabetic patient. Med Mycol 45:463–467
173. Heath CH, Lendrum JL, Wetherall BL et al (1997) *Phaeoacremonium parasiticum* infective endocarditis following liver transplantation. Clin Infect Dis 25:1251–1252
174. Padhye AA, Davis MS, Baer D et al (1998) Phaeohyphomycosis caused by *Phaeoacremonium inflatipes*. J Clin Microbiol 36:2763–2765
175. Crous PW, Gams W, Wingfield MJ et al (1996) *Phaeoacremonium* gen. nov., associated with wilt and decline diseases of woody hosts and human infections. Mycologia 88:786–796
176. Matsui T, Nishimoto K, Udagawa S et al (1999) Subcutaneous phaeohyphomycosis caused by *Phaeoacremonium rubrigenum* in an immunosuppressed patient. Nippon Ishinkin Gakkai Zasshi 40:99–102
177. Guarro J, Silvestre AM Jr, Verkley G et al (2006) Limitations of DNA sequencing for diagnosis of a mixed infection by two fungi, *Phaeoacremonium venezuelense* and a *Plectophomella* sp., in a transplant recipient. J Clin Microbiol 44:4279–4282
178. Baddley JW, Mostert L, Summerbell RC et al (2006) *Phaeoacremonium parasiticum* infections confirmed by beta-tubulin sequence analysis of case isolates. J Clin Microbiol 44:2207–2211

179. Summerbell RC, Krajden S, Levine R et al (2004) subcutaneous phaeohyphomycosis caused by *Lasiodiplodia theobromae* and successfully treated surgically. Med Mycol 42:543–547
180. Woo PC, Lau SK, Ngan AH et al (2008) *Lasiodiplodia theobromae* pneumonia in a liver transplant recipient. J Clin Microbiol 46:380–384
181. Rebell G, Forster RK (1976) *Lasiodiplodia theobromae* as a cause of keratomycoses. Sabouraudia 14:155–157
182. Slomovic AR, Forster RK, Gelender H (1985) *Lasiodiplodia theobromae* panophthalmitis. Can J Ophthalmol 20:225–228
183. Srinivasan A, Wickes BL, Romanelli AM et al (2009) Cutaneous infection caused by *Macrophomina phaseolina* in a child with acute myeloid leukemia. J Clin Microbiol 47:1969–1972
184. Moore MK (1986) *Hendersonula toruloidea* and *Scytalidium hyalinum* infections in London, England. J Med Vet Mycol 24:219–230
185. Moore MK (1992) The infection of human skin and nail by *Scytalidium* species. Curr Top Med Mycol 4:1–42
186. Mani RS, Chickabasaviah YT, Nagarathna SJ et al (2008) Cerebral phaeohyphomycosis caused by *Scytalidium dimidiatum*: a case report from India. Med Mycol 46:705–711
187. Benne CA, Neeleman C, Bruin M et al (1993) Disseminating infection with *Scytalidium dimidiatum* in a granulocytopenic child. Eur J Clin Microbiol Infect Dis 12:18–121
188. Kanj SS, Amr SS, Roberts GD (2001) *Ramichloridium mackenziei* brain abscess: report of two cases and review of the literature. Med Mycol 39:97–102
189. Badali H, Chander J, Bansal S et al (2010) First autochthonous case of *Rhinocladiella mackenziei* cerebral abscess outside the Middle East. J Clin Microbiol 48:646–649
190. Arango M, Jaramillo C, Cortés A et al (1998) Auricular chromoblastomycosis caused by *Rhinocladiella aquaspersa*. Med Mycol 36:43–45
191. Arzanlose M, Groenewald JZ, Gams W et al (2007) Phylogenetic and morphotaxonomic revision of *Ramichloridium* and allied genera. Stud Mycol 58:57–93
192. del Palacio-Hernanz A, Moore MK, Campbell CK et al (1989) Infection of the central nervous system by *Rhinocladiella atrovirens* in a patient with acquired immunodeficiency syndrome. J Med Vet Mycol 27:127–130
193. N'diaye B, Develoux M, Dieng MT et al (2000) Current report of mycetoma in Senegal: report of 109 cases. J Mycol Med 10:140–144
194. Cai Q, Lv GX, Jiang YQ et al (2013) The first case of phaeohyphomycosis caused by *Rhinocladiella basitona* in an immunocompetent child in China. Mycopathologia 176:101–105
195. Al-Mohsen IZ, Sutton DA, Sigler L et al (2000) *Acrophialophora fusispora* brain abscess in a child with acute lymphoblastic leukemia. Case history, clinical course, and literature review. J Clin Microbiol 38:4569–4576
196. Shukla PK, Khan ZA, Lal B et al (1983) Clinical and experimental keratitis caused by the *Colletotrichum* state of *Glomerella cingulata* and *Acrophialophora fusispora*. Sabouraudia 21:137–147
197. Osherov A, Schwammenthal E, Kuperstein R et al (2006) *Phialemonium curvatum* prosthetic valve endocarditis with an unusual echocardiographic presentation. Echocardiography 23:503–505
198. Guarro J, Nucci M, Akiti T et al (1999) Phialemonium. Fungemia: two documented nosocomial cases. J Clin Microbiol 37:2493–2497
199. Dan M, Yossepowitch O, Hendel D et al (2006) *Phialemonium curvatum* arthritis of the knee following intra-articular injection of a corticosteroid. Med Mycol 44:571–574
200. Gams W, McGinnis MR (1983) *Phialemonium*, a new anamorph genus intermediate between *Phialophora* and *Acremonium*. Mycologia 75:977–987
201. Weinberger M, Mahrshak I, Keller N et al (2006) Isolated endogenous endophthalmitis due to a sporodochial-forming *Phialemonium curvatum* acquired through intracavernous autoinjections. Med Mycol 44:253–259

202. Rivero M, Hidalgo A, Alastruey-Izquierdo A et al (2009) Infections due to *Phialemonium* species: case report and review. Med Mycol 47:766–774
203. Cortez KJ, Roilides E, Quiroz-Telles F et al (2008) Infections caused by *Scedosporium* spp. Clin Microbiol Rev 21:157–197
204. Tintelnot K, Just-Nubling G, Horre R et al (2009) A review of German *Scedosporium prolificans* cases from 1993 to 2007. Med Mycol 47:351–358
205. Berenguer J, Rodriguez-Tudela JL, Richard C et al (1997) Deep infections caused by *Scedosporium prolificans*. A report on 16 cases in Spain and a review of the literature. *Scedosporium prolificans* Spanish study group. Medicine (Baltimore) 76:256–265
206. Bouza E, Munoz P, Vega L et al (1996) Clinical resolution of *Scedosporium prolificans* fungemia associated with reversal of neutropenia following administration of granulocyte colony-stimulating factor. Clin Infect Dis 23:192–193
207. Wilson CM, Rourke EJO, McGinnis MR et al (1990) *Scedosporium inflatum*: clinical spectrum of a newly recognized pathogen. J Infect Dis 161:102–107
208. Wood GM, McCormack JG, Muir DB et al (1992) Clinical features of human infection with *Scedosporium inflatum*. Clin Infect Dis 14:1027–1033
209. Rodriguez-Tudela JL, Berenguer J, Guarro J et al (2009) Epidemiology and outcome of *Scedosporium prolificans* infection, a review of 162 cases. Med Mycol 47:359–370
210. Grenouillet F, Botterel F, Crouzet J et al (2009) *Scedosporium prolificans*: an emerging pathogen in France? Med Mycol 47:343–350
211. Gilgado F, Cano J, Gene J et al (2005) Molecular phylogeny of the *Pseudallescheria boydii* species complex: proposal of two new species. J Clin Microbiol 43:4930–4942
212. Gilgado F, Cano J, Gene J et al (2008) Molecular and phenotypic data supporting distinct species statuses for *Scedosporium apiospermum* and *Pseudallescheria boydii* and the proposed new species *Scedosporium dehoogii*. J Clin Microbiol 46:766–771
213. Leao AC, Weiss VA, Vicente VA et al (2017) Genome sequence of type strain *Fonsecaea multimorphosa* CBS 980.96(T), a causal agent of feline cerebral phaeohyphomycosis. Genome Announc 16:5–7
214. Najafzadeh MJ, Gueidan C, Badali H et al (2009) Genetic diversity and species delimitation in the opportunistic genus *Fonsecaea*. Med Mycol 47:17–25
215. Minotto R, Bernardi CD, Mallmann LF et al (2001) Chromoblastomycosis: a review of 100 cases in the state of Rio Grande do Sul, Brazil. J Am Acad Dermatol 44:585–592
216. Xi L, Lu C, Sun J et al (2009) Chromoblastomycosis caused by a meristematic mutant of *Fonsecaea monophora*. Med Mycol 47:27–33
217. Surash S, Tyagi A, de Hoog GS et al (2005) Cerebral phaeohyphomycosis caused by *Fonsecaea monophora*. Med Mycol 43:465–472
218. Sutton DA, Timm WD, Morgan-Jones G et al (1999) Human phaeohyphomycotic osteomyelitis caused by the coelomycete *Phomopsis saccardo* 1905: criteria for identification, case history, and therapy. J Clin Microbiol 37:807–811
219. de Hoog GS, Attili-Angelis D, Vicente VA et al (2004) Molecular ecology and pathogenic potential of *Fonsecaea* species. Med Mycol 42:405–416
220. Guarro J, Mayayo E, Tapiol J et al (1999) *Microsphaeropsis olivacea* as an etiological agent of human skin infection. Med Mycol 37:133–137
221. Shah CV, Jones DB, Holz ER (2001) *Microsphaeropsis olivacea* keratitis and consecutive endophthalmitis. Am J Ophthalmol 131:142–143
222. Kiehn TE, Polsky B, Punithalingam E et al (1987) Liver infection caused by *Coniothyrium fuckelii* in a patient with acute myelogenous leukemia. J Clin Microbiol 25:410–2412
223. Siu K, Izumi AK (2004) Phaeohyphomycosis caused by *Coniothyrium*. Cutis 73:127–130
224. Verkley GJM, da Silva M, Wicklow DT et al (2004) *Paraconiothyrium*, a new genus to accommodate the mycoparasite *Coniothyrium minitans*, anamorphs of *Paraphaeosphaeria*, and four new species. Stud Mycol 50:323–335
225. Ocampo MA, Kanitakis J, Bienvenu AL et al (2012) Phaeohyphomycosis caused by *Pyrenochaeta romeroi* mimicking a plantar wart in a kidney transplant recipient. Transpl Infect Dis 14:173–174

226. Urano S, Suzuki Y, Anzawa K et al (2014) Phaeomycotic cyst caused by *Exophiala xenobiotica* in a patient with rheumatoid arthritis and lung cancer. Med Mycol J 55:151–156
227. Welfringer A, Vuong V, Argy N et al (2017) A rare fungal infection: Phaeohyphomycosis due to *Veronaea botryosa* and review of literature. Med Mycol Case Rep 15:21–24
228. Kondo Y, Hiruma M, Matsushita A et al (2007) Cutaneous phaeohyphomycosis caused by *Veronaea botryosa* observed as sclerotic cells in tissue. Int J Dermatol 46:625–627
229. Xue Y, Chen H, Hu S et al (2011) Cutaneous phaeohyphomycosis on the auricle due to *Veronaea botryosa*. Eur J Dermatol 21:418–419
230. Sutton DA, Rinaldi MG, Kielhofner M (2004) First U.S. report of subcutaneous phaeohyphomycosis caused by *Veronaea botryosa* in a heart transplant recipient and review of the literature. J Clin Microbiol 42:2843–2846
231. Foulet F, Duvoux C, de Bievre C et al (1999) Cutaneous phaeohyphomycosis caused by *Veronaea botryosa* in a liver transplant recipient successfully treated with itraconazole. Clin Infect Dis 29:689–690
232. Soto E, Richey C, Reichley SR et al (2017) Diversity of *Veronaea botryosa* from different hosts and evaluation of laboratory challenge models for phaeohyphomycosis in *Acipenser transmontanus*. Dis Aquat Org 125:7–18
233. Hosoya T, Hanafusa Y, Kudo T et al (2015) First report of *Veronaea botryosa* as a causal agent of chromomycosis in frogs. Med Mycol 53:369–377
234. Engleberg NC, Johnson J IV, Bluestein J et al (1987) Phaeohyphomycotic cyst caused by a recently described species, *Phaeoannellomyces elegans*. J Clin Microbiol 25:605–608
235. Guarro J, DeVroey C, Gene J (1991) Concerning the implication of *Arthrobotrys oligospora* in a case of keratitis. J Med Vet Mycol 29:349–352
236. Thomas PA, Kuriakose T (1990) Keratitis due to *Arthrobotrys oligospora* Fres. 1850. J Med Vet Mycol 28:47–50
237. Naidu J, Singh SM, Pouraniik M (1991) Onychomycosis caused by *Scopulariopsis brumptii*: A case report and sensitivity studies. Mycopathologia 113:159–164
238. Markham WD, Key RD, Padhye AA et al (1990) Phaeohyphomycotic cyst caused by *Tetraploa aristata*. J Med Vet Mycol 28:147–150
239. Newmark E, Polack FM (1970) Tetraploa keratomycosis. Am J Ophthalmol 70:1013–1015
240. Lecso-Bronet M, Gueho E, Barbier-Boehm GE et al (1991) Prosthetic valve endocarditis due to *Thermomyces lanuginosus* Tsiklinsky–first case report. J Med Vet Mycol 29:205–209
241. Benoldi D, Alinovi A, Polonelli L et al (1991) *Botryomyces caespitosus* as an agent of cutaneous phaeohyphomycosis. J Med Vet Mycol 29:9–13
242. Anandi V, Jacob JT, Walter A (1989) Cerebral phaeohyphomycosis caused by *Chaetomium globosum* in a renal transplant recipient. J Clin Microbiol 27:2226–2229
243. Naidu J, Singh SM, Pouranik M (1991) Onychomycosis caused by *Chaetomium globosum Kunze*. Mycopathologia 113:31–34
244. Ajello L (1978) The black yeasts as disease agents: historical perspective. In: Proceedings fourth international conference on the mycoses. Scientific publication no. 356. Pan American Health Organization, Washington, DC, pp 9–16
245. Rodrigues MM, Laibson P, Kalpan W (1975) Exogenous corneal ulcer caused by *Tritirachium roseum*. Am J Opthalmol 80:804–806
246. Punithalingam E (1979) Sphaeropsidales in culture from humans. Nova Hedwigia 31:19–158
247. Bakerspigel A, Lowe D, Rostas A (1981) The isolation of *Phoma eupyrena* from a human lesion. Arch Dermatol 117:362–363
248. Baker JG, Salkin LF, Forgacs P et al (1987) First report of subcutaneous phaeohyphomycosis of the foot caused by *Phoma minutella*. J Clin Microbiol 25:2395–2397
249. Chen WT, Tu ME, Sun PL (2016) Superficial phaeohyphomycosis caused by *Aureobasidium melanogenum* mimicking tinea nigra in an immunocompetent patient and review of published reports. Mycopathologia 181:555–560
250. Blomqvist K, Salonen A (1969) *Oidiodendron cerealis* isolated from neurodermitis nuchae. Dermatologica 139:255–293
251. Dawson CO, Lepper AWD (1970) *Peyronellaea glomerata* infection of the ear pinna in goats. Sabouraudia 8:145–148

252. McGinnis MR, McKenzie RA, Connole MD (1985) *Phaeosclera dematioides*, a new etiologic agent of phaeohyphomycosis in cattle. J Med Vet Mycol 23:133–135
253. McGinnis MR, Padhye AA (1978) *Cladosporium castellanii* is a synonym of *Stenella araguata*. Mycotaxon 8:415–418
254. Pietrini P, Stewart WM (1977) Granulome perinarinaire du a *Taeniolella stilbospora* (Corda) Hughes. Bull Soc Fr Mycol Med 6:97–99
255. Hibbits J, Hughes GC, Sparker AK (1981) *Trichomaris invadens* gen. et sp. nov., an ascomycete parasite of the tanner crab (*Chionoecetes bairdi* Rathbun *Crustacea*; *Brachyura*). Can J Bot 59:2121–2178
256. Ajello L, Padhye AA, Payne M (1980) Phaeohyphomycosis in a dog caused by *Pseudomicrodochium suttonii* sp. Mycotaxon 12:131–136
257. Wood C, Russel-Bell B (1983) Characterization of pigmented fungi by melanin staining. Am J Dermatopathol 5:77–81
258. Lomax LG, Cole JR, Padhye AA et al (1986) Osteolytic phaeohyphomycosis in a German shepherd dog caused by *Phialemonium obovatum*. J Clin Microbiol 23:987–991
259. McGinnis MR, Sorrell DE, Miller RL et al (1981) Subcutaneous phaeohyphomycosis caused by *Exophiala moniliae*. Mycopathologia 73:69–72
260. Drouhet E, Dupont B (1983) Laboratory and clinical assessment of ketoconazole in deep-seated mycoses. Am J Med 74:30–47
261. de Hoog GS, Horre R (2002) Molecular taxonomy of the *Alternaria* and *Ulocladium* species from humans and their identification in the routine laboratory. Mycoses 45:259–276
262. Mouisdale MT, Harper JM, Thatcher GN (1981) Gal peritonitis: complication of continuous ambulatory peritoneal dialysis. Med J Aust 1:88
263. Drouhet E, de Bievre C, Wahab S (1985) *Drechslera longirostrata* (Subramanian) and other *Drechslera* species pathogenic to humans and animals. Proc Indiana Acad Sci 94:453–463
264. Barde AK (1983) Studies on human fungal infection of skin and nails with special reference to non dermatophytes at Balaghat (M.P.). PhD thesis, Rani Durgawati University, Jabalpur, India
265. Castanet J, Lacour JP, Toussaint-Gary M et al (1995) *Alternaria tenuissima* plurifocal cutaneous infection. Ann Dermatol Venereol 122:115–118
266. Magina SM, Lisboa C, Oliveira G et al (2000) Cutaneous alternariosis by *Alternaria chartarum* in a renal transplant patient. Br Assoc Dermatol 142:1234–1264
267. Estes SA, Merz WG, Maxwell LG (1977) Primary cutaneous phaeohyphomycosis caused by *Drechslera spicifera*. Arch Dermatol 113:813–815
268. Morais OO, Porto C, Coutinho AS et al (2011) Infection of the lymphatic system by *Aureobasidium pullulans* in a patient with erythema nodosum leprosum. Braz J Infect Dis 15:288–292
269. Chalet M, Howard DH, McGinnis MR et al (1986) Isolation of *Bipolaris australiensis* from a lesion of viral vesicular dermatitis on the scalp. J Med Vet Mycol 24:461–465
270. Zapatar RC, Albesi EJ, Garcia GH (1975) Mycotic keratitis by *Drechslera spicifera*. Sabouraudia: J Med Vet Mycol 13:295–298
271. Ge H, Pan M, Chen G et al (2017) The first case of cutaneous phaeohyphomycosis caused by *Bipolaris spicifera* in Northern China: A case report. Exp Ther Med 14:1875–1878
272. Amma SM, Paniker CKJ, Lype PT et al (1979) Phaeohyphomycosis caused by *Cladosporium bantianum* in Kerala (India). Sabouraudia 17:419–423
273. Noguchi H, Hiruma M, Matsumoto T et al (2017) Fungal melanonychia: ungual phaeohyphomycosis caused by *Botryosphaeria dothidea*. Acta Derm Venereol 9:765–766
274. Yu J, Yang S, Zhao Y et al (2006) A case of subcutaneous phaeohyphomycosis caused by *Chaetomium globosum* and the sequence analysis of *C. globosum*. Med Mycol 44:541–545
275. Lumlee Q, Meason-Smith C, Dieterly A et al (2017) Chaetomiaceae fungi, novel pathogens of equine neurotropic phaeohyphomycosis. Vet Pathol 54:813–819
276. Hironaga M, Watanabe S (1980) Cerebral phaeohyphomycosis caused by *Cladosporium bantianum*: a case in a female who had cutaneous alternariosis in her childhood. Sabouraudia 18:229–235

277. Wilson E (1982) Cerebral abscess caused by *Cladosporium bantianum*. Case Rep Pathol 14:91–96
278. Barde AK, Singh SM (1984) *Cladosporium carrionii* Trejos 1954 infection of human nail. Mycosen 27:366–369
279. Ma X, Gu Y, Liu X et al (2013) Phaeohyphomycotic dermatitis in a giant panda (*Ailuropoda melanoleuca*) caused by *Cladosporium cladosporioides*. Med Mycol Case Rep 6:119–121
280. Chen YT, Lu PL, Lee KM et al (2013) Acute meningitis caused by *Cladosporium sphaerospermum*. Am J Med Sci 346:523–525
281. Kim RC, Hodge CJ Jr, Lamberson HV Jr et al (1981) Traumatic intracerebral implantation of *Cladosporium trichoides*. Neurology 31:1145–1148
282. Bennet JE, Bonner H, Jennings AE et al (1973) Chronic meningitis caused by *Cladosporium trichoides*. Am J Clin Pathol 59:398–407
283. Seasworth BJ, Kwon-Chung KJ, Hamilton JD et al (1983) Brain abscess caused by a variety of *Cladosporium trichoides*. Am J Clin Pathol 79:747–752
284. Lanternier F, Barbati E, Meinzer U et al (2015) Inherited CARD 9 deficiency in 2 unrelated patients with invasive *Exophiala* infection. J Infect Dis 211:1241–1250
285. Kotwal A, Biswas D, Kakati B et al (2015) Non traumatic keratitis due to *Colletotrichum coccodes*: a case report. J Clin Diagn Res 9:1–2
286. Ogawa M, Reis V, Godoy P et al (2014) Phaeohyphomycosis caused by *Colletotrichum gloeosporio*ides and *Alternaría infectoria* in renal transplant recipient. Rev Chil Infectol 31:468–472
287. Llamos R, Al-Hatmi AM, Martínez G et al (2016) Non-traumatic keratitis due to *Colletotrichum truncatum*. JMM Case Rep 3:e005047
288. Lv GX, Ge YP, Shen YN et al (2011) Phaeohyphomycosis caused by a plant pathogen, *Corynespora cassiicola*. Med Mycol 49:657–661
289. Bay C, González T, Munoz G et al (2017) Nasal phaeohyphomycosis by *Curvularia spicifera* in pediatric patient with neutropenia and acute myeloid leukemia. Rev Chil Infectol 34:280–286
290. Bittencourt AL, Machado PR, Araujo MG (2002) Subcutaneous phaeohyphomycosis caused by *Cyphellophora pluriseptata*. Eur J Dermatol 12:103–106
291. Najafzadeh MJ, Suh MK, Lee MH et al (2013) Subcutaneous phaeohyphomycosis caused by *Exophiala equina*, with susceptibility to eight antifungal drugs. J Med Microbiol 62:797–800
292. Kohashi S, Toyama T, Hashimoto N et al (2017) Sinusitis caused by *Exserohilum rostratum* after cord blood transplantation for myelodysplastic syndrome: A case report and literature review. Transpl Infect Dis 8:10–15
293. Gupta AJ, Singh M, Yadav S et al (2017) Phaeohyphomycosis breast masquerading as fibroadenoma in a young teenage girl. Diagn Cytopathol 45:939–942
294. Chua JD, Gordon SM, Banbury J et al (2001) Relapsing *Exophiala jeanselmei* phaeohyphomycosis in a lung-transplant patient. Infect Dis 3:235–238
295. Tokuhisa Y, Hagiya Y, Hiruma M et al (2011) Phaeohyphomycosis of the face caused by *Exophiala oligosperma*. Mycoses 54:240–243
296. Tsujioka K, Tanaka R, Anzawa K et al (2015) A case of cutaneous phaeohyphomycosis caused by *Exophiala lecanii-corni* showing a seasonal fluctuation of skin lesions. Med Mycol J 56:117–121
297. Doymaz MZ, Seyithanoglu MF, Hakyemez I et al (2015) A case of cerebral phaeohyphomycosis caused by *Fonsecaea monophora*, a neurotropic dematiaceous fungus, and a review of the literature. Mycoses 58:187–192
298. Kindo AJ, Pramod C, Anita S et al (2010) Maxillary sinusitis caused by *Lasiodiplodia theobromae*. Indian J Med Microbiol 28:167–169
299. Fernández AL, Andres PO, Vecino CH et al (2017) Subcutaneous infection by *Graphium basitruncatum* in a heart transplant patient. Braz J Infect Dis 21:670–674
300. Crawford SJ, Chen SC, Halliday C et al (2015) *Microsphaeropsis arundinis* skin and soft tissue infection in renal transplant recipients: three case reports and a review of the literature. Transpl Infect Dis 17:915–920

301. Garinet S, Tourret J, Barete S et al (2015) Invasive cutaneous *Neoscytalidium* infections in renal transplant recipients: a series of five cases. BMC Infect Dis 19:15–53
302. Geramishoar M, Zomorodian K, Zaini F et al (2004) First case of cerebral phaeohyphomycosis caused by *Nattrassia mangiferae* in Iran. Jpn J Infect Dis 57:285–286
303. Boggild AK, Poutanen SM, Mohan S et al (2006) Disseminated phaeohyphomycosis due to *Ochroconis gallopavum* in the setting of advanced HIV infection. Med Mycol 44:777–782
304. Marques SA, Camargo RM, Summerbell RC et al (2006) Subcutaneous phaeohyphomycosis caused by *Phaeoacremonium parasiticum* in a renal transplant patient. Med Mycol 44:671–676
305. Gordon RA, Sutton DA, Thompson EH et al (2012) Cutaneous phaeohyphomycosis caused by *Paraconiothyrium cyclothyrioides*. J Clin Microbiol 50:3795–3798
306. Furudate S, Sasai S, Numata Y et al (2012) Phaeohyphomycosis caused by *Phaeoacremonium rubrigenum* in an immunosuppressive patient: A case report and review of the literature. Case Rep Dermatol 4:119–124
307. Slifkin M, Bowers HM (1975) *Phialophora mutabilis* endocarditis. Am J Clin Pathol 63:120–130
308. Tsang CC, Chan JF, Philip PCIp et al (2014) Subcutaneous phaeohyphomycotic nodule due to *Phialemoniopsis hongkongensis* sp. nov. J Clin Microbiol 52:3280–3289
309. Desoubeaux G, Garcia D, Bailly E et al (2014) Subcutaneous phaeohyphomycosis due to *Phialemoniopsis ocularis* successfully treated by voriconazole. Med Mycol Case Rep 10:54–58
310. Pracharktam R, Chongtrakool P, Sriurairatana S et al (2000) Mycetoma and phaeohyphomycosis caused by *Phialophora parasitica* in Thailand. J Med Assoc Thail 83:542–545
311. Hughart R, Merrick M, Adelaja OT et al (2016) Cutaneous phaeohyphomycosis caused by *Biatriospora mackinnonii* in a renal transplant recipient. JAAD Case Rep 2:230–232
312. Faisal M, Elsayed E, Fitzgerald SD et al (2007) Outbreaks of phaeohyphomycosis in the chinook salmon (*Oncorhynchus tshawytscha*) caused by *Phoma herbarum*. Mycopathologia 163:41–48
313. Hernández-Hernández F, Vargas-Arzola J, Ríos-Cruz OP et al (2018) First case of chromoblastomycosis due to *Phoma insulana*. Enferm Infecc Microbiol Clin 36:95–99
314. Amazan E, Desbois N, Fidelin G et al (2014) First case of phaeohyphomycosis due to *Pleurostoma ootheca* in a kidney transplant recipient in Martinique (French West Indies). Med Sante Trop 24:323–325
315. Ahmed SA, Desbois N, Quist D et al (2015) Phaeohyphomycosis caused by a novel species, *Pseudochaetosphaeronema martinelli*. J Clin Microbiol 53:2927–2934
316. Jennings JE (2016) Phaeohyphomycosis due to *Pyrenophora phaeocomes* and *Drechslera nobleae* in an Appaloosa mare. Can Vet J 57:431–433
317. Amr SS, Al-Tawfiq JA (2007) Aspiration cytology of brain abscess from a fatal case of cerebral phaeohyphomycosis due to *Ramichloridium mackenziei*. Diagn Cytopathol 35:695–699
318. Chander J, Singla N, Kundu R et al (2017) Phaeohyphomycosis Caused by *Rhytidhysteron rufulum* and Review of Literature. Mycopathologia 182:403–407
319. Noguchi H, Hiruma M, Matsumoto T et al (2017) Subcutaneous cystic phaeohyphomycosis due to *Pleurostomophora richardsiae*. J Dermatol 44:62–63
320. Larbcharoensub N, Chongtrakool P, Wirojtananugoon C et al (2013) Treatment of a brain abscess caused by *Scedosporium apiospermum* and *Phaeoacremonium parasiticum* in a renal transplant recipient. Southeast Asian J Trop Med Public Health 44:484–489
321. Badali H, Chander J, Gupta A et al (2011) Fatal cerebral phaeohyphomycosis in an immunocompetent individual due to *Thielavia subthermophila*. J Clin Microbiol 49:2336–2341
322. Jennings Z, Kable K, Halliday CL et al (2016) *Verruconis gallopava* cardiac and endovascular infection with dissemination after renal transplantation: Case report and lessons learned. Med Mycol Case Rep 15:5–8
323. Guarro J, Gugnani HC, Sood N et al (2008) Subcutaneous phaeohyphomycosis caused by *Wallemia sebi* in an immunocompetent host. J Clin Microbiol 46:1129–1131

Endemic Mycoses in Americas

6

Sandra de Moraes Gimenes Bosco, Giselle Souza da Paz, Jéssica Luana Chechi, Alana Lucena Oliveira, Ana Carolina do Prado, Danielle Hamae Yamauchi, Hans Garcia Garces, and Eduardo Bagagli

Abstract

American continent concentrates important endemic mycoses, both systemic and subcutaneous, such as coccidioidomycosis, blastomycosis, histoplasmosis, paracoccidioidomycosis and sporotrichosis. The common factor among these mycoses is the dimorphic nature of their etiologic agents, like *Coccidioides* spp., *Blastomyces* spp., *Histoplasma* spp., *Paracoccidioides* spp. and *Sporothrix* spp., respectively. Human and animal coccidioidomycoses are found in the USA (states of California, Texas, Arizona and New Mexico), Brazil (states of Piauí and Ceará) and Argentina (states of Catamarca, Santiago del Estero, San Luis and Córdoba). Blastomycosis is restricted to North America, mainly in the valleys of the Mississippi and Ohio Rivers of the USA, and is predominantly observed in dogs. Cases of classical histoplasmosis in humans and animals occur mainly in the USA (Mississippi, Missouri, Ohio River valley), Canada (Ontario and Montreal) and in some regions of Latin America (Colombia and Brazil). Paracoccidioidomycosis is the most important systemic mycosis in Latin American countries, highlighting Brazil, Colombia, Venezuela and Argentina. It

S. de Moraes Gimenes Bosco (✉) · G. S. da Paz · J. L. Chechi
Laboratory of Medical Mycology, Department of Microbiology and Immunology, Institute of Biosciences, Universidade Estadual Paulista, UNESP/Botucatu, Botucatu, SP, Brazil
e-mail: sandra.bosco@unesp.br

A. L. Oliveira · A. C. do Prado
Laboratory of Medical Mycology, Department of Microbiology and Immunology, Institute of Biosciences, Universidade Estadual Paulista, UNESP/Botucatu, Botucatu, SP, Brazil

D. H. Yamauchi · H. G. Garces · E. Bagagli
Laboratory of Fungal Biology, Department of Microbiology and Immunology, Institute of Biosciences, Universidade Estadual Paulista, UNESP/Botucatu, Botucatu, SP, Brazil
e-mail: eduardo.bagagli@unesp.br

K. Singh, N. Srivastava (eds.), *Recent Trends in Human and Animal Mycology*,
https://doi.org/10.1007/978-981-13-9435-5_6

predominantly affects humans, despite some cases of animal disease being recently reported. In Brazil, there was a special situation concerning sporotrichosis, where thousands of cases of human disease were transmitted due to cat scratches, characterizing a great epidemic of zoonotic transmission. The most prevalent species reported was *Sporothrix brasiliensis* which is considered to be one of the most virulent species of *Sporothrix* species complex.

Keywords
Systemic mycosis · Blastomycosis · Coccidioidomycosis · Histoplasmosis · Paracoccidioidomycosis · Sporotrichosis

6.1 General Aspect

The endemic systemic mycoses of American continent such as blastomycosis, coccidioidomycosis, histoplasmosis and paracoccidioidomycosis share the dimorphic nature of their etiological agents (which have a saprophytic mycelial phase in environment that infect hosts by airborne route through the inhalation of conidia). Another aspect that has to be emphasized is the taxonomic classification of these fungi: phylum *Ascomycota*, order *Onygenales* [1–3].

The order *Onygenales* comprises several important fungi in Medical Mycology. It is important to highlight at least three families: *Arthrodermataceae* (includes causative agents of dermatophytosis), *Onygenaceae* (includes *Coccidioides immitis* and *Coccidioides posadasii*, etiological agents of coccidioidomycosis) and the *Ajellomycethaceae*, which includes *Blastomyces dermatitidis*, *Blastomyces gilchristii*, *Histoplasma capsulatum*, *Paracoccidioides brasiliensis*, *Paracoccidioides lutzii*, *Emmonsia parva*, *Emmonsia crescens* and *Lacazia loboi* [4]. *Emmonsia parva* and *E. crescens* cause adiaspiromycosis, a disease found mainly in rodents, acquired by inhalation of the conida present in soil. Once in the lung, the conidia undergo an enlargement in volume (up to a million-fold) to form the adiaspores (parasitic form in tissue), around 2–4 μm to 40–500 μm in diameter [5]. The last decade has witnessed an increase in reports of a novel *Emmonsia*-like fungus as causative agent of systemic mycosis in immunocompromised patients worldwide, frequently leading to death. These new species differ from *Emmonsia* by producing a yeast-like phase rather than adiaspores in tissues [6]. *Emergomyces pasteurianus*, formerly called *Emmonsia pasteuriana* [7] was reported as causative agent of disease in HIV patients in Italy [7, 8], Spain [9] and India [10]. In China, there are two reports in non-HIV patients, but in both cases there was a significant immunosuppression due to renal transplant [11] and high dose of corticosteroids [12]. *Emergomyces africanus* is the main species that cause emergomycosis in South African patients with high case-fatality ratio; however, the actual habitat of fungus in nature is unknown; it is believed that soil is the main source of infection through inhalation of conidia [7, 13].

It is important to mention that new members of *Ajellomycethaceae*, viz., *Emmonsiellopsis coralliformis*, *Emmonsiellopsis terrestris*, *Helicocarpus griseus* and *Polytolypa hystricis* have been described. They are found in environmental samples and do not appear to undergo a thermally regulated dimorphic transition in animals [14].

Despite the emerging importance of *Emmonsia* spp. and *Emergomyces* spp. in Medical Mycology field, the present chapter is focused on coccidioidomycosis, blastomycosis, histoplasmosis, paracoccidioidomycosis and sporotrichosis, as these are the common endemic mycoses in American continent.

The etiological agents of coccidioidomycosis, blastomycosis, histoplasmosis and paracoccidioidomycosis have specific geographical distribution, for example, blastomycosis is mostly reported in North America, while paracoccidioidomycosis corresponds to the major systemic mycosis from Latin American countries [15, 16]. Coccidioidomycosis is observed in arid regions of the USA, Mexico, Brazil and Argentina [17]. Histoplasmosis is autochthonous in more than 60 countries, being not restricted only to American continent [18].

Historically, the first systemic mycosis described was coccidioidomycosis in 1892, by a medical student, Alejandro Posadas, and his advisor (the pathologist Robert Wernick), in an Argentine soldier who had disseminated skin lesions for over 3 years [19]. In 1894, Thomas Casper Gilchrist described blastomycosis in a male patient having a skin lesion in the hand [20]. Samuel Taylor Darling, a pathologist from the Ancon Hospital, Panama, described systemic histoplasmosis in a patient who had hepatosplenomegaly and anemia [21]. Later in 1908, the first case of paracoccidioidomycosis was described by Adolfo Lutz, in São Paulo state, Brazil, in a male rural worker patient presenting lesions in the mucocutaneous junction in the mouth [22].

Sporotrichosis, an important and emerging endemic subcutaneous mycosis in Brazil, is caused by a non-*Onygenales* dimorphic fungus, and the disease has two main forms of transmission: classical by scratches through vegetal material and zoonotic through cats [23, 24]. Historically, sporotrichosis was described in 1898 at The Johns Hopkins Hospital, in Baltimore, USA, by Robert Benjamin Schenckii, in a patient with skin lesions [25].

6.2 Systemic Endemic Mycoses

6.2.1 Coccidioidomycosis

6.2.1.1 Definition and Etiology

Coccidioidomycosis is a systemic granulomatous mycosis of humans and animals, characterized mainly by involvement of the lungs and other tissues, such as the skin, bones, joints and meninges. It is caused by the dimorphic fungus *Coccidioides immitis* or *Coccidioides posadasii*. The teleomorph phase of both species is still unknown.

Coccidioides sp. is a dimorphic fungus, but unlike other dimorphic fungi, its dimorphism is not totally temperature dependent. The dimorphism of *Coccidioides* spp. is mainly related to parasitic condition and may be achieved in vitro by using a special culture medium for reversion and incubation with microaerophilia

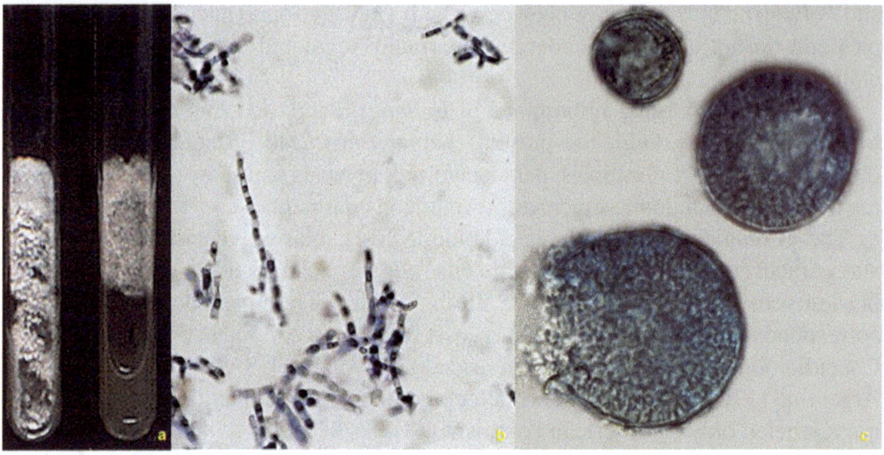

Fig. 6.1 Morphology of *Coccidioides posadasii.* (**a**) Filamentous colonies obtained on Sabouraud's agar with 10 days of incubation; (**b**) microscopy of filamentous culture, evidencing intercalary arthroconidia (lactophenol cotton blue, 400 ×); (**c**) spherules in direct mycological examination (lactophenol cotton blue, 400 ×). "Copyright of Gen National Publishing Group (Rio de Janeiro, Brazil), Book: Doenças Infecciosas em Animais de Produção e Companhia, Chapter of Coccidioidomicose (page 870), reprinted with permission"

conditions [26]. The mycelial phase is characterized by a colony of humid and glabrous aspect, with a membranous appearance and a grayish color, later becoming cottony with color that varies from white to cream (Fig. 6.1a). Microscopically, hyaline and septate hyphae are observed, with numerous quadrangular alternating barrel-shaped thick-walled arthroconidia, measuring about 2.5–6 μm in diameter (Fig. 6.1b). The spherule (sporangium) is equivalent to yeast phase and is observed in host tissues. The spherules are characterized by thick cell walls and spherical or oval in shape and measure about 10–60 μm, reaching up to 100 μm in diameter (Fig. 6.1c). The cytoplasmatic content inside the spherule began progressive cleavages to form the endospores, which measure about 2–5 μm. The spherule breaks its cell wall and promotes the release of endospores. Once released, the endospores are spread by hematogenic route to other tissues and begin to form a new spherule [26].

Until the early 2000s, it was believed that *Coccidioides immitis* was the unique species to cause coccidioidomycosis. Fisher and co-workers (2001, 2002) demonstrated that *Coccidioides immitis* is a cryptic species, since they found genetic diversity among isolates of *C. immitis* from California and other regions (Texas, Mexico and South America). The authors separated these clades in Californian (CA) and non-Californian (non-CA) isolates and later named the non-CA isolates as *Coccidioides posadasii*, to honor Alejandro Posadas [27, 28]. Genetic diversity has also been found in isolates among both species from different endemic regions [29–31].

C. immitis is considered as the most virulent fungus, both for humans and animals. The inhalation of few arthroconidia is sufficient for the infection in lung tissue. It is the only fungal pathogen that is classified into risk group 3, and its handling is restricted to laboratories with biosafety level 3 [32]. It is also considered the only eukaryotic organism with potential use for bioterrorism, having been included among the biological agents that are controlled by antiterrorist legal statutes to transport and receive selected agents, a process regulated by the American Anti-Terrorism Act and Effective Death Penalty, according to the US Act of 1996.

6.2.1.2 Epidemiology

Coccidioides spp. may be found in semi-arid and arid soil conditions, between the parallels 40° N (USA) and 40° S (Argentina) [33]. Characteristically, these regions present sandy soil, alkaline pH, summers with high temperatures and mild winters, in addition to long periods of drought, low rainfall indices and xerophytic vegetation. The soil favorable to development of *Coccidioides* spp. presents high concentrations of organic matter, mainly based on carbon compounds and high concentrations of salts based on calcium sulfate, boron and sodium chloride [34]. The fungus may be generally found in soils at a depth of about 20 cm or more, where competition with other microorganisms is lower. In endemic areas, the distribution of fungus in the soil is focal, commonly associated with animal burrows (mainly rodents and marsupials in the USA and armadillos in Brazil) and archeological sites, constituting an important occupational disease for archeologists, geologists and other professionals dealing with soil in endemic areas [35–39].

Around 40% of coccidioidomycosis infection in humans evolves to clinical disease, which resembles an influenza-like disease (also called Valley fever) [40]. Several outbreaks have been reported in literature, and a recent review revealed 47 outbreaks during 1940–2015, which resulted in 1,464 cases of the disease in humans. The majority of these outbreaks were related to occupational activities and resulted from an earthquake and a large dust storm [41]. In Brazil, outbreaks of coccidioidomycosis have been reported in human patients and dogs after armadillos hunting (Fig. 6.2). Generally, after an incubation period of 10 days (after the hunt), both hunters and dogs presented with fever and respiratory manifestations; some even led to death. The agent has already been isolated from sputum and other clinical samples from patients, lungs of dogs and tissues of armadillos and soil samples collected from burrows of armadillos [35]. In the USA, Arizona and California are considered as the highest endemic areas. Benedict and co-workers warned about the severity of the disease in residents of non-highly endemic states as the less severe cases generally remained undiagnosed or unreported [42]. Besides Arizona and California states that register around 60% of coccidioidomycosis in the USA, the disease has also been reported to be endemic in some regions of Mexico, El Salvador, Honduras, Guatemala, Venezuela, Colombia, Bolivia, Paraguay, Argentina and Brazil [17, 29, 43–46].

The morbidity of disease is not high in animals. However, the lethality depends on the species involved. In cattle and sheep, the disease is usually benign or evolves to a localized form. For the canine and feline species, the lethality can reach 100% in animals with clinical manifestations of the disease. Epidemiological surveys conducted in endemic areas of the USA, using skin tests with coccidioidin (antigen), showed that a large percentage of dogs are infected, but without clinical manifestation of the disease. In Arizona, the prevalence of coccidioidomycosis in dogs and cats, between 2009 and 2015, was 23 and 17, respectively. The average age of these animals was 7 years for dogs and 9 years for cats [47].

Fig. 6.2 Armadillo hunting: situation observed in Brazilian rural communities. Hunter accompanied with his dog. Image courtesy of Dr. Jael Soares Batista. "Copyright of Gen National Publishing Group (Rio de Janeiro, Brazil), Book: Doenças Infecciosas Em animais de Produção e Companhia, Chapter of Coccidioidomicose (page 872), reprinted with permission"

In horses, Ziemer and co-workers reported 15 cases between 1975 and 1984 in California and Arizona [48]. A survey of skin test with coccidioidin was conducted in 11,643 cattles in Arizona, between 1954 and 1959, and a total of 2,859 (24.6%) were found positive [49]. Infection of *Coccidioides* spp. and coccidioidomycosis were also reported in wild animals, such as llama, alpacas, rhesus macaque, koala, black rhinoceros, snake and bats [50–57]. Marine mammals such as sea lion and bottlenose dolphin have also been affected by coccidioidomycosis [58, 59]. A survey in marine mammals, during the years 1998–2012 in marine coast of California, USA, revealed that among the 41 animals infected by fungi, 36 were infected by *Coccidioides immitis* (20 sea otters, 15 sea lions and 1 harbor seal) [60].

6.2.1.3 Clinical Aspects

The high prevalence of positive intradermal tests in endemic areas showed that subclinical infection is more common in dogs, which are considered sentinels for disease and a way to map the disease in endemic regions [61, 62].

The disease may begin firstly with respiratory sign like cough, which can be dry (similar to that observed in cases of tracheobronchitis or "kennel cough") or productive. Dry cough is a result of hilar lymphadenomegaly or diffuse pulmonary

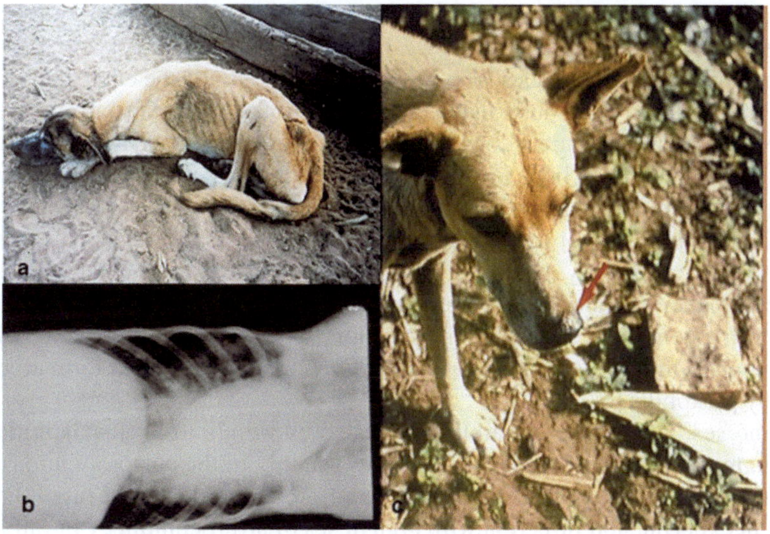

Fig. 6.3 Coccidioidomycosis in dogs: hunting dogs with coccidioidomycosis, from the municipality of Oeiras, State of Piauí, Brazil. (**a**) Cachexia; (**b**) radiographic appearance of the lung of the same dog evidencing bilateral pneumonia with hilar adenopathy; (**c**) another dog showing snout injury (arrow). "Copyright Mycopathologia (2000) 148:57. https://doi.org/10.1023/A:1007183022761, and Gen National Publishing Group (Rio de Janeiro, Brazil), Book: Doenças Infecciosas Em Animais de Produção e Companhia, Chapter of Coccidioidomicose (page 874), reprinted with permission"

interstitial disease; productive cough occurs due to alveolar involvement. This pulmonary condition may resolve or evolve into a severe generalized pneumonia (Fig. 6.3b) with worsening of respiratory signs or even disseminate to other organs [62, 63]. The clinical signs resulting from the spread of the disease are associated with the organ in which the fungus is lodged. In addition to respiratory signs, claudication, peripheral localized lymphadenopathy, exudation of cutaneous lesions and ocular signs (keratitis, uveitis, chorioretinitis, panophthalmitis and even blindness) are observed. Signs of impairment of the digestive system, as well as peripheral generalized lymphadenopathy, are rare. Left congestive heart failure may occur. Other cardiac manifestations may vary from disturbances of the blood circulation to changes in the musculature, making it difficult to contract and conduct impulses. Changes in the pericardium are also reported. The main manifestations of the central nervous system like ataxia, behavioral changes, convulsions, strokes and coma are rarely observed. Claudication is the most evident clinical sign of bone lesions, which is extremely painful and commonly seen in long bones. Joint lesions are uncommon, although immune-mediated polyarthritis has been described in infected dogs. Most of the skin lesions result from dissemination of the fungus; however, they may also result from the involvement of the underlying bone. Clinically, tegumentary lesions range from small wounds to abscesses and ulcers that drain purulent contents (Fig. 6.3c). Nonspecific signs associated with disseminated disease are represented by constant or intermittent fever, anorexia, weight loss leading to cachexia (Fig. 6.3a), depression and weakness [62, 64, 65].

Coccidioidomycosis in cats is less frequent than dogs and the clinical manifestations are similar to those observed in dogs. It seems that the cutaneous involvement is more frequent in cats than dogs. It is common for cats to have cutaneous lesions without any underlying bone involvement, as commonly seen in dogs. Other manifestations commonly observed with skin lesions are fever, anorexia and weight loss. Respiratory signs are not frequent in cats, possibly because these animals have less intense physical activity than dogs. Although bone lesions are similar, in cats they are less frequent. Ocular lesions occur in a manner similar to that observed in dogs both in frequency and in clinical presentation [64, 65]. In wild felids, the disease was reported in mountain lions (*Felis concolor*) with disseminated form accompanied by peritonitis [62, 66, 67].

In horses, the disease is mainly manifested by weight loss, emaciation, fever, cough, muscular pain, cutaneous abscesses (mainly in the pectoral region), anemia and colic. It seems that cattles may be refractory to development of disease, since in experimental infection with high doses of arthroconidia by intratracheal route, no signs of disease were observed [68]. In endemic areas of coccidioidomycosis in Arizona, skin tests showed prevalence of infection in cattles quite similar to humans [62, 69]. Lesions of coccidioidomycosis and comprehend granuloma in thoracic lymph nodes have been observed in cattle from slaughter houses [68]. Coccidioidomycosis in sheep is similar to that observed in cattles of slaughter houses, i.e. granulomatous lesions in mediastinal and bronchoalveolar lymph nodes [62].

6.2.2 Blastomycosis

6.2.2.1 Definition and Etiology

Blastomycosis, a systemic granulomatous mycosis of humans and animals, is caused by the dimorphic fungus *Blastomyces dermatitidis*. The disease is characterized mainly by pulmonary involvement, but other tissues are also affected. The *B. dermatitidis* corresponds to the anamorph (asexual) stage. Its teleomorph phase (sexual) has already been described and is named *Ajellomyces dermatitidis* [70, 71]. A new species, named *Blastomyces gilchristii*, was described as other etiological agent of blastomycosis [72]. Recently a new species, *Blastomyces percursus*, was reported in Israel and South Africa [73].

The mycelial form has a white-brown coloration, smooth or grooved topography, raised center and generally cottony appearance (Fig. 6.4a). Microscopically, the mycelial form has hyaline and thin hyphae with conidiophores having rounded or globular conidia at their ends (Fig. 6.4b). The yeast form is characterized by the cerebriform and glabrous appearance, creamy consistency and coloration that vary in tones from beige to light yellow (Fig. 6.5a). Microscopically, the yeast has a diameter of 5–20 μm and a thick and birefringent cell wall. Cell budding is unique, and the daughter cell attaches to the mother cell by a broad base binding (Fig. 6.5b).

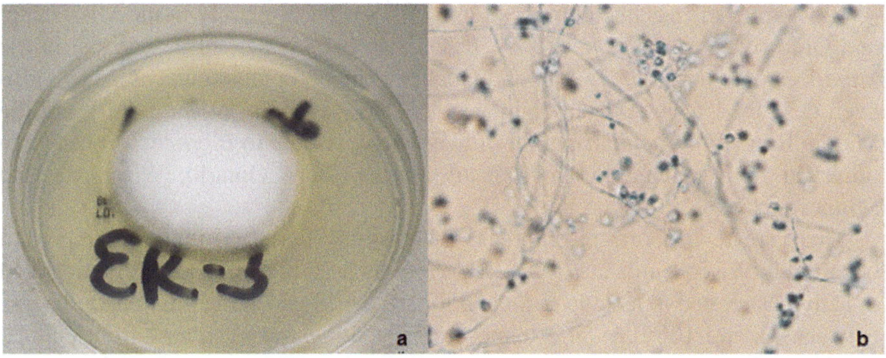

Fig. 6.4 Filamentous colony of *Blastomyces dermatitidis*. (**a**) Cottony appearance and whitish coloring; (**b**) Microscopy (lactophenol cotton blue, 400 ×). Image courtesy of Dr. Dennis Baumgardner (Family Practice Center, Aurora St. Luke's Medical Center, Milwaukee, WI, USA)."Copyright of Gen National Publishing Group (Rio de Janeiro, Brazil), Book: Doenças Infecciosas Em Animais de Produção e Companhia, Chapter of Blastomicose (page 862), reprinted with permission"

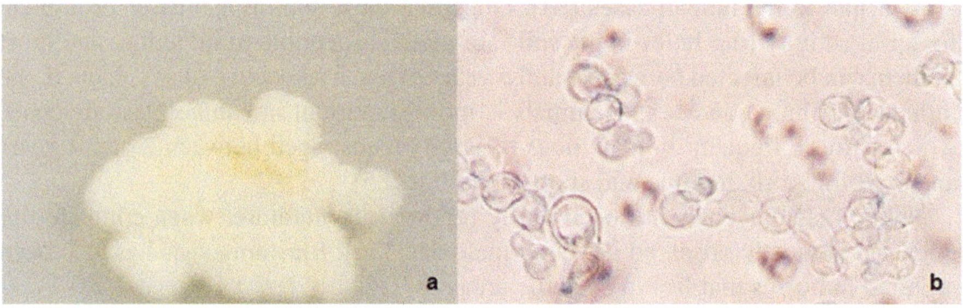

Fig. 6.5 Yeast colony of *Blastomyces dermatitidis*. (**a**) Macroscopic aspect; (**b**) microscopy of the single- and broad-based buds (lactophenol cotton blue, 400 ×). Image courtesy of Dr. Dennis Baumgardner (Family Practice Center, Aurora St. Luke's Medical Center, Milwaukee, WI, USA). "Copyright of Gen National Publishing Group (Rio de Janeiro, Brazil), Book Doenças Infecciosas Em Animais de Produção e Companhia, Chapter of Blastomicose (page 862), reprinted with permission"

6.2.2.2 Epidemiology

It is believed that *B. dermatitidis* has the soil as the natural reservoir, although its habitat is not known till now and the fungus was rarely isolated from soil especially near to watercourses and from riparian forests with high organic matter, enriched with animal excreta and with low pH [74–76]. The infection is acquired through the inhalation of the conidia from environment. Once inhaled, the conidia reach the pulmonary alveoli and convert to yeast form. The risks of infection increase when rains and intense dew occur, which facilitate the release of infecting spores of the fungus [76].

Blastomycosis has been observed in North America, mainly in the Mississippi and Ohio river valleys in the USA. It has also been observed in Canada, in the provinces of Manitoba, Ontario and Quebec. The autochthonous cases have also been described from African countries [15, 77, 78], parts of India [79] and the Middle East [80]. The distribution of blastomycosis due to *B. gilchristii* has been observed in hyperendemic regions of Northwestern Ontario, Wisconsin and Minnesota [72, 81].

There are few reports of outbreaks of blastomycosis, and the most important occurred in Wisconsin, state of the USA, in which 26 people who had participated in activities related to a camp in a natural park developed acute pulmonary disease between 21 and 106 days after exposure. The fungus was isolated from clinical specimens from patients and also from environment [82].

The principal animal species affected by blastomycosis is the dog, in which the annual incidence of the disease is estimated in 1,420 for every 100,000 dogs. This incidence is considered around tenfolds higher when compared to human cases of the disease, a fact that reinforces the concept that canine blastomycosis can be a harbinger of disease in humans [83]. It is estimated that one-third of patients had one of their dogs diagnosed with blastomycosis at least 6 months before the symptoms began in the human cases [84]. Dogs that have hunting habits or prominent fossorial activity (dig holes in the soil) are more susceptible to infection, and most of them can be infected between 1 and 5 years of age irrespective of sex. Indeed, the influence of breed and sex of animals is more important in canine blastomycosis, which is more frequent in male dogs of large breeds, aged between 2 and 4 years and used for work, hunting and sports activities [85].

Feline blastomycosis has been rarely reported in literature, when compared to dogs and is mostly observed in young male cats [86]. Blastomycosis has also been reported in horses that live in endemic areas [87].

Among wild animals, there are some reports in wild canides, such as red foxes (*Vulpes vulpes*), grey wolves (*Canis lupus*), kinkajou (*Potos flavus*), ferret, redruffed lemur (*Varecia rubra*), rhesus monkey (*Macaca mulatta*), lions (*Panthera leo*), Siberian tiger (*Panthera tigris*), cheetah (*Acinonyx jubatus*), snow leopard (*Panthera uncia*), American black bear (*Ursus americanus*) and marine animals such as sea lion (*Zalophus californianus*) and Atlantic bottlenose dolphin (*Tursiops truncatus*) [88–95].

In humans, around 55–65% of the cases of blastomycosis are observed in men with age range between 30 and 60 years. The majority of cases associated to men are due to exposition to *B. dermatitidis* in some risk activities, such as hunting, fishing, or forestry work [96]. Human-to-human transmission of blastomycosis is rare and usually occurred by sexual contact [97, 98] and intrauterine transmission [99, 100]. Primary cutaneous infection is caused by dog bite, projectile injury during yard work, dog-contaminated necropsy, human-contaminated autopsy, cat scratches, gardening and laboratory contamination [101].

6.2.2.3 Clinical Aspects

Among domestic animals, blastomycosis is undoubtedly much more frequent in dogs followed by cats and horses less frequently [86]. The most evident clinical signs in dogs are anorexia, weight loss, fever, cough, lymphadenomegaly, dyspnoea and cutaneous lesions. In some dogs, the disease tends to stabilize showing mild clinical signs for weeks to months and sudden deteriorate. Majority of cases of canine blastomycosis present pulmonary involvement, which, when discreet, manifests as intolerance to physical exercise, dyspnoea and nasal discharges. In severe cases, dyspnoea is observed at rest. Ocular manifestations and nasal secretion may be the first clinical signs of canine blastomycosis. Ocular lesions represent about 40% of the cases in dogs, with predominance of uveitis (initially, lacrimation, hyperemia, myositis, blepharospasm, keratitis, conjunctivitis and photophobia), retinal detachment and granulomas, intense vitreous hemorrhage, corneal edema (which hinders ocular examination), glaucoma secondary to fungal obstruction, panophthalmitis, chorioretinitis, periorbital cellulitis and involvement of the nictitating membrane. Often, the ocular manifestations of blastomycosis can lead to blindness [102–104]. Cutaneous lesions of blastomycosis are the result of fungal spread from the primary pulmonary focus and in many cases, represent an advanced stage of the disease. They correspond to about 20 to 50% of clinical cases and are characterized by ulcers that drain serosanguinolent or purulent exudate. Other lesions may be proliferative, granulomatous and fleshy in appearance. Although cutaneous lesions can occur anywhere, nasal plane, facial region and plantar cushion have been predominant [86, 103, 104]. Figure 6.6a, c show skin lesions of blastomycosis in dogs, and Fig. 6.6d shows pulmonary involvement in canine blastomycosis. Bone involvement is also a manifestation of blastomycosis in dogs, accounting for about 30% of cases. The lesion has an osteolytic character, with proliferation of the periosteum. Claudication is the main clinical manifestation. Genital organs (testis, epididymis and prostate) of male dog can be affected and become swollen and painful. When the urinary tract (kidneys and bladder) is affected, the fungus can also be found in the urine. Involvement of central nervous system can occur in 5% of dogs, manifesting focal deficits, ataxia, behavioral changes or seizures. Cardiac disorders in dogs are uncommon and involve signs and symptoms of myocarditis, arrhythmias, syncope, endocarditis, cardiac block and mass lesions. Occasionally, involvement of the larynx, peritoneum, mammary gland, intestines and joints has also been reported [86, 104, 105].

In domestic cats, the lesions are similar to those observed in dogs, including dyspnoea, weight loss, lymphadenitis, ocular (uveitis), cutaneous lesions and central nervous system disease (paralysis). However, because of the low frequency of cases, it is not possible to determine the main characteristic of feline blastomycosis. Database of veterinary hospitals of the USA documented 324 cases of canine blastomycosis and only three cases of feline disease for the same period of evaluation [86]. In an outbreak of blastomycosis in domestic cats in a suburban area of Chicago (Illinois, USA), skin lesions (in the hind limbs and head) and lethargy were observed in five animals. Four of the five cats had anorexia and fever, whereas three had dyspnoea. But all had nasal congestion, ataxia and uveitis [106]. In another registry also in Illinois, all eight cats had radiological evidence of lung disease (nodules, masses

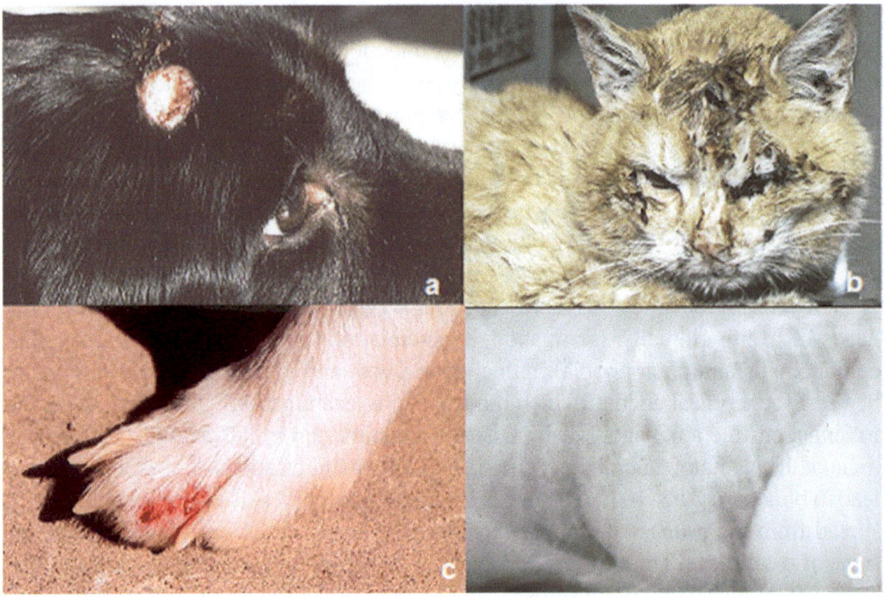

Fig. 6.6 Canine and feline blastomycosis. (**a**) Circular lesion in frontal region in dog; (**b**) facial lesion with exudation in cat; (**c**) lesion in interdigital space in posterior limb of dog; (**d**) chest X-ray with evidence of diffuse pulmonary infiltrate in canine blastomycosis. Images courtesy of Dr. Dennis Baumgardner and Dr. Daniel Paretsky. "Copyright of Gen National Publishing Group (Rio de Janeiro, Brazil), Book: Doenças Infecciosas Em Animais de Produção e Companhia, Chapter of Blastomicose (page 865), reprinted with permission"

or alveolar consolidation) and skin lesions [107]. Figure 6.6b shows cutaneous lesion with exudation in a cat.

There are few cases of blastomycosis in horses reported in the literature, particularly from the USA [87]. In one report, subcutaneous abscesses on the perineum, perivulvar region, udder and abdomen were described. Likewise, in another case, history of 6 months of chronic subcutaneous infection in cervical and pectoral regions together with dysphagia was reported in a pony. Necropsy of this animal revealed piogranulomatous lesions in the lungs and kidney, peritonitis and multiple abscesses in organs. Two other horses had fatal disseminated blastomycosis, presenting lethargy, cough and weight loss, with no signs of skin lesion. Necropsy of the animals revealed pyogranulomas in the lungs and liver, peritonitis and pleuritis [108].

6.2.3 Histoplasmosis

6.2.3.1 Definition and Etiology

Histoplasmosis is a systemic granulomatous mycosis of humans and animals, characterized mainly by involvement of lungs and other tissues, caused by the dimorphic fungus *Histoplasma capsulatum*. The term *H. capsulatum* corresponds to the

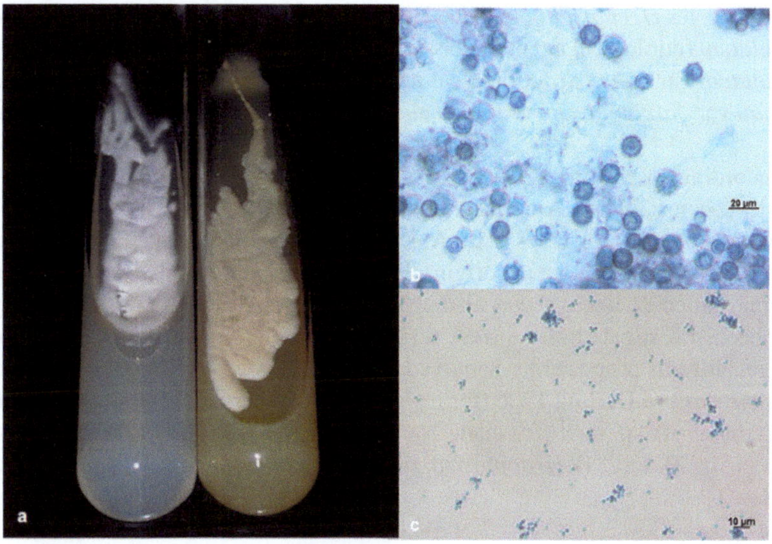

Fig. 6.7 Macro- and microscopic aspects of *Histoplasma capsulatum* dimorphism. (**a**) Filamentous colony, whitish color and cottony texture in potato dextrose agar (left); beige colouration, discreetly wrinkled and creamy texture on Sabouraud dextrose agar (right); (**b**) microscopic appearance of the mycelium showing a large amount of tuberculated macroconidia (lactophenol cotton blue, 400 ×); (**c**) microscopic appearance of the yeast cells with single buddings (lactophenol cotton blue, 200 ×). Images courtesy of Dr. Rosely Maria Zancopé Oliveira (FIOCRUZ/Rio de Janeiro state, Brazil). "Copyright of Gen National Publishing Group (Rio de Janeiro, Brazil), Book: Doenças Infecciosas Em animais de Produção e Companhia, Chapter of Histoplasmose (page 930), reprinted with permission"

anamorph (asexual) stage, and the teleomorph (sexual) phase has already been described and is named *Ajellomyces capsulatus* [109].

H. capsulatum cultured at 25 °C produces whitish or cream-colored colonies in 2–4 weeks, with cottonous hyphae, which become greyish or brown with the aging of the cultures (Fig. 6.7a, at left). The fungus presents two types of conidia, macro- and microconidia. Macroconidia measure about 8–15 µm in diameter and are produced laterally or at the terminal ends of hyphae. Morphologically, macroconidia are spherical, having smooth cell wall initially; however, with the aging of the colonies, they develop digitiform extensions in the outer layer of the cell wall, which gives them the appearance of tuberculate macroconidia. Microconidia measure 2–5 µm in diameter and are produced at the ends of the short conidiophores that are at right angles with the hyphae into which they are inserted. They are oval and have smooth cell wall (Fig. 6.7b). The yeast form of *H. capsulatum* has a smooth or slightly wrinkled surface and humid and shiny appearance, with white to beige color (Fig. 6.7a, at right). Microscopically, they are characterized by oval aspect and average size of 2 to 3 × 3 to 4 µm. They multiply by single buddings, which arise in the narrower portion of the mother cell and are connected by very narrow attachments (Fig. 6.7c).

The species *H. capsulatum*, indeed, comprises three varieties: *H. capsulatum* var. *capsulatum* (etiological agent of classic histoplasmosis, worldwide distribution), *H. capsulatum* var. *duboisii* (etiological agent of African histoplasmosis) and *H. capsulatum* var. *farciminosum* (etiological agent of epizootic lymphangitis in equides) [3].

According to Kasuga and co-workers (2003), *H. capsulatum* comprises seven cryptic species: North American population (clades NAm1 and NAm2), Latin American population (clades LAm A, LAm B), African, Eurasian, Australia and the Netherlands (from Indonesian origin). Besides these genetic clades, an isolate from Panama, called 81 lineage, was also observed [110]. Although these cryptic species have been investigated, there were no changes in the nomenclature of the fungus until recently, as proposed by Sepúlveda and co-workers: (1) *Histoplasma capsulatum sensu stricto* Darling 1906 (for isolates from Panama lineage), (2) *Histoplasma mississippiensis* sp. nov. (for clade Nam 1), (3) *Histoplasma ohiense* sp. nov. (for clade NAm 2) and (4) *Histoplasma suramericanum* sp. nov. (for clade LAm A) [111].

6.2.3.2 Epidemiology

Histoplasmosis caused by *H. capsulatum* var. *capsulatum* has a worldwide distribution, considered autochthonous in more than 60 countries [18]. In the USA, the infection is frequently found in the regions of Mississippi and Ohio River valleys [21]. It predominates in river valleys between latitudes 45° N and 30° S, in localities with high temperatures, between 22 °C and 29 °C, with annual average rainfall ranging from 800 to 1200 mm and humidity between 67 and 87% [112].

The infections by *H. capsulatum* var. *capsulatum* and *H. capsulatum* var. *duboisii* are acquired mainly by inhalation of microconidia. But, *H. capsulatum* var. *farciminosum* disseminated through injured skin caused by contaminated objects like grooming tools, feeding and watering utensils, harnesses and wound dressings especially in equides [113]. This species also causes infections other than equids such as dogs in Japan and cats in France and Switzerland [114–119]. Concerning *H. capsulatum* var. *duboisii*, the infection was observed in baboons (*Papio cynocephalus*), bats (*Nycteris hispida*, *Tadarida pumila*) and aardvarks (*Orycteropus afer*) [118, 119].

Mycelial form of *H. capsulatum* is found predominantly associated to environments rich in guano of bats or birds, since guano has nitrogen-rich content, acting as source of nutrients for fungal growth as well as by reducing the microbial competition in environment [120–123]. The fungus has been repeatedly isolated and molecularly detected from such environmental conditions around the world [124–127]. Besides environment, *H. capsulatum* has already been isolated in culture and molecularly detected from bat tissues worldwide as well [128–137].

Several outbreaks of histoplasmosis linked to inhalation of the fungus from environment of bat caves are reported in literature. Indeed, the great majority of the outbreaks of histoplasmosis are linked to bats [138–140]. Benedict and co-workers reviewed the outbreaks of histoplasmosis in the USA between 1938 and 2013 and found a total of 105 outbreaks which involved 2,850 individuals in 26 states and also in the territory of Puerto Rico. A total of 77% of these outbreaks were related to birds' or bats' droppings, and workplace exposure corresponded to 41% [141]. It is interesting to point out that the outbreaks related to workplace activity were indirectly involved with bat or bird droppings, as reported by O'Keefe and co-workers [142]. According to this report, the workers might have acquired infection from bat guano during cleaning the sheltered areas of a camp (such as raking leaves, cleaning picnic tables, digging fire pits and moving firewood). The workers did not wear any personal protective equipment while cleaning [142].

In recent reviews, histoplasmosis is considered the most important fungal infection acquired during adventure activities, such as spelunking and also during travelling [143–145]. Cottle and co-workers (2013) reported a multinational outbreak of acute histoplasmosis after a field trip in a forest in Uganda. Thirteen of 24 biology students from 10 different countries entered a hollow trunk of a large tree infested with bats and developed signs of acute histoplasmosis 2 weeks after this trip [139].

It is natural to associate the infection of *H. capsulatum* due to entry in the protected environments containing guano of bats or birds; indeed, the majority of the outbreaks reinforce this idea. However, Jülg and co-workers (2008) reported that the infection may also occur at the entrances of a bat cave. In this report, the researchers who stayed at a distance of 1 m of the cave entrance for 1.5 h to record the data on bat behavior developed acute histoplasmosis after 5 days. The researchers did not wear any personal protective equipment, because it was not necessary as they worked outside the cave. During observation, about 100 to 300 bats per second flew closely to them. The authors concluded that a high pathogen burden can also be dispersed by bats during flying [146].

Among the systemic mycoses, histoplasmosis is most frequently reported in HIV patients [147, 148].

6.2.3.3 Clinical Aspects

The clinical manifestations of the disease depend on the variety of the fungus, the animal species affected, the infective load of the microrganism and the susceptibility of the animal. In immunocompetent dogs, initial infection is usually self-limiting; however, the hosts that are immunocompromised or received large doses of microconidia have a greater risk of developing the disease [148]. The incubation period varies between 7 and 14 days, and clinical disease has been observed mainly in horses, dogs and cats and secondarily in swine, cattle and camelids. In companion animals, cases of histoplasmosis are generally reported; however, in equidae, outbreaks are observed in endemic regions only [118, 149].

In dogs, nonspecific clinical signs such as weight loss, inappetence and persistent fever are observed. When the lesion is predominantly pulmonary, dyspnoea, cough and abnormality of sounds are observed in pulmonary auscultation [150]. However, in most dogs, the disease spreads from the gastrointestinal tract [151]. Enteric signs include watery diarrhoea, tenesmus, and faeces with mucus and fresh blood. Pallor of the mucous membranes is observed in dogs due to blood loss through the gastrointestinal tract or bone marrow involvement. Hepatosplenomegaly, lymphadenomegaly, jaundice and ascites are commonly associated with canine

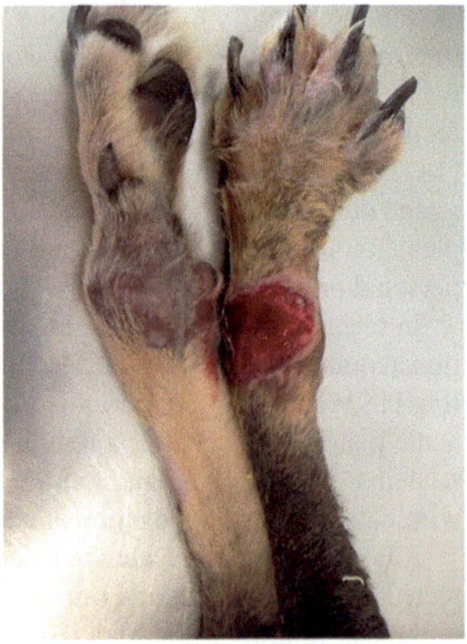

Fig. 6.8 Anterior limbs of dog with disseminated histoplasmosis from São Paulo state, Brazil. Ulcerated lesion with regular border and serosanguinolent exudate (right), swollen of the carpal joint due to bone dissemination (left). Image courtesy of Dr. CláudiaValéria Seulner Brandão (School of Veterinary Medicine and Animal Science/UNESP/Botucatu, São Paulo state, Brazil)

histoplasmosis. Ocular, encephalic, cutaneous and bony lesions (Fig. 6.8) may also be observed [150].

The disease may affect cats of different ages, with a mean age of 4 years. Domestic felines in general show unspecific signs of anemia, weight loss, lethargy, fever, anorexia, lymphadenomegaly, hepatosplenomegaly, depression and jaundice. Respiratory signs such as dyspnoea and tachypnoea may be observed in around 40% of infected cats. Skin lesions, joint pain, ocular signs (such as chorioretinitis, anterior uveitis, conjunctivitis, optic neuritis and retinal detachment) and encephalic and bone lesions may also be observed in cats with histoplasmosis [150].

The main clinical manifestation in horses is the painless nodules or ulcers on the skin along the lymphatic vessels (lymphangitis) and lymph nodes (lymphadenitis), which drain purulent material. The lesions are mainly observed in the skin of limbs, neck, head and chest regions. Lesions are rarely observed in the mucous membranes (ocular, oral and nasal/respiratory). It is very common that the affected horses may be restless due to the numerous flies attracted by the exudates of cutaneous lesions. As the disease progresses, the horses present loss of appetite and worsening of the general condition, leading the animal to death [118].

6.2.4 Paracoccidioidomycosis

6.2.4.1 Definition and Etiology

Paracoccidioidomycosis (PCM) is a systemic granulomatous mycosis that affects primarily the lungs and may be disseminated to other tissues by lymphohematogenic route. The disease is caused by the thermo-dimporphic fungus *Paracoccidioides* spp. whose teleomorph phase is still unknown [76, 152, 153].

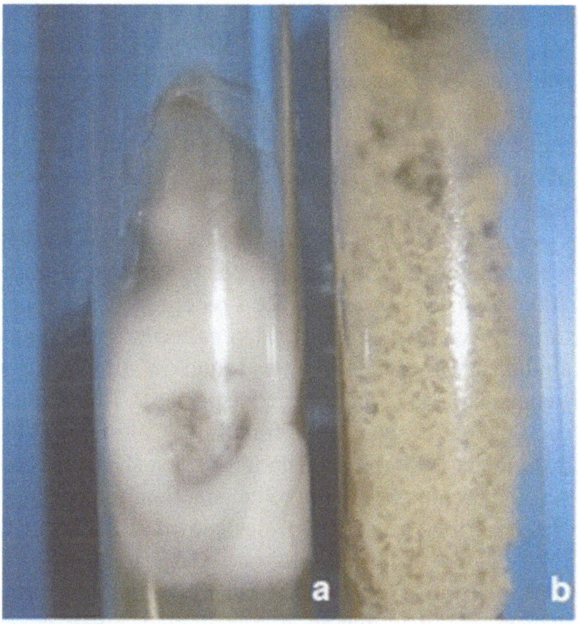

Fig. 6.9 Macroscopic aspects of *P. brasiliensis* cultured at Sabouraud. (**a**) Mycelial form showing cottonous white colony with center containing small elevation and fissures; (**b**) yeast form demonstrating typical beige color and cerebriform aspect. "Copyright of Gen National Publishing Group (Rio de Janeiro, Brazil), Book: Doenças Infecciosas Em Animais de Produção e Companhia, Chapter of Paracoccidioidomicose (page 936), reprinted with permission"

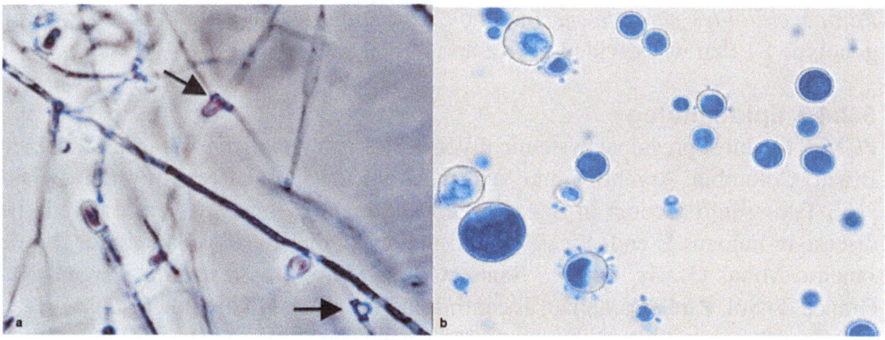

Fig. 6.10 Microscopic aspects of *P. brasiliensis.* (**a**) Mycelial phase, evidencing the arthroconidia (arrows). (**b**) Yeast phase, showing rounded cells, thick cell wall, with several buddings and "pilot wheel" appearance (lactophenol cotton blue, 400 ×). "Copyright of Gen National Publishing Group (Rio de Janeiro, Brazil), Book: Doenças Infecciosas Em animais de Produção e Companhia, Chapter of Paracoccidioidomicose (page 937), reprinted with permission"

The mycelial colony of *Paracoccidioides* spp. presents slow growth, in general 15 to 20 days at 25°C, in culture media such as Sabouraud, Mycosel® and Potato Dextrose agar. The colony initially appears white and cottony and become wrinkled in aspect with the presence of fissures of brown color, resembling popcorn (Fig. 6.9a). Microscopically, it is observed septate and thin hyphae containing the infective conidia (called arthroconidia, arthroaleurioconidia or planoconidia [154]), which vary in size from 3.6 to 4.6 μm in length (Fig. 6.10a). The morphology of yeast colony shows the characteristic cerebriform aspect and beige color (Fig. 6.9b). Micromorphology reveals big and rounded yeast with thick and birefringent cell wall (measuring around 6–30 μm in diameter) surrounded by several medium-sized (2–10 μm in diameter) budding cells (Fig. 6.10b). This aspect of yeast cell surrounded by several buddings is called "pilot wheel". Another typical characteristic of yeast is the "Mickey mouse cap", also shown in culture and histopathological slides [155, 156].

Paracoccidioides brasiliensis was named by Floriano Almeida in 1930 and was thought to be the single species as the etiologic agent of PCM. In the year 2006, *P. brasiliensis* has been recognized as a species complex that comprises three different genotypes (also called cryptic species): S1 (species 1, the most abundant genotype, mainly found in Brazil, Argentina, Paraguay, Peru and Venezuela); PS2 (phylogenetic species 2, found in Brazil and Venezuela) and PS3 (phylogenetic species 3, found in Colombia) [156]. A fourth genotype was described; PS4 (phylogenetic species 4) was reported from Venezuela [152]. Another genotype proved to be genetically more distant than the previous one and was described as a new species named *Paracoccidioides lutzii* in honor to Adolfo Lutz who described the disease [153, 157]. Recently, Turissini and co-workers (2017) proposed that each genotype should be designated as separated species; thereby, S1 was named *P. brasiliensis sensu stricto*, PS2 as *Paracoccidioides americana* sp. nov., PS3 as *Paracoccidioides restrepiensis* sp. nov. and PS4 as *Paracoccidioides venezuelensis* sp. nov. The authors also called the attention to genotype S1 that was divided in S1a and S1b [158].

6.2.4.2 Epidemiology

PCM is the most prevalent systemic mycosis in Latin American countries, in which Brazil, Colombia, Argentina and Venezuela are the most affected by the disease [16]. Taking into account in Brazil, where the majority of cases are observed, the disease in humans is endemically found mainly in the states of São Paulo, Rio de Janeiro, Minas Gerais, Espírito Santo, Goiás, Mato Grosso do Sul, Paraná, Rio Grande do Sul, Pará, Maranhão, Tocantins and Rondônia. In the states of Northeast, a semi-arid region, no autochthonous cases have been reported (Fig. 6.11) [159].

The infection is acquired through inhalation of conidia present in the environment. The exact place of the fungus in nature is still unknown, but there are strong evidences that the fungus lives as saprobe in soil, since it has been isolated and molecularly detected from soil and the disease is predominantly found in male agricultural workers [76, 159–162]. *P. brasiliensis* and *P. lutzii* have also been detected in soil and in aerosol samples from different endemic areas in Brazil [162–164]. There are some sporadic isolation of *P. brasiliensis* from faeces of frugivorous bat (*Artibeus lituratus*) in Colombia and penguin (*Pygoscelis adeliae*) in Antarctica and dog food contaminated with soil in Brazil [165] and also (molecular detection) in faeces of armadillos [164]. In contrast to other systemic mycoses, outbreaks of PCM have never been reported. However, do Valle and co-workers (2017), described a series of acute cases of PCM after deforestation and massive earth removal for constructing a highway in the metropolitan region of Baixada Fluminense, state of

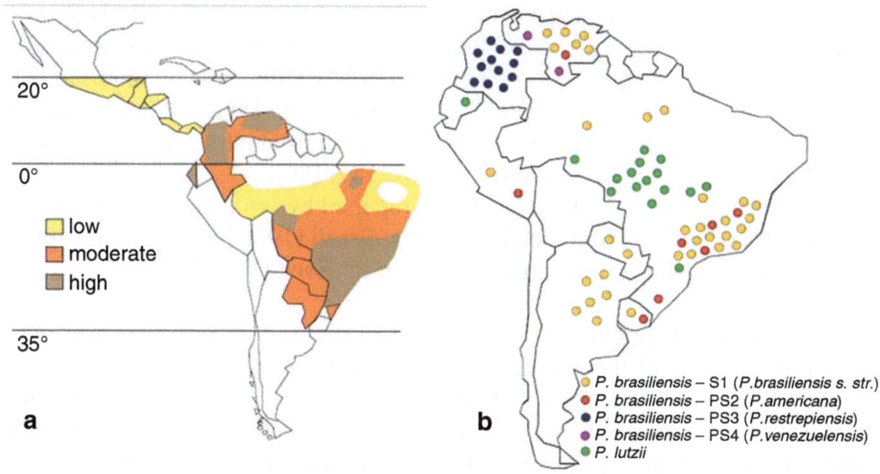

Fig. 6.11 (**a**) Endemic areas of paracoccidioidomycosis, according to Shikanai-Yasuda and co-workers (167); (**b**) geographic distribution of the cryptic species of *P. brasiliensis* and *P. lutzii*, adapted from Theodoro and co-workers [157]. "Copyright Springer International Publishing AG, Book: Emerging and Epizootic Fungal Infections in Animals, Chapter of Paracoccidioidomycosis (page 131)"

Rio de Janeiro, Brazil. During the period from December 2015 to December 2016, the authors observed an increasing rate around 5.7 times higher (8 cases/year) than the one observed during 1988 to 2015 in the same region (1.4 cases/year) [166].

PCM infection in humans is divided into asymptomatic (corresponding to subjects that respond positively to delayed hypersensitivity tests with paracoccidioidin and do not exhibit any clinical signs of disease) and the symptomatic form (acute/juvenile or chronic/adult disease) [167]. The natural infection in domestic and wild animals was studied mainly in Brazil by intradermal tests, serological surveys, histopathological analysis, molecular detection and culture of *P. brasiliensis* from tissue samples. Among wild animals, intradermal tests with paracoccidioidin showed that terrestrial animals (coatimundis, *Nasua nasua*) had highest positivity in comparison to arboreal animals (weeping capuchins and marmosets), while in domestic animals, horses showed the highest positivity than cattle and sheep [168, 169]. Concerning serological surveys, the most employed test was ELISA, and it was carried out in different species, such as dogs, cats, dairy cattle and goats, horses, sheep, pigs, rabbits, chickens and capuchin monkeys (*Cebus* sp.) and golden howler monkeys (*Alouatta caraya*) [170–177]. Dogs were the most studied species in serological surveys, in which the most significant parameter for positivity was their rural origin [178–180]. Dogs have also proved to be susceptible to experimental PCM disease [181]. Richini-Pereira and co-workers (2008), using nested PCR, detected the infection in wild road-killed animals, such as armadillos (*Dasypus novemcinctus*, *Dasypus septemcinctus*), guinea pigs (*Cavia aperea*), porcupines (*Sphiggurus spinosus*), grisons (*Galictis vittata*) and raccoons (*Procyon cancrivorus*) [182].

An important finding that opened new avenues for understanding the ecology of *P. brasiliensis* was the systematic fungal culturing from tissues of armadillos, *Dasypus novemcinctus* [183–187]. The origin of armadillos was in South America in the same geographic region where PCM is found [186, 188]. Besides armadillos, the infection of *P. brasiliensis* has also been reported from other members of super-order Xenarthra like anteaters and sloths [182, 189, 190].

The infection in animals by *P. lutzii* was reported by serological survey in dogs, horses and wild animals from South Region of Brazil [191], but the isolation of the fungus from armadillo's tissues has never been achieved [163, 164].

6.2.4.3 Clinical Aspects

The great majority of the reports of PCM in animals correspond to subclinical infection [187]. In humans, two main forms are observed: (1) acute-subacute, juvenile type or infantojuvenil, which affects mainly children and young adults under 30 years of age (Fig. 6.12a) and (2) chronic, adult type, which usually affects patients over 30 years of age (Fig. 6.12a, c). Acute-subacute PCM (juvenile type) affects both sexes and represents about 10% of the general casuistry of the disease. Of these cases, about 5% occur in children between 10 and 14 years of age. Eventually, the mycosis occurs in very young children, situations that are usually very serious and potentially fatal. The juvenile form can be subdivided into moderate or severe, according to the degree of dissemination. The clinical manifestations are caused by the rapid and progressive involvement of the mononuclear phagocytic system with diffuse, superficial and deep lymphadenomegaly, hepatosplenomegaly and bone marrow dysfunction in the most severe cases. Cutaneous manifestations and bone lesions may be noticed in this form. Fever and weight loss may also be observed, which quickly leads to the patient's overall commitment. Chronic adult PCM accounts for the majority of the cases in humans (about 90%), mainly observed in male adults between 30 and 60 years of age, who have the agricultural work or frequent contact with soils as the main risk activity. In women the disease is less frequent, in the average proportion of one woman to

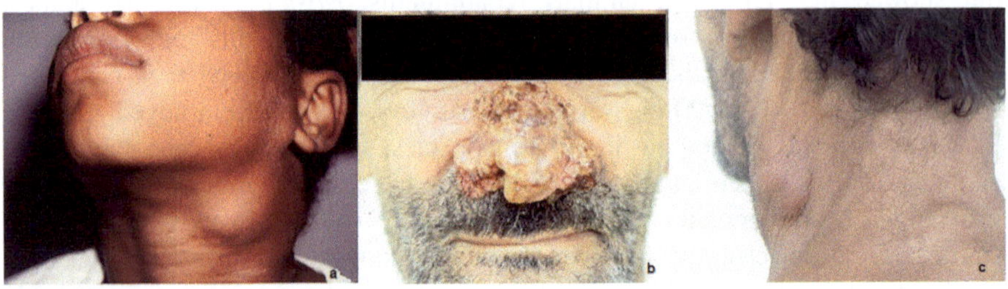

Fig. 6.12 Clinical manifestations of paracoccidioidomycosis in humans. (**a**) Subacute disseminated form in children, with generalized lymphadenomegaly; (**b** and **c**) chronic form of the adult, with involvement of the lungs, mucosa and nasal dorsum and cervical lymph node."Copyright of Gen National Publishing Group (Rio de Janeiro, Brazil), Book: Doenças Infecciosas Em animais de Produção e Companhia, Chapter of Paracoccidioidomicose (page 940), reprinted with permission"

ten men with PCM, since the female hormone (17-β-estradiol) confers protection by preventing the conversion of inhaled conidia to yeast in the lung parenchyma [192]. Chronic PCM can be divided into unifocal and multifocal. Latter is subdivided into mild, moderate and serious [16, 155, 159, 167, 193]. It is important to note that this classification was broadly based on the observation of clinical disease in humans [193], since animal disease is rare.

In animals, PCM disease has been reported in dogs and sloth. The first natural case in dogs was reported by Ricci and co-workers (2004) in a female Doberman presenting poor general conditions, generalized lymphadenomegaly and hepatosplenomegaly. The animal was treated with ketoconazole, with total remission of the symptoms, but after 18 months there was a relapse, and the animal was euthanized [194]. The second report was also in a female Doberman with the same clinical findings; however, the dog was treated with itraconazole, and remission of the symptoms was totally observed after 2 years (Fig. 6.13) [195]. It is interesting to note that both cases in dogs had the same clinical aspects and were similar to the ones observed in acute/juvenile form of disease in humans. Similarly, third case was reported in a female Labrador presenting lymph node enlargement and skin lesion in the superior left lip, providing the first report of skin lesion in dog is caused by *P. brasiliensis* (Fig. 6.14) [196].

Despite several reports of fungal infection, the disease was not observed in armadillos. There are some evidences of granuloma formation in the lungs and liver that may indicate a possibly active PCM disease [186, 197]. A generalized PCM was

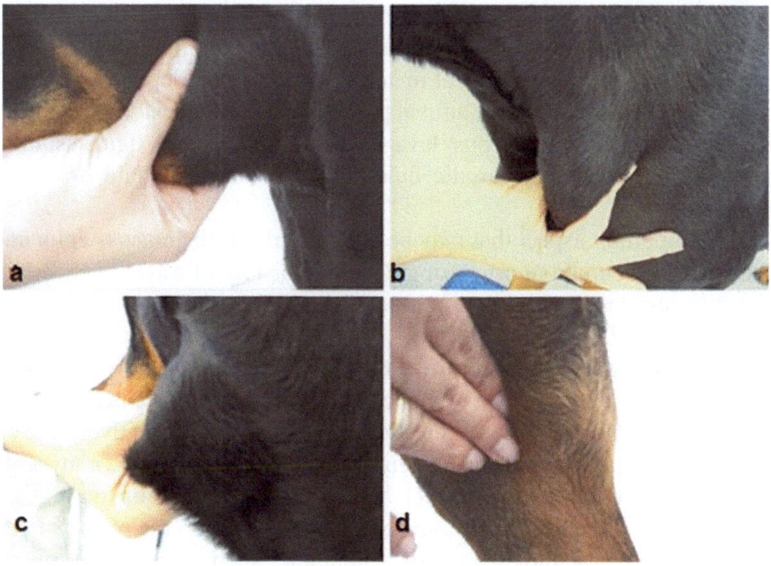

Fig. 6.13 Generalized lymphadenomegaly in a 6-year-old female Doberman. (**a**) Submandibular lymph node; (**b**) prescapular lymph node; (**c**) inguinal lymph node; (**d**) popliteal lymph node. "Copyright Mycopathologia 172(2):147–152, 2011, reprinted with permission"

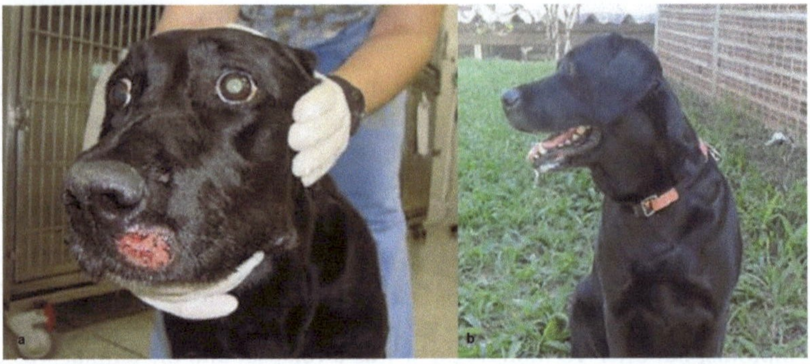

Fig. 6.14 First report of skin lesion in canine paracoccidioidomycosis. (**a**) Before treatment; (**b**) after treatment. "Copyright Mycopathologia 182(3–4):425–434, 2017, reprinted with permission"

reported in a two-tailed sloth, which presented anorexia, lethargy and dehydratation, leading to death [190].

6.2.5 Diagnosis of Systemic Mycoses

The procedures for diagnosing the four systemic mycoses are the same and comprise direct mycological examination, culture, histopathology, serology, and more recently molecular biology procedures. The isolation of fungi still remains as the gold standard method; however, such procedure may be susceptible to contamination, which depends on the viability of the fungi in clinical samples and require long time for fungal growth. The clinical material for diagnosis depends on the type of injury that manifests in human or animal patients. Generally, lesion scrapings, respiratory secretions, bronchoalveolar lavage, pus, cerebrospinal fluid, aspiration of lymph nodes, synovial fluid, ascitic fluid and other biopsied fluid or tissue fragments are used.

Another important aspect that may be considered for diagnosis is epidemiological data such as occupation, origin of the patient, history of travel to endemic areas and also non-responsiveness to previous antibacterial treatments.

6.2.5.1 Direct Mycological Examination

It is one of the first approaches for diagnosis. The clinical sample is assembled with a drop of KOH (10%) placed between slide and coverslip. Alternatively, the clinical samples may be stained with lactophenol cotton blue, Giemsa, Gram and rapid hematology (Diff-Quik) stains. Direct mycological examination may be helpful mainly for diagnosis of coccidioidomycosis and PCM, since the nature of etiological agents in clinical samples are pathognomonic, such as spherules with endospores (Fig. 6.15) and multibudding yeasts (Fig. 6.16), respectively. Identification of *H. capsulatum* by direct microscopy is usually difficult, since the yeasts are too

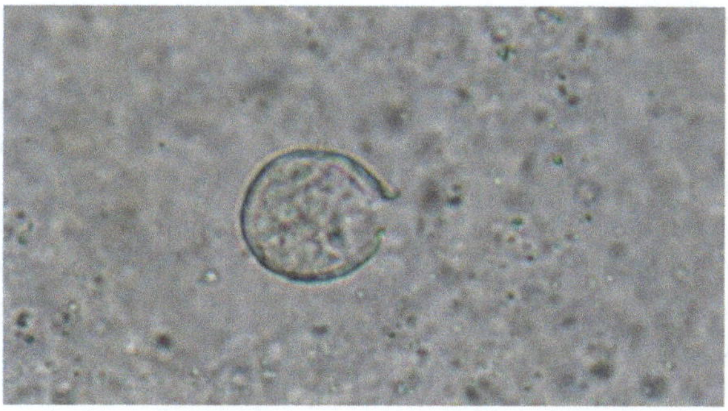

Fig. 6.15 Direct mycological examination of a sputum sample from patient with pulmonary coccidioidomycosis from Piauí state, Brazil, showing a spherule with disrupted cell wall. Image courtesy of Dr. Liline Maria Soares Martins (University Hospital of Federal University of Piauí, Piauí state, Brazil)

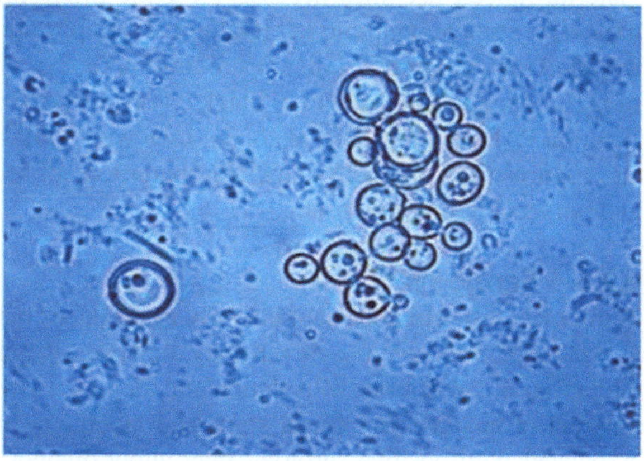

Fig. 6.16 Direct mycological examination of a skin scraping from patient with cutaneous paracoccidioidomycosis from São Paulo state, Brazil, showing rounded multi-budding yeasts. Image courtesy of Dr. Silvio Alencar Marques (Botucatu Medical School/UNESP/Botucatu, São Paulo state, Brazil)

small and are found inside macrophages. Youssef and co-workers (2011) demonstrated a negative image of *B. dermatitidis* yeasts in cytological smears of pleural fluid obtained from a 79-year-old male patient by fine-needle aspiration. Such approach may help clinicians to rapidly identify the etiological agent [198].

6.2.5.2 Culture

It is recommended to culture clinical samples in temperatures 25°C and 37°C, to demonstrate the dimorphism of the etiological agents. The cultures may be kept under aerobic condition, and the culture media employed are Sabouraud Dextrose agar without cycloheximide, Brain Heart Infusion agar, Potato Dextrose agar and Mycosel® agar, the last one better appropriate for *Paracoccidioides* spp. Due to high virulence, and extraordinary aerial dispersion capacity, cultivation and handling of *Coccidioides* spp., cultures should be restricted to the laboratories having biosafety level 3. Moreover, specially trained technicians are also required. The other dimorphic fungi can be handled at biosafety level 2 conditions.

6.2.5.3 Histopathology

The most important characteristic of histopathological analysis is the observation of granulomatous lesions with the fungus in the center. Staining techniques such as hematoxylin-eosin (HE) reveal (in coccidioidomycosis) a pyogranulomatous reaction. The fungus is not differentiated, however, the cell wall of the spherule and endospores may be both basophilic and eosinophilic. In canine and feline blastomycosis, purulent or pyogranulomatous lesions in infected tissues have been observed. Histologically, the agent is surrounded by neutrophils, macrophages and multinucleated giant cells. Histological sections of histoplasmosis stained with HE show slightly basophilic yeasts with spherical or oval morphology, surrounded by a clear halo delimited by a thin cell wall within phagocytes. Diagnosis of PCM in histological sections needs special attention as only a few small and unibroting yeast elements can be seen which may be very small forms of *P. brasiliensis* and usually indistinguishable from *H. capsulatum*, small forms of uncapsulated *Cryptococcus* spp. and endospores of *Coccidioides* spp. Thus, in such situation, the culture of the material indeed becomes necessary for the differential diagnosis. The best techniques for evidencing fungi in histological sections are PAS (periodic acid of Schiff) and the Gomori-Grocott (silver) staining methods, which stain the cell wall of the fungi in magenta and brownish-black colors, respectively. It is important to emphasize that PAS may not stain *H.capsulatum* well, making silver stain more appropriate. Figures 6.17 and 6.18 show histopathological sections of coccidioidomycosis and paracoccidioidomycosis.

6.2.5.4 Serology

Serological tests are important not only in diagnosis but in follow-up of the treatment as well. The exclusive use of serology, however, is not definitive in establishing the diagnosis. In the absence of isolation and identification of the etiological agent, combination of anamnesis, clinical signs and laboratory and radiographic examimation associated with positive serology may allow diagnosis. The most commonly used serological techniques are agar gel immunodiffusion (AGID) and enzyme-linked immunosorbent assay (ELISA), which have good sensitivity and specificity. AGID is

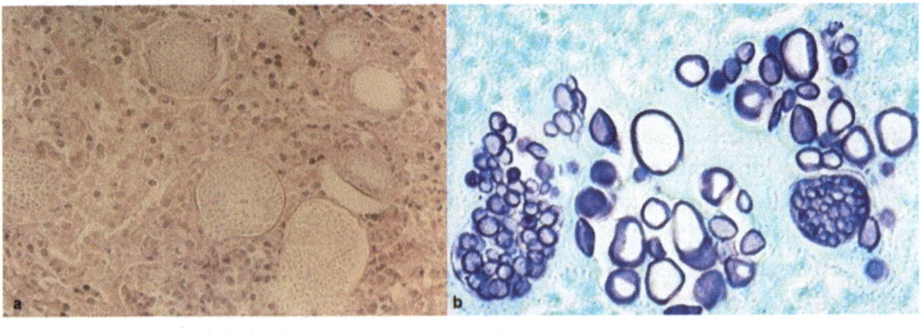

Fig. 6.17 Histopathological sections of pulmonary coccidioidomycosis showing different stages of spherules. (**a**) Hematoxilin-eosin; (**b**) Gomori-Grocott (400 ×). Images courtesy of Dr. Kelsen Eulálio Dantas (Federal University of Piauí, Piauí state, Brazil)

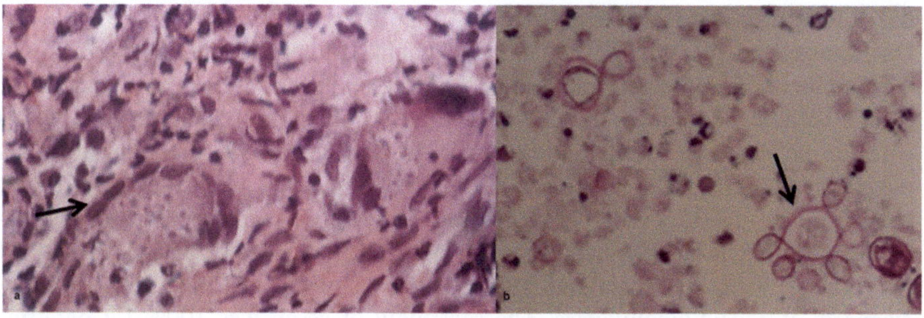

Fig. 6.18 Microscopic aspects of *P. brasiliensis* in histological sections. (**a**) Arrow shows granuloma with giant cells containing yeast elements suggestive of the fungus (hematoxilin-eosin, 400 ×); (**b**) multi-budding yeast (PAS, 400 ×). "Copyright of Gen National Publishing Group (Rio de Janeiro, Brazil), Book: Doenças Infecciosas Em Animais de Produção e Companhia, Chapter of Paracoccidioidomicose (page 942), reprinted with permission"

technically a simple assay to be performed; however the results can be time-consuming and can take even 3 days for a negative result to be confirmed. Serodiagnosis of human as well as canine coccidioidomycosis occurs at a similar rate. About 70% of infections are asymptomatic [62, 199]. Chow and co-workers developed an enzyme immunoassay (EIA) for detecting IgG antibodies in coccidioidomycosis of canine and other mammalian species. The EIA assay was employed in different endemic regions of coccidioidomycosis in the USA, using sera from infected and noninfected dogs, and the authors observed sensitivity and specificity of 94.6% and 96%, respectively [200]. Cryptic species in *Paracoccidioides* may have an impact on serology of PCM. In southeastern Brazil, all human patients suffering from PCM were found serologically positive, when tested with B339 antigen of *P. brasiliensis*. But when the same sera were tested using an antigen preparation from *P. lutzii* 510B isolate, positivity decreased to 41%. Conversely, patients from the Midwest of Brazil tested 92% positive with antigens from *P. lutzii* and 26% with *P. brasiliensis* antigen [201, 202]. Antigenemia may be an useful approach for complementary diagnosis, especially at the beginning of infection and prior to seroconversion or in the case of immunosuppressed individuals who are unable to develop a serological response [203].

Antigen detection tests are now widely used for the diagnosis of human histoplasmosis, both in sera and urine. Cunningham and co-workers standardized an EIA for detection of *H. capsulatum* antigens in urine samples of dogs. The authors showed a sensitivity and specificity of 89.47% and 100%, respectively [204]. Hage and co-workers (2011) combined both methods bronchoalveolar lavage (BAL) antigen detection and cytopathology, for rapid diagnosis, and they found a sensitivity of 96.8% [205]. Hanzlicek and co-workers (2016) evaluated the correlation of decreasing antigen concentration in urine and serum samples of cats undergoing treatment for histoplasmosis. They observed a significant positive linear correlation between time to clinical remission and low concentration of antigen in urine and serum at baseline. The serum and urine samples of cats were found positive at the time of disease relapse [206].

6.2.5.5 Molecular Identification

Diagnosis of endemic mycoses remains challenging. Molecular identification of fungi is now playing a significant role in this. Such techniques have been widely used for diagnosis, both in cultures and clinical samples. Polymerase chain reaction (PCR) with universal primers from ribosomal DNA (ITS1 and ITS4; ITS4 and ITS5) is mostly being used for the identification of fungi [207, 208] followed by nested PCR with specific primers. Nested PCR increases the sensitivity and specificity of diagnosis. Real-time PCR has also been tested for a sensitive and rapid diagnosis. Gago and co-workers (2014) developed and validated a quantitative real-time PCR for early diagnosis of coccidioidomycosis. The sensitivity of assay was 100% for paraffin-embedded tissues as well as for samples from patients proving that the assay is effective for the diagnosis and monitoring *Coccidioides* infection. It is also useful in preventing health hazards and in rapid identification of cultures in clinical setting [209]. Mitchell and co-workers (2015) reported that the diagnosis of coccidioidomycosis may be achieved in 4 h with a sensitivity equivalent to culture. The identification at genus and species levels is critical in order to provide antifungal timely [210]. In this sense, Morjaria and co-workers (2015) highlighted the importance of molecular diagnostics for a fast and definitive diagnosis of disseminated blastomycosis in a diabetic patient presenting as a brain mass initially thought to be tumoral mass [211]. A fatal acute respiratory distress syndrome due to *B. gilchristii* infection was confirmed by PCR and sequence analysis [212]. Bialek and co-workers (2002) evaluated nested PCR assays for detecting *H. capsulatum* in human tissues targeting rDNA (18S) and Hc100 and concluded that despite being single copy, the detection of Hc100 gene did not provide false-positive results compared to 18S rDNA [213]. Similarly, a real-time PCR for diagnosis of *H. capsulatum* has been standardized on the detection of Hc 100 gene. The authors concluded that such approach may be advantageous for a quick diagnosis of histoplasmosis with 6 pg/μl of (*H. capsulatum*) DNA analytical sensitivity, 88.9% of clinical sensitivity and 100% of analytic specificity [214].

Regarding molecular diagnosis of PCM, one of the first studies employed the detection of a 43 KDa protein nucleotide sequence, a single copy gene that codifies for the glycoprotein gp43 for cell wall of *P. brasiliensis* [215]. On the basis of its presence, Ricci and co-workers (2004) confirmed the first naturally infected PCM disease in dogs [194]. The majority of the reports for detecting *Paracoccidioides* spp. in tissue samples employ the nested PCR based on the amplification of complete ITS1-5.8S-ITS2 using the panfungal ITS4 and ITS5 primers [207] and specific inner primers [161, 162, 164, 182]. A real-time PCR approach was also evaluated for diagnosis of PCM. Buitrago and co-workers (2009) detected 1 fg of fungal DNA per μl of sample (sputum and tissue biopsies) and concluded that this technique may represent a sensitive method for rapid diagnosis of paracoccidioidomycosis and could also help monitoring patients in treatment [216].

6.2.6 Treatment of Systemic Mycosis

In general, the treatment of systemic mycoses involves the use of antifungals for a long period of time (months or years) and may produce socioeconomic impact due to the high costs of some antifungals. There are limited options of antifungal agents, and majority of them are either ergosterol synthetase inhibitors (azolics, triazolics, and alilamines) or bind directly to ergosterol in fungal cell membrane (polyenes). Due to the biological similarity between fungi, animals and humans, the treatment frequently leads to side effects, especially hepatic and renal damages [217]. In the beginning of the 2000s, a new class of antifungal represented by echinocandins was described. Echinocandins have fewer side effects due to the selective toxicity. They have limited activity against the dimorphic fungi, being more effective for *Candida* spp., *Aspergillus* spp. and *Pneumocystis* spp. [218].

Special attention must be given to drug interaction with azoles. The use of antacids, cimetidine, diphenylhydantoin and rifampicin may decrease their absorption. Concomitant use of azoles prolongs the effect or even increases the toxicity of the following drug groups: benzodiazepines (midazolam), glucocorticoids, antihistamines, quinidine, cyclosporine A, clarithromycin, telithromycin, everolimus, warfarin, terfenadine, astemizole, phenytoin, nifedipine, drospirenone, tetrahydrocannabinol, cannabidiol, sulfonylurea, digoxin, rosiglitazone, isavuconazole, riociguat and vincristine. Coadministration of negative inotropic drugs may enhance the risk of congestive heart failure while plasma levels of calcium antagonists are enhanced by itraconazole [219].

6.2.6.1 Coccidioidomycosis

Due to the lack of controlled studies and a wide variety of clinical manifestations in coccidioidomycosis in domestic animals, there is still no standard treatment protocol. The recommended antifungals in the treatment of coccidioidomycosis are, mainly, the azole derivatives (ketoconazole, fluconazole and itraconazole). New azole derivatives have recently been introduced for treatment of coccidioidomycosis, such as posaconazole. Herrin and co-workers (2005) treated a chimpanzee (*Pan troglodytes*), presenting ascites and neurological impairment with 10 mg/kg fluconazole PO/SID during 6 months and did not observe any improvement. The treatment was changed to 50 mg/kg posaconazole PO/SID with clinical remission of symptoms and negative serum titres after 24 months of therapy [220]. Posaconazole and voriconazole showed the best in vitro inhibition of strains of *C. immitis* and *C. posadasii* from endemic and non-endemic areas of coccidioidomycosis [221]. A retrospective study in 50 cats with different presentations of coccidioidomycosis from Arizona and EUA showed that the majority of cats were treated only with 50 mg/kg of fluconazole PO/BID, but some cats needed therapy combined with other antifungals, such as itraconazole, amphotericin B, posaconazole and terbinafine [222]. In a recent review, Davidson and co-workers (2019) suggested the use of fluconazole (5–10 mg/kg PO/BID), ketoconazole (5–20 mg/kg PO/BID), itraconazole (5 10 mg/kg PO/SID), posaconazole (5 mg/kg PO/ BID), voriconazole (4–5 mg/kg PO/BID) and lipid complex amphotericin B (1–3 mg/kg IV by every other day 3 times/week) for 6 to 12 months to treat canine coccidioidomycosis [223]. Foley and Legendre (1992) successfully treated bone lesions in a foal with itraconazole (2.6 mg/kg PO/BID) in 6 months [224]. Chitin synthase inhibitors such as lufenuron and nikkomycin Z have been suggested for the treatment of coccidioidomycosis; however no controlled studies and placebo were included in such studies. Besides, no in vivo efficacy was observed in experimentally infected mice [225–227].

6.2.6.2 Blastomycosis

A survey carried out during 1998–2008, in the USA, compared the period of treatment, hepatotoxicity, occurrence of relapses or deaths and costs between treatment with fluconazole and itraconazole in 144 dogs having systemic blastomycosis. No significant differences in efficacy, relapse/death rates and level of hepatic enzymes were observed between both groups. Treatment duration was considerably longer for fluconazole (median 183 days) than for itraconazole (138 days); however, the cost of fluconazole treatment was lower than itraconazole (median US$ 1223 versus US$ 3717) [228].

The following dosages for dogs and cats were recommended for at least 60 days of treatment: ketoconazole, 10 or 20 mg/kg, PO/BID for dogs, and 50 mg/kg, PO/SID for cats; fluconazole, 5 mg/kg, PO/BID for dogs, and 50 mg/kg, PO/TID for cats; itraconazole, 5 mg/kg, PO/SID for dogs, and 5 mg/kg, PO/BID for cats; and amphotericin B deoxycholate 0.5 mg/kg IV, each 48 h for dogs, and 0.25 mg/kg IV, each 48 h for cats [86].

6.2.6.3 Histoplasmosis

Itraconazole is the treatment of choice for dogs and cats. The recommended dose is 5–10 mg/kg PO, every 12 or 24 h, for at least 4 to 6 months. Oral solutions are better absorbed than capsules. Ketoconazole is advised VO for dogs, at the dose of 10 mg/kg, every 12 h, for 4 to 6 months. Fluconazole is prescribed for animals with neurological and/or ocular signs resulting from diffusion into the central nervous system. It is recommended to maintain treatment with these drugs for an additional 2 months for the complete clinical recovery of the animal. Amphotericin B is another alternative drug for dogs and cats, although it is nephrotoxic and requires monitoring of the renal function of the animals being treated. It is indicated at a dose of 0.25–0.5 mg/kg IV every 48 h until remission of clinical signs, although it should not exceed the cumulative dose of 5–10 mg/kg in dogs and 4–8 mg/kg in cats. In dogs with intestinal involvement, supportive treatment should include high digestibility diet, control of secondary bacterial proliferation and diarrhoea. Serial listing of biochemical and haematological examinations as well as diagnostic imaging tests is recommended, in all treated animals [150]. Treatment of epizootic lymphangitis consists of intravenous injection of 100 ml of sodium iodide of a 10% solution, repeated weekly for 4 weeks [113]. Gabal (1984) studied the effect of amphotericin B, nystatin (polyenes) and 5-fluorocytosine in vitro against five isolates of *H. capsulatum* var. *farciminosum*, and best inhibition was found against polyenes. The author suggested the use of intravenous injection of amphotericin B in combination with topical application of nystatin. Other options for treatment include surgical removal of the lesion and fire cauterization of the lesion [229].

6.2.6.4 Paracoccidioidomycosis

Human PCM has been treated with polyenes (amphotericin B), azole derivatives (ketoconazole, fluconazole or itraconazole) and the combination of sulfonamides and trimethoprim [167]. Concerning animal PCM, there are only three reports of naturally infected dogs. The first case was treated with ketoconazole, leading to total regression of the lymphadenomegaly; however clinical recurrence was observed after 18 months, and the dog was euthanized [194]. In the second report, dog was treated with itraconazole, 10 mg/kg/day/PO for 24 months, and showed total remission of clinical signs [195]. The third dog was also treated with itraconazole 10 mg/kg/day/PO/BID for at least 18 months [196]. Despite these few reports, it may be possible to conclude that itraconazole is the first antifungal of choice for the treatment of PCM in dogs.

6.2.7 Prophylaxis, Control and Public Health Concerns

Inspite of their non-transmissible nature, systemic mycoses may be classified as sapronoses (saprozoonoses), since the transmission is related to an abiotic element (usually soil). Therefore, eradication of these fungi from the contaminated sites would be an impractical task. Apart from this, vaccines are not available for preventing new infections.

Several attempts to develop vaccines for coccidioidomycosis, blastomycosis, histoplasmosis and paracoccidioidomycosis were carried out; however, it still remains a great challenge. Development of therapeutic vaccines (immuno-boosters) may reduce the time of treatment and prevention of relapses [230].

It is important to avoid the interaction of animals to places of risk in endemic areas, especially hunting of animals that build burrows in the soil, such as armadillos, as this practice favors the production of aerosols that may contain the fungus in its infective form. This was observed particularly in Brazil predominantly in coccidioidomycosis and paracoccidioidomycosis. Likewise, hunting dogs from endemic regions of blastomycosis should be monitored with the aim to early diagnosis of disease.

The use of appropriate personal protective equipments when entering caves, roofs or other protected places inhabited by bats or birds should be considered to avoid histoplasmosis. NIOSH (National Institute for Occupational Safety and Health)-approved respirator is recommended for the protection from fungal aerosols. Precaution must be taken while dealing with cultures and clinical materials for diagnosis of systemic mycoses. This should be handled only by trained people and in biosafety conditions accordingly to the fungus involved, e.g. *Coccidioides* spp. level 3, *Blastomyces* spp., *Histoplasma* spp. and *Paracoccidioides* spp. level 2. Campaigns aiming for educating public about the risks and potential means of acquiring the infection should be strongly recommended.

6.3 Sporotrichosis

6.3.1 Definition and Etiology

Sporotrichosis is a subcutaneous mycosis of humans and animals, characterized by ulcerated or nodular granulomatous lesions in the skin. The lesions may be unique or disseminated in the skin and subcutaneous tissue. Systemic dissemination is observed when the host is immunocompromised. The disease is endemic and also called as "implantation mycoses", since it is acquired by transcutaneous trauma through which the fungal conidia enter the host [231].

The etiological agent of sporotrichosis is *Sporothrix schenckii*, a thermal dimorphic fungus belonging to phylum *Ascomycota*, class *Pyrenomycetes*, order *Ophistomatales* and family *Ophistomataceae*. The sexual form of *S. schenckii* is *Ophiostoma stenoceras* [232, 233]. *Sporothrix schenckii* was believed to be the unique etiologic agent; however, Kwon-Chung (1979) observed some differences concerning the ability of fungal growth at 37 °C and virulence in animal models among isolates obtained from patients with lymphocutaneous and fixed cutaneous forms of sporotrichosis [234]. Marimon and co-workers (2006, 2008) based on sequence analysis of chitin synthase, ß-tubulin and calmodulin genes and also by phenotypic characteristics (morphology of conidia and growth at 30 °C, 35 °C and 37 °C and assimilation of sucrose, raffinose and ribitol) described three clinically relevant (*S. brasiliensis*, *S. globosa*, *S. luriei*) and two environmental species (*S. mexicana* and *S. albicans*) [235, 236]. The original *S. schenckii* now is known as *S. schenckii sensu stricto* [237]. A new species, *S. chilensis*, was also described based on morphological and molecular methods [238].

Sporothrix spp. is a thermal dimorphic fungus found in environment as filamentous form, mainly associated to decay vegetation, wood, sphagnum moss, hay and soil. It is interesting to emphasize that, despite the dimorphic nature of *Sporothrix* spp. and the endemicity of sporotrichosis, some differences may be highlighted between sporotrichosis and other systemic fungal infections caused by *Onygenales* fungi: *Sporothrix* sp. is widely distributed in environment, and sporotrichosis has widespread endemicity; the main route of infection is traumatogenic and seems to have a substantial direct zoonotic transmission [237].

The classical route of sporotrichosis involves the perforation of skin or mucosa due to scratches with soil and/or vegetation contaminated with conidia of filamentous form. In human and/or animal host, the fungus changes into yeast form. Macroscopically, the filamentous colony of *Sporothrix* spp. grows as colorless to beige in color and later becomes dark brown to black due to the production of melanin. The colony has a glabrous and membranous aspect, and microscopically, thin hyphae with erect conidiophores bearing several hyaline single-celled conidia, disposed in a flower-like arrangement, are seen (Fig. 6.19a, c). The yeast colony of *Sporothrix* spp. is similar to colonies of *Candida* spp. and *Saccharomyces cerevisiae*, characterized by colorless to beige color colony with smooth surface, and microscopically it appear as a single-bud yeast (Fig. 6.19a, b).

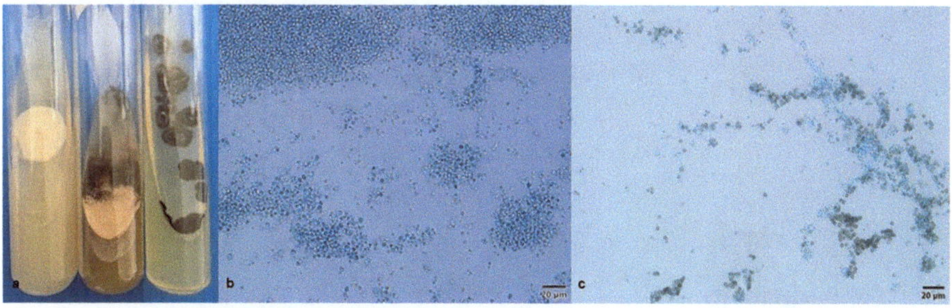

Fig. 6.19 Macro- and microscopic aspects of *Sporothrix* spp. (**a**) Yeast colony of *S. schenckii* cultured at Sabouraud agar (left), mycelial growth of *S. schenckii* showing a membranous aspect and the production of melanin (middle), mycelial growth of *S. brasiliensis* showing great production of melanin (right); (**b**) microscopic aspect of *S. brasiliensis* yeasts (lactophenol cotton blue, 400 ×); (**c**) microscopic aspect of hyphae and conidia of *S. brasiliensis*; note the presence of dark conidia, corresponding to melanin deposition in the cell wall (lactophenol cotton blue, 400 ×)

6.3.2 Epidemiology

Sporotrichosis can affect anyone regardless of age or gender; it depends on exposure. Occupational and recreational habits specific to different populations increase the risk of infection. For example, in Northeast India and Japan, the disease is mostly found in women due to their greater engagement in agricultural activities [239, 240]. In South Africa, males are more affected due to being frequently engaged in outdoor and mining activities [241]. In Uruguay, sporotrichosis has a higher prevalence among males and armadillo hunters; the infection occurs due to scratches received during armadillo hunting [242].

Several outbreaks have been reported in literature. The most famous occurred in South Africa, during the 1940s, affecting more than 3000 gold mine workers [243]. The great majority of outbreaks of sporotrichosis have been related to sphagnum moss contact, as observed in Florida by Hajjeh and co-workers [244], in Wisconsin by Grotte and Younger [245]. Besides sphagnum moss, hay has been the source of infection for outbreaks of sporotrichosis, as described by Dooley and co-workers [246] in a Halloween haunted house in the USA and by Feeney and co-workers [247] in 11 patients from Australia. Attention is drawn to the fact that these previously reported outbreaks were associated with plant sources of infection and *S. schenckii* as causative agent of the disease.

In Brazil, since the 1990s, a zoonotic outbreak of sporotrichosis has been reported associated to cat scratches and/or bites due to *S. brasiliensis*, mainly in south and southeast regions [248, 249]. The fungus may also be transmitted to human through contamination of preexisting skin wounds, due to high number of yeasts in the lesions of cats [250]. Particularly in the state of Rio de Janeiro, thousands of cases were observed, the great majority occurring in metropolitan region with low sanitation rates [249]. *S. brasiliensis* has been considered as the most virulent species among the species complex of *Sporothrix* [251]. Zoonotic transmission due to cat scratches was also reported in the USA [252, 253], Mexico [254], Malaysia [255,

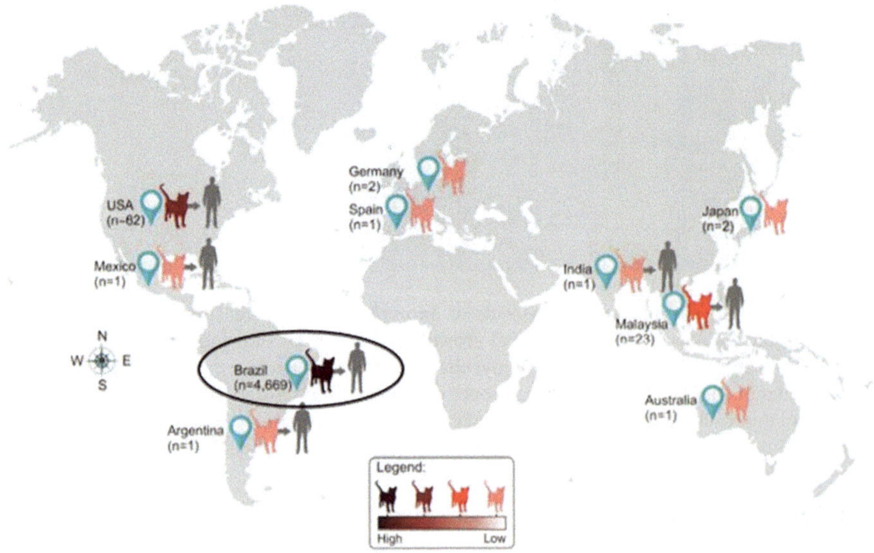

Fig. 6.20 Zoonotic sporotrichosis due to cat scratches around the world highlighting the largest absolute number of cases in southeast region of Brazil. (Source: Gremião and co-workers [249])

256] and India [257]. It is important to emphasize that these zoonotic reports were due to *S. schenckii s. str.* In Argentina, between 2011 and 2014, four cases of zoonotic transmission of human sporotrichosis followed by cat scratches were reported [258]. Later, the etiological agent of these reports was identified as *S. brasiliensis* [259]. These reports confirmed the existence of *S. brasiliensis* in an area of Argentina with proximity to the southern region of Brazil, where the disease is endemic in cats [249]. While in the countries mentioned above the number of zoonotic transmissions of sporotrichosis may vary from 1 to 62 cases, Brazil drew worldwide attention by reporting 4699 zoonotic cases (Fig. 6.20) [249].

An outbreak of 53 cases over 3 years in fishermen that was reported around Lake Ayarza in Guatemala was the result of injuries caused by contaminated fishes [260]. Haddad and co-workers reported sporotrichosis in fishermen from southeast region of Brazil due to injury with spines of the dorsal fin of fishes [261]. The fungus has also been isolated from nails and oral cavity of cats with and without clinical manifestations of sporotrichosis in Brazil [262]. In Thailand, sporotrichosis caused by *S. schenckii* was first time reported in a stray cat from Bangkok, with a poor prognosis [263].

6.3.3 Clinical Aspects

Clinical manifestations of sporotrichosis may be influenced mainly by the load and depth of the inoculum, pathogenicity, thermal tolerance of the strain and

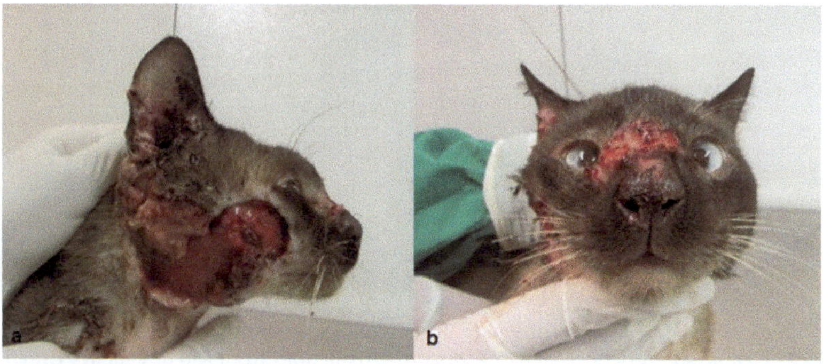

Fig. 6.21 Cephalic lesions in feline sporotrichosis due to *S. brasiliensis* in a cat from Rio de Janeiro, Brazil. (**a**) Side view of ulcerated and exudative lesion; (**b**) frontal view of ulcerated lesion in nasal region. Images courtesy of Dra. Isabella Dib Ferreira Gremião (FIOCRUZ/Rio de Janeiro state, Brazil)

immunological status of the host. Sporotrichosis is clinically observed in cats, dogs and horses, while cats are the most frequently affected animal species.

The cats show broad range of clinical manifestations, including subclinical infection, single lesions with spontaneous remission and fatal disseminated disease [264]. The disease is mainly found in adult male, mongrel and unneutered cats [265]. The lesions in general present with ulcerated appearance, which drain sero-sanguinolent or purulent exudates. A firm and nodular subcutaneous mass may also be observed commonly on the head, especially on the nose (Fig. 6.21). In majority of cases, cats present multiple skin lesions with mucosal involvement, especially mucous membranes of the respiratory tract [264]. Most skin lesions are observed in the head, the most affected area during fights. However, mucosal involvement may also occur due to cat habits such as licking, which can transfer a considerable amount of yeast cells from the skin lesions to the oral cavity, as well as to other distant body parts [266]. Respiratory signs, such as sneezing, dyspnoea and nasal discharge, are frequently observed in cats, and such clinical condition is associated with treatment failure and death [267]. A significant correlation among feline immunodeficiency virus (FIV) and/or feline leukemia virus (FeLV) in coinfected and non-coinfected cats has not been observed [264].

In dogs, fixed cutaneous and disseminated cutaneous lesions are the most frequent clinical forms observed. Figure 6.22 shows fixed cutaneous lesion in dogs.

6.3.4 Diagnosis

Sporotrichosis may be diagnosed using classical to molecular methodologies [231]. The classical methodologies comprise direct mycological examination, culture, serology and histopathology.

For direct mycological examination, the clinical specimen is placed on a glass slide with a drop of 10% potassium hydroxide (KOH) and covered with a coverslip.

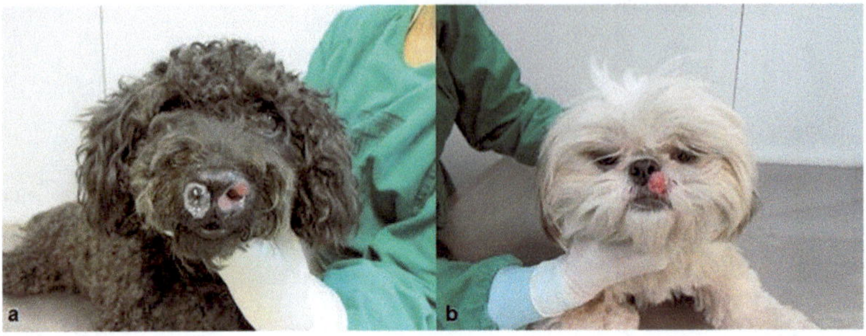

Fig. 6.22 Canine sporotrichosis due to *S. brasilliensis* in dogs from Rio de Janeiro, Brazil, showing the commitment of nasal region. (**a**) Nasal discharge of left nostril and pale and ulcerated mucosa of right nostril in a Poodle; (**b**) ulcerated legion at the lip in a Shitzu. Images courtesy of Dra. Isabella Dib Ferreira Gremião (FIOCRUZ/Rio de Janeiro state, Brazil)

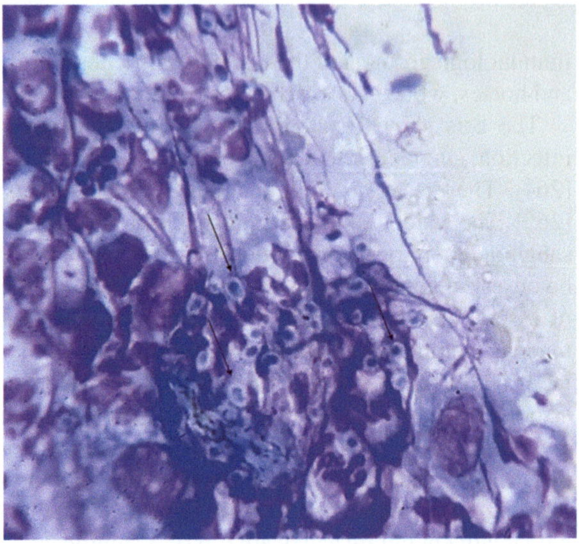

Fig. 6.23 Cytological examination of a feline lesion showing several yeasts (arrows) of *S. brasiliensis* (Diff-Quick, 400 ×). Image courtesy of Alana Lucena Oliveira (School of Veterinary Medicine and Animal Science/UNESP/Botucatu, São Paulo state, Brazil)

The yeasts are small (2–6 μm in diameter) and scarce in the lesions of humans and dogs in contrast to lesions in cats, which contain high amount of yeasts. Therefore, such approach is not very useful for diagnosis in both species. [268]. Cells stained by Diff-Quick method is very useful to diagnose sporotrichosis in cats, which reveals yeasts in cigar-shaped form (Fig. 6.23). This technique is rapid and convenient and has a high sensitivity and low cost.

Culturing *Sporothrix* spp. is the gold standard method; however, attention must be given to contamination and viability of the sample. *Sporothrix* sp. is resistant to the presence of cycloheximide, such as Mycosel® which is good for primary isolation from contaminated skin lesions. Cultures must be incubated aerobically, preferably at 25°C aiming to recognize the melanin production macroscopically [269, 270].

Among several serological tests, ELISA has been extensively employed for diagnosis as well as to monitoring therapy by serological follow-up of patients. Antigens preparations for serological assays include mycelium or yeast culture filtrates as well as partially purified molecules. Mendoza and co-workers (2002) reported the production of exoantigens from mycelia phase of *S. schenckii* which had no cross-reaction with serum samples from patients with coccidioidomycosis, histoplasmosis or paracoccidioidomycosis [271]. Fernandes and co-workers (2011) standardized an ELISA for diagnosis of feline sporotrichosis using two types of antigen: crude exoantigen preparation of *S. schenckii* and Concanavalin A-binding fraction (SsCBF). The authors tested 30 sera samples of cats with proven sporotrichosis and found 90% sensitivity with SsCBF, and the specificity was 96%, while crude exoantigens demonstrated 96% sensitivity and 98% specificity [272].

The presence of yeasts in histopathological sections is better seen with special stains, such as periodic acid-Schiff (PAS) or Gomori methenamine silver (GMS). Important to emphasize that yeasts cells of *S. schenckii* are difficult to visualize due to its paucity in lesions of humans and dogs [273, 274]. In general, a mixed suppurative and granulomatous inflammatory reaction in the dermis and subcutaneous tissue, frequently accompanied by microabscess and fibrosis, is observed. Cutaneous infections may also exhibit hyperkeratosis, parakeratosis and pseudo-epitheliomatous hyperplasia. Foreign bodies of vegetal origin related to the traumatic inoculation of the agent may also be encountered [231]. Figure 6.24 shows histological section of cat sporotrichosis stained with PAS, showing a high load of yeasts cells.

Molecular-based methods are not routinely used for diagnosis of sporotrichosis. Indeed, such approach may be more widely used for identification of species complex and also for phylogenetic studies. The ITS regions of rDNA are particularly useful to differentiate among *S. brasiliensis*, *S. schenckii*, *S. globosa* and *S. luriei* [275]. Molecular methods are sensitive and specific for the detection and identification of all *Sporothrix* species of clinical interest, require small amounts of sample and take less time than the traditional methods [266, 276].

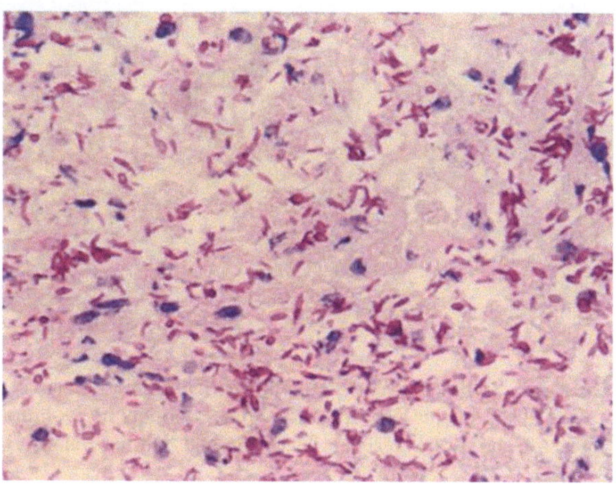

Fig. 6.24 Histopathological section of sporotrichosis stained with PAS evidencing high amounts of yeast with cigar-shaped format (440 ×)

6.3.5 Treatment

Despite the fact that exact mechanism of action of potassium iodide remains unknown till date, it has been used for treatment of sporotrichosis since the early twentieth century with satisfactory results. It is believed that potassium iodide acts on the resolution of granulomas through increased proteolysis and phagocytosis [277]. Due to adverse side effects related to potassium iodide, azolic compounds were introduced in the 1990s, highlighting itraconazole, which is currently the first-choice treatment, both for humans and animals (cats and dogs) [278]. Treatment of sporotrichosis usually requires long-term administration of itraconazole, potassium iodide or amphotericin B, depending on the severity and location of the lesions [249]. Successful treatment outcomes will also rely on rapid and accurate diagnosis. Therefore, differential diagnosis is important in cats. Unfortunately, the treatment of sporotrichosis in cats begins with antibacterials, even without a mycological diagnosis. Due to the characteristics of the lesions, which drain serosanguineous or purulent exudates, many veterinarians think that the lesions are due to bacterial infection. The lesions in general worsen a lot, and the veterinarians change the active principle of antibacterial without considering fungal etiology. Figure 6.25 shows clinical sporotrichosis in a Persian cat from Brazil with 7 months of evolution. When the correct diagnosis was achieved, the cat died due to dissemination of the pathogen as well as septic bacteremia.

The major side effects of itraconazole are anorexia, vomiting and diarrhoea associated with elevations of alanine aminotransferase, which requires its monthly measurement. The administration of itraconazole with meals of cats and dogs is recommended [267]. Schubach and co-workers evaluated different therapeutic regimens in 266 diseased cats. Clinical cure was achieved in 68 (25.4%), and treatment duration ranged from 16 to 80 weeks (median = 36 weeks). The most observed adverse effects were anorexia, vomiting and diarrhoea [264].

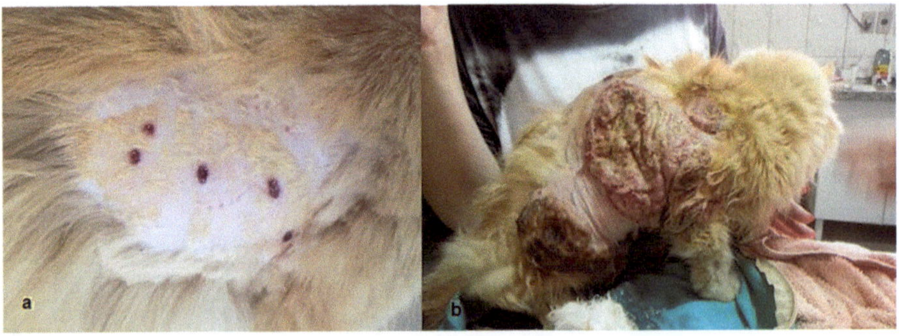

Fig. 6.25 Sporotrichosis due to *S. brasiliensis* in a Persian cat, from São Paulo state, Brazil. (**a**) Lesion at the beginning; (**b**) lesion after 7 months of evolution. Images courtesy of Dr. Rodrigo Trolezi (School of Veterinary Medicine and Animal Science/UNESP/Botucatu, São Paulo state, Brazil)

Important to consider is the increasing report of amphotericin B and itraconazole-insensitive strains of *Sporothrix* spp. over time [279]. The association of itraconazole and potassium iodide has been an effective option for the treatment of feline sporotrichosis. Reis and co-workers in 2016 reported that 25 (96.15%) cats were cured with itraconazole and potassium iodide, with a mean dose of 26.3 mg/kg (19.6 to 33.3 mg/kg) and 3.1 mg/kg (2.5–5.4 mg/kg), respectively. The median time of treatment until cure was 14 weeks (8–30 weeks); however, cats with mucosal involvement presented a longer median time (23 weeks) when compared to cats with skin lesion only [280]. It is recommended to keep treatment for at least 4–8 weeks after clinical cure of lesions to avoid relapse. A recurrent situation in Brazil is the abandonment of cats during the treatment period, a fact that may have contributed to the resistance to itraconazole

Alternatively, terbinafine, a fungicidal allylamine, may be used for feline sporotrichosis in cases of low response, intolerance or resistance to itraconazole. Viana and co-workers (2018) reported two cases of sporotrichosis in dogs successfully treated with terbinafine at doses of 25 and 30 mg/kg, once a day with food, during a period of 12–19 weeks, respectively, without side effects and relevant alterations in full blood examination and serum biochemical analysis [281]. Cutaneous sporotrichosis in dogs, despite requiring a long period of therapy, usually presents a favorable prognosis, responding to therapy with iodides or with itraconazole. However, disseminated cutaneous or extra-cutaneous sporotrichosis in dogs is usually associated with systemic immunocompromised hosts and, therefore, presents a reserved prognosis.

6.3.6 Public Health and Control

Sporotrichosis has two main forms of transmission, classical and zoonotic, and both of them include traumatisme that causes injury to the skin leading to inoculation of the fungus. It is recommended to wear gloves and long sleeves, as well as heavy boots, during activities like pruning of roses and the handling of sphagnum moss, wires, bushes, hay bales, conifer (pine) seedlings, or other materials that may facilitate the exposure to the fungus. As already mentioned by Hajjeh and co-workers (1997), the risk of sporotrichosis increased significantly in conditions like working with sphagnum moss, filling of topiaries and particularly with less gardening experience [244].

Sporotrichosis is a neglected disease, and since 2011 the cases in cats and dogs become notifiable to health authorities in Rio de Janeiro, Brazil (Joint Technical Note number 03/2011). Human cases of disease were mandatorily notified in 2013, according to the Resolution of the State Secretariat of Health (Resolution SES number 674, July 12, 2013).

It has been a great challenge to control sporotrichosis among cat population. The disease has been mostly observed among male non-neutered cats (around 73.1%) in Rio de Janeiro, Brazil [265]. Cats have some behavioral characteristics that favor the infection, such as toileting habits in contact with soil, sharpening the nails in environment, frequent disputes over females during estrus and territorial disputes [282]. Associated to their natural behavior, we must consider the fact that cats are susceptible to the fungus [249]. It is important to note that severe cases of sporotrichosis in cats may develop independently of retrovirus (FIV and/or FeLV) coinfections, which are immunosuppressive. Cats are usually allowed to roam outdoors, and most of them are neither vaccinated nor neutered and do not receive regular prophylactic deworming in endemic regions of disease. These features cannot be ruled out for contributing to the susceptibility of disease [249].

As preventive measures for controlling disease in cats, health authorities in Rio de Janeiro has promoted campaigns for castration without costs to owners, as well as the donation of itraconazole for treatment. Even then, the abandonment rate of these animals is high [265]. Precautions during the administration of oral medication must be performed, since the scratches occur during the physical restraint of diseased cats. It is strongly recommended that diseased animals should be kept separately from other animals, since exudates of lesions are rich in yeast form of fungus [248].

Experimental immunization with monoclonal antibodies against 70 kDa protein (putative adhesin) induced a significant reduction of fungal load in liver and spleen of infected mice [283, 284]. Portuondo and co-workers (2015) have evaluated protective properties of an aluminium hydroxide-adsorbed *S. schenckii* cell wall protein-based vaccine in experimentally infected mouse and found an increase in ex vivo release of IL-12, IFN-γ, IL-4 and IL-17. The authors observed that the immunization was able to facilitate in vivo protection in a subsequent challenge with *S. schenckii*, becoming a viable vaccine candidate for further testing. However, the use of adjuvants should be carefully reviewed in future investigations in cats, since this compound is associated with feline vaccine-induced sarcomas [285].

As observed, controlling zoonotic transmission of sporotrichosis has been a difficult task, and besides all the considerations discussed above, health education should be strongly recommended for population in general.

References

1. Rippon JW (1988) Medical mycology: The pathogenic fungi and the pathogenic actinomycetes. Saunders, Philadelphia
2. Alexopoulos CJ, Mims CW, Blackwell M (1996) Introductory Mycology, 4th edn. Wiley, New York
3. de Hoog GS, Guarro J, Gene J et al (2009) Atlas of clinical fungi: the ultimate bench tool for diagnostics. Centraalbureau voor Schimmelcultures, KNAW fungal. Universitat Rovira I Virgili, Utrecht
4. Untereiner WA, Scott JA, Naveau FA et al (2004) The Ajellomycetaceae, a new family of vertebrate-associated Onygenales. Mycologia 96:812–821
5. Emmons CW, Jellison WL (1960) *Emmonsia crescens* sp. nov. and adiaspiromycosis (haplomycosis) in mammals. Ann N Y Acad Sci 89:91–101
6. Schwartz IS, Kenyon C, Feng P et al (2015) 50 years of *Emmonsia* disease in humans: The dramatic emergence of a cluster of novel fungal pathogens. PLoS Pathog 11:e1005198

7. Drouhet E, Guého E, Gori S et al (1998) Mycological, ultrastructural and experimental aspects of a new dimorphic fungus *Emmonsia pasteuriana* sp. nov. isolated from a cutaneous disseminated mycosis in AIDS. J Mycol Med 8:64–77
8. Gori S, Drouhet E, Guého E et al (1998) Cutaneous disseminated mycosis in a patient with AIDS due to a new dimorphic fungus. J Mycol Med 8:57–63
9. Pelegrín I, Ayats J, Xiol X et al (2011) Disseminated adiaspiromycosis: case report of a liver transplant patient with human immunodeficiency infection, and literature review. Transpl Infect Dis 13:507–514
10. Malik R, Capoor MR, Vanidassane I et al (2016) Disseminated *Emmonsia pasteuriana* infection in India: a case report and a review. Mycoses 59:127–132
11. Feng P, Yin S, Zhu G et al (2015) Disseminated infection caused by *Emmonsia pasteuriana* in a renal transplant recipient. J Dermatol 42:1179–1182
12. Tang XH, Zhou H, Zhang XQ et al (2015) Cutaneous disseminated emmonsiosis due to *Emmonsia pasteuriana* in a patient with cytomegalovirusenteritis. JAMA Dermatol 151:1263–1264
13. Schwartz IS, Govender NP, Corcoran C et al (2015) Clinical characteristics, diagnosis, management and outcomes of disseminated emmonsiosis: a retrospective case series. Clin Infect Dis 61:1004–1012
14. Marin-Felix Y, Stchigel AM, Cano-Lira JF et al (2015) *Emmonsiellopsis*, a new genus related to the thermally dimorphic fungi of the family Ajellomycetaceae. Mycoses 58:451–460
15. Baumgardner DJ, Laundre B (2001) Studies on the molecular ecology of *Blastomyces dermatitidis*. Mycopathologia 152:51–58
16. Wanke B, Londero AT (1994) Epidemiology and Paracoccidioidomycosis Infection. In: Restrepo-Moreno A, Negro GD (eds) Franco M, Lacaz C da S. Paracoccidioidomycosis. CRC Press, Boca Raton, pp 109–119
17. Negroni R (2008) Evolución de losconocimientos sobre aspectos clínico-epidemiológicosde la Coccidioidomycosisenlas Américas. Rev Argent Microbiol 40:246–256
18. Antinori S (2014) *Histoplasma capsulatum*: more widespread than previously thought. Am J Trop Med Hyg 90:982–983
19. Hirschmann JV (2007) The early history of coccidioidomycosis: 1892–1945. Clin Infect Dis 44:1202–1207
20. Al-Doory Y (1992) Introduction. In: Al-Doory Y, Di Salvo AF (eds) Blastomycosis. Springer, Plenum Publishing Corporation, New York
21. Schwarz J, Baum GL (1957) The history of histoplasmosis, 1906 to 1956. New England J Med 246:253–258
22. Marques SA (2008) Paracoccidioidomycosis: a century from the first case report. An Bras Dermatol 83:271–273
23. Conti Diaz IA (1989) Epidemiology of sporotrichosis in Latin America. Mycopathologia 108:113–116
24. Barros MB, Schubach AO, do Valle AC et al (2004) Cat transmitted sporotrichosis epidemic in Rio de Janeiro, Brazil: description of a series of cases. Clin Infect Dis 38:529–535
25. Schenck BR (1898) On refractory subcutaneous abscess caused by a fungus possibly related to the Sporotricha. Bull Johns Hopkins Hosp 9:286–290
26. Lubarsky R, Plunkett OA (1955) *In vitro* production of the spherule phase of *Coccidioides immitis*. J Bacteriol 70:182–186
27. Fisher MC, Koenig GL, White TJ et al (2001) Biogeographic range expansion into South America by *Coccidioides immitis* mirrors new world patterns of human migration. Proc Natl Acad Sci USA 98:4558–4562
28. Fisher MC, Koenig GL, White TJ et al (2002) Molecular and phenotypic description of *Coccidioides posadasii* sp. nov. previously recognized as the non-California population of *Coccidioides immitis*. Mycologia 94:73–84
29. Brilhante RS, de Lima RA, Ribeiro JF et al (2013) Genetic diversity of *Coccidioides posadasii* from Brazil. Med Mycol 51:432–437

30. Duarte-Escalante E, Zúñiga G, Frías-De-León MG et al (2013) AFLP analysis reveals high genetic diversity but low population structure in *Coccidioides posadasii* isolates from Mexico and Argentina. BMC Infect Dis 3:411
31. Teixeira MM, Barker BM (2016) Use of population genetics to assess the ecology, evolution, and population Structure of *Coccidioides*. Emerg Infect Dis 22:1022–1030
32. Stevens DA, Clemons KV, Levine HB et al (2009) Expert opinion: what to do when there is *Coccidioides* exposure in a laboratory. Clin Infect Dis 49:919–923
33. Galgiani JN, Ampel NM, Blair JE et al (2016) Infectious Diseases Society of America (IDSA) clinical practice guideline for the treatment of coccidioidomycosis. Clin Infect Dis 63:e112–e146
34. Elconin AF, Egeberg RO, Egeberg MC (1964) Significance of soil salinity on the ecology of *Coccidioides immitis*. J Bacteriol 87:500–503
35. Eulálio KD, de Macedo RL, Cavalcanti MA et al (2000) *Coccidioides immitis* isolated from armadillos (*Dasypus novemcinctus*) in the state of Piauí, northeast Brazil. Mycopathologia 149:57–61
36. Macêdo RC, Rosado AS, Mota FF et al (2011) Molecular identification of *Coccidioides* spp. in soil samples from Brazil. BMC Microbiol 16:108
37. Reyes-Montes M del R, Pérez-Huitrón MA, Ocaña-Monroy JL et al (2016) The habitat of *Coccidioides* spp. and the role of animals as reservoirs and disseminators in nature. BMC Infect Dis 16:550
38. Coopersmith EJ, Bell JE, Benedict K et al (2017) Relating coccidioidomycosis (valley fever) incidence to soil moisture conditions. Geohealth 17:51–63
39. Bowers JR, Parise KL, Kelley EJ et al (2019) Direct detection of *Coccidioides* from Arizona soils using CocciENV, a highly sensitive and specific real-time PCR assay. Med Mycol 57:246–255
40. Brown J, Benedict K, Park BJ et al (2013) Coccidioidomycosis: epidemiology. Clin Epidemiol 5:185–197
41. Freedman M, Jackson BR, McCotter O et al (2018) Coccidioidomycosis outbreaks, United States and worldwide, 1940–2015. Emerg Infect Dis 24:417–424
42. Benedict K, Ireland M, Weinberg MP et al (2018) Enhanced Surveillance for Coccidioidomycosis, 14 US States. Emerg Infect Dis 24:1444–1452
43. Baptista Rosas RC, Riquelme M (2007) Epidemiologia de la coccidioidomicosis en México. Rev Iberoam Micol 24:100–105
44. Centers for Disease Control and Prevention (CDC) (2009) Increase in coccidioidomycosis-California, 2000–2007. MMWR Morb Mortal Wkly Rep 58:105–109
45. Canteros CE, Toranzo A, Ibarra-Camou B et al (2010) La coccidioidomicosis en Argentina, 1892–2009. Rev Arg Microbiol 42:261–268
46. Canteros CE, Vélez HA, Toranzo AI et al (2015) Molecular identification of *Coccidioides immitis* in formalin-fixed, paraffin-embedded (FFPE) tissues from a Colombian patient. Med Mycol 53:520–527
47. Simões DM, Dial SM, Coyner KS et al (2016) Retrospective analysis of cutaneous lesions in 23 canineand 17 feline cases of coccidioidomycosis seen in Arizona, USA (2009–2015). Vet Dermatol 27:346–e87
48. Ziemer EL, Pappagianis D, Madigan JE et al (1992) Coccidioidomycosis in horses: 15 cases (1975-1984). J Am Vet Med Assoc 201:910–916
49. Maddy KT, Crecelius HG, Cornell RG (1960) Distribution of *Coccidioides immitis* determined by testing cattle. Public Health Rep 75:955–962
50. Coster ME, Ramos-Vara JA, Vemulapalli R et al (2010) *Coccidioides posadasii* keratouveitis in a llama (*Lama glama*). Vet Ophthal 1:53–57
51. Fernandez JA, Hidalgo MN, Hodzic E et al (2018) Pathology of coccidioidomycosis in llamas and alpacas. J Vet Diag Invest 30:560–564
52. Diab S, Johnson SM, Garcia J et al (2013) Case report: Abortion and disseminated infection by *Coccidioides posadasii* in an alpaca (*Vicugna pacos*) fetus in Southern California. Med Mycol Case Rep 2:159–162

53. Kundu MC, Ringenberg MA, d'Epagnier DL et al (2017) Coccidioidomycosis in an indoor-housed rhesus macaque (*Macaca mulatta*). Comparative Medicine 67:452–455
54. Wallace RS, Clyde VL, Steinberg H (2009) Coccidioidomycosis in a black rhinoceros (*Diceros bicornis*). J Zoo Wildl Med 40:365–368
55. Burgdorf-Moisuk A, Stalis IH, Pye GW (2012) Disseminated coccidoidomycosis in a koala (*Phascolarctos cinereus*). J Zoo Wildl Med 43:197–199
56. Churgin SM, Garner MM, Swenson J et al (2013) Intestinal coccidioidomycosis in a red coachwhip snake (*Masticophis flagellum piceus*). J Zoo Wildl Med 44:1094–1097
57. Cordeiro RA, de Castro e Silva KR, Brilhante RSN et al (2012) *Coccidioides posadasii* infection in bats, Brazil. Emerg Infect Dis 18:668–670
58. Reed RE, Migaki G, Cummings JA (1976) Coccidioidomycosis in a California sea lion (*Zalophus californianus*). J Wildl Dis 12:372–375
59. Reidarson TH, Griner LA, Pappagianis D et al (1998) Coccidioidomycosis in a bottlenose dolphin. J Wildl Dis 34:629–631
60. Huckabone SE, Gulland FMD, Johnson SM et al (2015) Coccidioidomycosis and other systemic mycoses of marine mammals stranding along the central California, USA coast:1998–2012. J Wildl Dis 51:295–308
61. Gautam R, Srinath I, Clavijo A et al (2013) Identifying areas of high risk of human exposure to coccidioidomycosis in Texas using serology data from dogs. Zoonoses Public Health 60:174–181
62. Barker BM (2018) Coccidioidomycosis in Animals. In: Guillot J, Verweij PE (eds) Seyedmousavi S, deHoog GS. Emerging and Epizootic Fungal Infections in Animals, Springer Nature, pp 81–114
63. Graupmann-Kuzma A, Valentine BA, Shubitz LF et al (2008) Coccidioidomycosis in dogs and cats: a review. J Am Anim Hosp Assoc 44:226–235
64. Greene RT (2012) Coccidioidomycosis and Paracoccidioidomycosis. In: Greene CE (ed) Infectious diseases of the dog and cat, 4th edn. St. Louis, Elsevier Saunders, pp 634–645
65. Bosco SMG, Bagagli E, Wanke B (2016) Coccidioidomicose. In: Megid J, Ribeiro MG, Paes AC (eds) Doenças infecciosas em animais de produção e companhia. Roca, Rio de Janeiro, Brazil, pp 869–877
66. Clyde VL, Kollias GV, Roelke ME et al (1990) Disseminated coccidioidomycosis in a western cougar (*Felis concolor*). J Zoo Wildl Med 21:200–205
67. Adaska JM (1999) Peritoneal coccidioidomycosis in a mountain lion in California. J Wildl Dis 35:75–77
68. Maddy KT, Reed RE, Trautman RJ et al (1960) Experimental bovine coccidioidomycosis. Am J Vet Res 21:748–752
69. Maddy KT (1954) Coccidioidomycosis of cattle in the southwestern United States. J Am Vet Med Assoc 124:456–464
70. McDonough ES, Lewis AL (1967) *Blastomyces dermatitidis*: production of sexual stage. Science 156:528–529
71. McDonough ES, Lewis AL (1968) The ascigerous stage of *Blastomyces dermatitidis*. Mycologia 60:76–83
72. Brown EM, McTaggart LR, Zhang SX et al (2013) Phylogenetic analysis reveals a cryptic species *Blastomyces gilchristii*, sp. nov. within the human pathogenic fungus *Blastomyces dermatitidis*. PLoS One 8:e59237 (*Erratum in*: PLoS One 2016 Dec 9;11:e0168018)
73. Dukik K, Muñoz JF, Jiang Y et al (2017) Novel taxa of thermally dimorphic systemic pathogens in the Ajellomycetaceae (Onygenales). Mycoses 60:296–309
74. Denton JF, McDonough ES, Ajello L et al (1961) Isolation of *Blastomyces dermatitidis* from soil. Science 133:1126–1127
75. Denton JF, Di Salvo AF (1964) Isolation of *Blastomyces dermatitidis* from natural sites at Augusta, Georgia. Am J Trop Med Hyg 13:716–722
76. Restrepo A, Baumgardner DJ, Bagagli E et al (2000) Clues to the presence of pathogenic fungi in certain environments. Med Mycol 38:67–77

77. Emmons CW, Murray IG, Lurie HI et al (1964) North American blastomycosis: two autochthonous cases from Africa. Sabouraudia 3:306–311
78. Litvinov IV, St-Germain G, Pelletier R et al (2013) Endemic human blastomycosis in Quebec, Canada, 1988–2011. Epidemiol Infect 141:1143–1147
79. Randhawa HS, Khan ZU, Gaur SN (1983) *Blastomyces dermatitidis* in India: first report of its isolation from clinical material. Sabouraudia 21:215–221
80. Kuttin ES, Beemer AM, Levij J et al (1978) Occurrence of *Blastomyces dermatitidis* in Israel. First autochthonous Middle Eastern case. Am J Trop Med Hyg 27:1203–1205
81. McTaggart LR, Brown EM, Richardson SE (2016) Phylogeographic analysis of *Blastomyces dermatitidis* and *Blastomyces gilchristii* reveals an association with North American freshwater drainage basins. PLoS ONE 11:e0159396
82. Baumgardner DJ, Burdick JS (1991) An outbreak of human and canine blastomycosis. Rev Infect Dis 13:898–905
83. Sarosi GA, Eckman MR, Davies SF et al (1979) Canine blastomycosis as a harbinger of human disease. Ann Intern Med 91:733–735
84. Saccente M, Woods GL (2010) Clinical and laboratory update on blastomycosis. Clin Microbiol Rev 23:367–381
85. Baumgardner DJ, Paretsky DP, Yopp AC (1995) The epidemiology of blastomycosis in dogs: north central Wisconsin EUA. J Med Vet Mycol 33:171–176
86. Legendre AM (2012) Blastomycosis. In: Greene CE (ed) Infectious diseases of the dog and cat, 4th edn. Elsevier, Amsterdam, pp 606–614
87. Wilson JH, Olson EJ, Haugen EW et al (2006) Systemic blastomycosis in a horse. J Vet Diagn Investig 18:615–619
88. Cates MB, Kaufman L, Grabau JH et al (1986) Blastomycosis in an Atlantic bottlenose dolphin. J Am Vet Med Assoc 189:1148–1150
89. Wilkinson LM, Wallace JM, Cline JM (1999) Disseminated blastomycosis in a rhesus monkey (*Macaca mulatta*). Vet Pathol 36:460–462
90. Zwick LS, Briggs MB, Tunev SS et al (2000) Disseminated blastomycosis in two California sea lions (*Zalophus californianus*). J Zoo Wildl Med 31:211–214
91. Storms TN, Clyde VL, Munson L et al (2003) Blastomycosis in non domestic felids. J Zoo Wildl Med 34:231–238
92. Harris JR, Blaney DD, Lindsley MD et al (2011) Blastomycosis in man after kinkajou bite. Emerg Infect Dis 17:268–270
93. Dykstra JA, Rogers LL, Mansfield SA et al (2012) Fatal disseminated blastomycosis in a free ranging American black bear (*Ursus americanus*). J Vet Diagn Investig 24:1125–1128
94. Nemeth NM, Campbell GD, Oesterle PT et al (2016) Red fox as sentinel for *Blastomyces dermatitidis*, Ontario, Canada. Emerg Infect Dis 22:1275–1277
95. Rosser MF, Lindemann DM, Barger AM et al (2016) Systemic blastomycosis in a captive red ruffed lemur (*Varecia rubra*). J Zoo Wildl Med 47:912–916
96. Crampton TL, Light RB, Berg GM et al (2002) Epidemiology and clinical spectrum of blastomycosis diagnosed at Manitoba hospitals. Clin Infect Dis 34:1310–1316
97. Farber ER, Leahy MS, Meadows TR (1968) Endometrial blastomycosis acquired by sexual contact. Obstet Gynecol 32:195–199
98. Craig MW, Davey WN, Green RA (1970) Conjugal blastomycosis. Am Rev Respir Dis 102:86–90
99. Watts EA, Gard PD Jr, Tuthill SW (1983) First reported case of intrauterine transmission of blastomycosis. Pediatr Infect Dis 2:308–310
100. Tuthill SW (1985) Disseminated blastomycosis with intrauterine transmission. South Med J 78:1526–1527
101. Gray NA, Baddour LM (2002) Cutaneous inoculation blastomycosis. Clin Infect Dis 34:E44–E49
102. Bloom JD, Hamor RE, Gerding PA (1996) Ocular blastomycosis in dogs: 73 cases, 108 eyes (1985–1993). J Am Vet Med Assoc 209:1271–1274

103. Arceneaux KA, Taboada J, Hosgood G (1998) Blastomycosis in dogs: 115 cases (1980–1995). J Am Vet Med Assoc 213:658–664
104. Schwartz IS (2018) Blastomycosis in mammals. In: Guillot J, Verweij PE (eds) Seyedmousavi S, deHoog GS. Emerging and Epizootic Fungal Infections in Animals, Springer Nature, pp 159–176
105. Bosco SMG, Bagagli E, Baumgardner DJ (2016) Blastomicose. In: Megid J, Ribeiro MG, Paes AC (eds) Doenças infecciosas em animais de produção e companhia. Roca, Rio de Janeiro, Brazil, pp 861–868
106. Blondin N, Baumgardner DJ, Moore GE et al (2007) Blastomycosis in indoor cats: Suburban Chicago, Illinois, EUA. Mycopathologia 163:59–66
107. Gilor C, Graves TK, Barger AM et al (2006) Clinical aspects of natural infection with *Blastomyces dermatitidis* in cats: 8 cases (1991-2005). J Am Vet Med Assoc 229:96–99
108. Kohn C (2007) Miscellaneous Fungal Diseases. Blastomycosis. In: Sellon DB, Long MT (eds) Equine infectious diseases. St. Louis, Saunders, Elsevier, pp 431–434
109. Kwon-Chung KJ (1972) Sexual stage of *Histoplasma capsulatum*. Science 175:326
110. Kasuga T, White TJ, Koenig G et al (2003) Phylogeography of the fungal pathogen *Histoplasma capsulatum*. Molecular Ecology 12:3383–3401
111. Sepúlveda VE, Márquez R, Turissini DA et al (2017) Genome sequences reveal cryptic speciation in the human pathogen *Histoplasma capsulatum*. mBio 8:e01339-17
112. Bonifaz A, Vázquez-González D, Perusquía-Ortiz AM (2011) Endemic systemic mycoses:coccidioidomycosis, histoplasmosis, paracoccidioidomycosis and blastomycosis. J Dtsch Dermatol Ges 9:705–714
113. Al-Ani FK (1999) Epizootic lymphangitis in horses: a review of the literature. Rev Sci Tech Off Int Epiz 18:691–699
114. Murata Y, Sano A, Ueda Y et al (2007) Molecular epidemiology of canine histoplasmosis in Japan. Med Mycol 45:233–247
115. Fantini O, Guillot J, Gier S et al (2014) A case of histoplasmosis in a cat limited to the skin. Proceedings of the annual congress of the European Society of Veterinary Dermatology, Salzburg
116. Guillot J, Pin D, Chabé M et al (2015) Feline histoplasmosis due to *Histoplasma capsulatum* var. *farciminosum* in eastern France. In: Proceedings of the annual congress of the International Society of Human and Animal Mycology, Melbourne
117. Fischer NM, Favrot C, Monod M et al (2013) A case in Europe of feline histoplasmosis apparently limited to the skin. Vet Dermatol 24:635–638
118. Guillot J, Guérin C, Chermette R (2018) Histoplasmosis in animals. In: Guillot J, Verweij PE (eds) Seyedmousavi S, deHoog GS. Emerging and Epizootic Fungal Infections in Animals, Springer Nature, pp 115–128
119. Chermette R, Guillot J (2010) Mycoses due to dimorphic fungi. In: Lefevre PC, Blancou J, Chermette R, Uilenberg G (eds) Infectious and parasitic diseases of livestock. Lavoisier, Paris, pp 1385–1394
120. Emmons CW (1961) Isolation of *Histoplasma capsulatum* from soil in Washington D.C. Public Health Rep 76:591–595
121. Emmons CW, Klite PD, Baer GM et al (1966) Isolation of *Histoplasma capsulatum* from bats in the United States. Am J Epidemiol 84:103–109
122. Klite PD, Diercks FH (1965) *Histoplasma capsulatum* in fecal contents and organs of bats in the Canal Zone. Am J Trop Med Hyg 14:433–439
123. Hoff GL, Bigler WJ (1981) The role of bats in the propagation and spread of histoplasmosis: a review. J Wildl Dis 17:191–196
124. Taylor RL, Shacklette MH, Kelley HB (1962) Isolation of *Histoplasma capsulatum* and *Microsporum gypseum* from soil and bat guano in Panama and the Canal Zone. Am J Trop Med Hyg 11:790–795
125. McMurray DN, Russel LH (1982) Contribution of bats to the maintenance of *Histoplasma capsulatum* in a cave microfocus. Am J Trop Med Hyg 31:527–531

126. Gugnani HC, Muotoe-Okafor FA, Kaufman L et al (1994) A natural focus of *Histoplasma capsulatum* var. *duboisii* is a bat cave. Mycopathologia 127:151–157
127. Norkaew T, Ohno H, Sriburee P et al (2013) Detection of environmental sources of *Histoplasma capsulatum* in Chiang Mai, Thailand, by nested PCR. Mycopathologia 176:395–402
128. Diercks FH, Shacklette MH, Kelley HB Jr et al (1965) Naturally occurring histoplasmosis among 935 bats collected in Panama and the Canal Zone, July 1961-February. Am J Trop Med Hyg 14:1069–1072
129. Ajello L, Hosty TS, Palmer J (1967) Bat histoplasmosis in Alabama. Am J Trop Med Hyg 16:329–331
130. Tesh RB, Arata AA, Schneidau JD Jr (1968) Histoplasmosis in Colombian bats, with a consideration of some of the factors influencing the prevalence of natural infection in chiroptera. Am J Trop Med Hyg 17:102–106
131. Fernández ACM (1988) Isolation of *Histoplasma capsulatum* in bats in Cuba. Rev Cubana Med Trop 40:36–43
132. Canteros CE, Iachini RH, Rivas MC et al (2005) First isolation of *Histoplasma capsulatum* from the urban bat *Eumops bonariensis*. Rev Argent Microbiol 37:46–56
133. Taylor ML, Chávez-Tapia CB, Rojas-Martínez A et al (2005) Geographical distribution of genetic polymorphism of the pathogen *Histoplasma capsulatum* isolated from infected bats, captured in a central zone of Mexico. FEMS Immunol Med Microbiol 45:451–458
134. Dias MA, Oliveira RM, Giudice MC et al (2011) Isolation of *Histoplasma capsulatum* from bats in the urban area of São Paulo State, Brazil. Epidemiol Infect 139:1642–1644
135. González-González AE, Ramírez JA, Aliouat-Denis CM et al (2013) Molecular detection of *Histoplasma capsulatum* in the lung of a free-ranging common noctule (*Nyctalus noctula*) from France using the Hcp100 gene. J Zoo Wildl Med 44:15–20
136. Dos Santos B, Langoni H, da Silva RC et al (2017) Molecular detection of *Histoplasma capsulatum* in insectivorous and frugivorous bats in Southeastern Brazil. Med Mycol 56:937–940
137. Paz GS, Adorno BMV, Richini-Pereira VB et al (2018) Infection by *Histoplasma capsulatum*, *Cryptococcus* spp. and *Paracoccidioides brasiliensis* in bats collected in urban areas. Transbound Emerg Dis 65:1797–1805
138. Negroni R, Duré R, Ortiz Nareto A et al (2010) Histoplasmosis outbreak in Morón, Buenos Aires Province, Argentina. Rev Argent Microbiol 42:254–260
139. Cottle LE, Gkrania-Klotsas E, Williams HJ et al (2013) A multinational outbreak of histoplasmosis following a biology field trip in the Ugandan rainforest. J Travel Med 20:83–87
140. Rocha-Silva F, Figueiredo SM, Silveira TT et al (2013) Histoplasmosis outbreak in Tamboril cave-Minas Gerais state, Brazil. Med Mycol Case Rep 4:1–4
141. Benedict K, Mody RK (2016) Epidemiology of histoplasmosis outbreaks, United States, 1938–2013. Emerg Infect Dis 22:370–378
142. O'Keefe A, Frederick J, Harmon B et al (2012) Histoplasmosis outbreak among day camp attendees-Nebraska, June 2012. MMWR Morb Mortal Wkly Rep 61:747–748
143. Alonso D, Muñoz J, Letang E et al (2007) Imported acute histoplasmosis with rheumatologic manifestations in Spanish travelers. J Travel Med 14:338–342
144. Gundacker ND, Rolfe RJ, Rodriguez JM (2017) Infections associated with adventure travel: a systematic review. Travel Med Infect Dis 16:3–10
145. Salzer HJF, Stoney RJ, Angelo KM et al (2018) Epidemiological aspects of travel-related systemic endemic mycoses: a Geo Sentinel analysis, 1997–2017. J Travel Med 25
146. Jülg B, Elias J, Zahn A et al (2008) Bat-associated histoplasmosis can be transmitted at entrances of bat caves and not only inside the caves. J Travel Med 15:133–136
147. Samayoa B, Roy M, Cleveland AA et al (2017) High mortality and coinfection in a prospective cohort of Human Immunodeficiency Virus/Acquired Immune Deficiency Syndrome patients with histoplasmosis in Guatemala. Am J Trop Med Hyg 97:42–48
148. Woods JP (2003) Knocking on the right door and making a comfortable home: *Histoplasma capsulatum* intracellular pathogenesis. Curr Opin Microbiol 6:327–331
149. Silva AV, Bosco SMG (2016) Histoplasmose. In: Megid J, Ribeiro MG, Paes AC (eds) Doenças infecciosas em animais de produção e companhia. Roca, Rio de Janeiro, pp 929–935

150. Brömel C, Greene CE (2012) Histoplasmosis. In: Greene CE (ed) Infectious diseases of the dog and cat, 4th edn. Elsevier, St. Louis, pp 614–621
151. Schumacher LL, Love BC, Ferrell M et al (2013) Canine intestinal histoplasmosis containing hyphal forms. J Vet Diag Invest 25:304–307
152. Teixeira MM, Theodoro RC, Nino-Vega G et al (2014) *Paracoccidioides* species complex: ecology, phylogeny, sexual reproduction, and virulence. PloS Pathog 10:e1004397
153. Teixeira MM, Theodoro RC, Oliveira FF et al (2014) *Paracoccidioides lutzii* sp. nov.: biological and clinical implications. Med Mycol 52:19–28
154. Bustamante-Simon B, McEwen JG, Tabares AM et al (1985) Characteristics of the conidia produced by the mycelial form of *Paracoccidioides brasiliensis*. Sabouraudia 23:407–414
155. Montenegro MR, Franco MF (1994) Pathology. In: Restrepo-Moreno A, Negro GD (eds) Franco M, Lacaz C da S. Paracoccidioidomycosis. CRC Press, Boca Raton, pp 131–150
156. Matute DR, McEwen JG, Puccia R et al (2006) Cryptic speciation and recombination in the fungus *Paracoccidioides brasiliensis* as revealed by gene genealogies. Mol Biol Evol 23:65–73
157. Theodoro RC, Teixeira MdeM, Felipe MSS et al (2012) Genus *Paracoccidioides*: Species recognition and biogeographic aspects. PloS ONE 7:e37694
158. Turissini DA, Gomez OM, Teixeira MM et al (2017) Species boundaries in the human pathogen *Paracoccidioides*. Fungal Genet Biol 106:9–25
159. Martinez R (2015) Epidemiology of paracoccidioidomycosis. Rev Inst Med Trop São Paulo 57:11–20
160. Franco M, Bagagli E, Scapolio S et al (2000) A critical analysis of isolation of *Paracoccidioides brasiliensis* from soil. Med Mycol 38:185–191
161. Theodoro RC, Candeias JMG, Araújo JP et al (2005) Molecular detection of *Paracoccidioides brasiliensis* in soil. Med Mycol 43:725–729
162. Arantes TD, Theodoro RC, Macoris SADG et al (2013) Detection of *Paracoccidioides* spp. in environmental aerosol samples. Med Mycol 51:83–92
163. Arantes TD, Theodoro RC, Teixeira MM et al (2016) Environmental mapping of *Paracoccidioides* spp. in Brazil reveals new clues into genetic diversity, biogeography and wild host association. PLoS Negl Trop Dis 10:e0004606
164. Hrycyk MF, Garces HG, Bosco SMG et al (2018) Ecology of *Paracoccidioides brasiliensis*, *P. lutzii* and related species: infection in armadillos, soil occurrence and mycological aspects. Med Mycol 56:950–962
165. Restrepo A, McEwen JG, Castañeda E (2001) The habitat of *Paracoccidioides brasiliensis*: how far from solving the riddle? Med Mycol 39:233–241
166. de Valle ACF, Macedo PM, Almeida-Paes R et al (2017) Paracoccidioidomycosis after highway construction, Rio de Janeiro, Brazil. Emerg Infect Dis 23:1917–1919
167. Shikanai-Yasuda MA, Telles Filho Fde Q, Mendes RP et al (2006) Guidelines in paracoccidioidomycosis. Rev Soc Bras Med Trop 39:297–310
168. Costa EO, Diniz LS, Netto CF (1995) The prevalence of positive intradermal reactions to paracoccidioidin in domestic and wild animals in São Paulo, Brazil. Vet Res Commun 19:127–130
169. Costa EO, Diniz LS, Netto CF et al (1995) Delayed hypersensitivity test with paracoccidioidin in captive Latin American wild mammals. J Med Vet Mycol 33:39–42
170. Corte AC, Svoboda WK, Navarro IT et al (2007) Paracoccidioidomycosis in wild monkeys from Paraná State, Brazil. Mycopathologia 164:225–228
171. Silveira LH, Paes RCS, Medeiros EV et al (2008) Occurrence of antibodies to *Paracoccidioides brasiliensis* in dairy cattle from Mato Grosso do Sul, Brazil. Mycopathologia 165:367–371
172. Oliveira GG, Silveira LH, Itano EN et al (2011) Serological evidence of *Paracoccidioides brasiliensis* infection in chickens from Paraná and Mato Grosso do Sul States, Brazil. Mycopathologia 171:197–202
173. Oliveira GG, Navarro IT, Freire RL et al (2012) Serological survey of paracoccidioidomycosis in sheep. Mycopathologia 173:63–68

174. Oliveira GG, Belitardo DR, Balarin MR et al (2013) Serological survey of paracoccidioidomycosis in cats. Mycopathologia 176:299–302
175. Ferreira JB, Navarro IT, Freire RL et al (2013) Evaluation of *Paracoccidioides brasiliensis* Infection in dairy goats. Mycopathologia 176:95–99
176. Belitardo DR, Calefi AS, Borges IK et al (2014) Detection of antibodies against *Paracoccidioides brasiliensis* in free-range domestic pigs (*Sus scrofa*). Mycopathologia 177:91–95
177. Belitardo DR, Calefi AS, Sbeghen MR et al (2014) *Paracoccidioides brasiliensis* infection in domestic rabbits (*Oryctolagus cuniculus*). Mycoses 57:222–227
178. Ono MA, Bracarense AP, Morais HS et al (2001) Canine paracoccidioidomycosis: a sero epidemiologic study. Med Mycol 39:277–282
179. Fontana FF, dos Santos CT, Esteves FM et al (2010) Seroepidemiological survey of paracoccidioidomycosis infection among urban and rural dogs from Uberaba, Minas Gerais, Brazil. Mycopathologia 169:159–165
180. Teles AJ, Klafke GB, Cabana ÂL et al (2016) Serological investigation into *Paracoccidioides brasiliensis* infection in dogs from Southern Rio Grande do Sul, Brazil. Mycopathologia 181(3–4):323–328
181. Ono MA, Kishima MO, Itano EN et al (2003) Experimental paracoccidioidomycosis in dogs. Med Mycol 41:265–268
182. Richini-Pereira VB, Bosco SdeM, Griese J et al (2008) Molecular detection of *Paracoccidioides brasiliensis* in road-killed wild animals. Med Mycol 46:35–40
183. Naiff RD, Ferreira LC, Barrett TV et al (1986) Enzootic paracoccidioidomycosis in armadillos (*Dasypus novemcinctus*) in the State of Pará. Rev Inst Med Trop São Paulo 28:19–27
184. Bagagli E, Sano A, Coelho KI et al (1998) Isolation of *Paracoccidioides brasiliensis* from armadillos (*Dasypus noveminctus*) captured in an endemic area of paracoccidioidomycosis. Am J Trop Med Hyg 58:505–512
185. Bagagli E, Franco M, Bosco Sde M et al (2003) High frequency of *Paracoccidioides brasiliensis* infection in armadillos (*Dasypus novemcinctus*): an ecological study. Med Mycol 41:217–223
186. Bagagli E, Bosco SMG (2008) Armadillos and dimorphic pathogenic fungi: ecological and evolutionary aspects. In: Lowghry WJ, Viscaino SF (eds) The biology of the Xenarthra. University Press of Florida, Gainesville, pp 103–110
187. Bosco SMG, Bagagli E (2018) Paracoccidioidomycosis in Animals and Humans. In: Guillot J, Verweij PE (eds) Seyedmousavi S, deHoog GS. Emerging and Epizootic Fungal Infections in Animals, Springer Nature, pp 129–145
188. Bagagli E, Bosco SMG, Theodoro RC et al (2006) Phylogenetic and evolutionary aspects of *Paracoccidioides brasiliensis* reveal a long coexistence with animal hosts that explain several biological features of the pathogen. Infect Genet Evol 6:344–351
189. Richini-Pereira VB, Bosco SMG, Theodoro RC et al (2009) Importance of xenarthrans in the eco-epidemiology of *Paracoccidioides brasiliensis*. BMC Res Notes 2:228
190. Trejo-Chávez A, Ramírez-Romero R, Ancer-Rodríguez J et al (2011) Disseminated paracoccidioidomycosis in a Southern two-toed sloth (*Choloepus didactylus*). J Comp Pathol 144:231–234
191. Mendes JF, Klafke GB, Albano APN et al (2017) Paracoccidioidomycosis infection in domestic and wild mammals by *Paracoccidioides lutzii*. Mycoses 60:402–406
192. Salazar ME, Restrepo A, Stevens DA (1988) Inhibition by estrogens of conidium-to-yeast conversion in the fungus *Paracoccidioides brasiliensis*. Infect Immun 56:711–713
193. Franco M, Montenegro MR, Mendes RP et al (1987) Paracoccidioidomycosis: a recently proposed classification of its clinical forms. Rev Soc Bras Med Trop 20:129–132
194. Ricci G, Mota FT, Wakamatsu A et al (2004) Canine paracoccidioidomycosis. Med Mycol 42:379–383
195. Farias MR, Condas LA, Ribeiro MG et al (2011) Paracoccidioidomycosis in a dog: case report of generalized lymphadenomegaly. Mycopathologia 172:147–152

196. Headley SA, Pretto-Giordano LG, Di Santis GW et al (2017) *Paracoccidioides brasiliensis*-associated dermatitis and lymphadenitis in a dog. Mycopathologia 182:425–434
197. Silva-Vergara ML, Martinez R (1999) Role of the armadillo *Dasypus novemcinctus* in the epidemiology of paracoccidioidomycosis. Mycopathologia 144:131–133
198. Youssef D, Shams W, Ganote CE et al (2011) Negative Image of *Blastomyces* on Diff-Quik Stain. Acta Cytologica 55:377–381
199. Shubitz LF, Dial SM (2005) Coccidioidomycosis: a diagnostic challenge. Clin Tech Small Anim Pract 20:220–226
200. Chow NA, Lindsley MD, McCotter OZ et al (2017) Development of an enzyme immunoassay for detection of antibodies against *Coccidioides* in dogs and other mammalian species. PLoS ONE 12:e0175081
201. Batista J, de Camargo ZP, Fernandes GF et al (2010) Is the geographical origin of a *Paracoccidioides brasiliensis* isolate important for antigen production for regional diagnosis of paracoccidioidomycosis? Mycoses 53:176–180
202. Gegembauer G, Araujo LM, Pereira EF et al (2014) Serology of paracoccidioidomycosis due to *Paracoccidioides lutzii*. PLoS Negl Trop Dis 8:e2986
203. Klotz SA, Penn RL, George RB (1986) Antigen detection in the diagnosis of fungal respiratory infections. Semin Respir Infect 1:16–21
204. Cunningham L, Cook A, Hanzlicek A et al (2015) Sensitivity and specificity of *Histoplasma* antigen detection by Enzyme Immunoassay. J Am Anim Hosp Assoc 51:306–310
205. Hage CA, Davis TE, Fuller D et al (2010) Diagnosis of histoplasmosis by antigen detection in BAL fluid. Chest 137:623–628
206. Hanzlicek AS, Meinkoth JH, Renschler JS et al (2016) Antigen concentrations as an indicator of clinical remission and disease relapse in cats with histoplasmosis. J Vet Intern Med 30:1065–1073
207. White TJ, Burns T, Lee S et al (1990) Amplification and direct sequencing of fungal ribosomal RNA genes for phylogenetics. In: Innis MA, Gelfand DH, Sninsky JJ, White TJ (eds) PCR protocols: a guide to methods and applications. Academic Press, New York, pp 315–322
208. Iwen PC, Hinrichs SH, Rupp ME (2002) Utilization of the internal transcribed spacer regions as molecular targets to detect and identify human fungal pathogens. Med Mycol 40:87–109
209. Gago S, Buitrago MJ, Clemons KV et al (2014) Development and validation of a quantitative real-time PCR assay for the early diagnosis of coccidioidomycosis. Diagn Microbiol Infect Dis 79:214–221
210. Mitchell M, Dizon D, Libke R et al (2015) Development of a real-time PCR Assay for identification of *Coccidioides immitis* by use of the BD Max system. J Clin Microbiol 53:926–929
211. Morjaria S, Otto C, Moreira A et al (2015) Ribosomal RNA gene sequencing for early diagnosis of *Blastomyces dermatitidis* infection. Int J Infect Dis 37:122–124
212. Dalcin D, Rothstein A, Spinato J et al (2016) *Blastomyces gilchristii* as cause of fatal acute respiratory distress syndrome. Emerg Infect Dis 22:306–308
213. Bialek R, Feucht A, Aepinus C et al (2002) Evaluation of two nested PCR assays for detection of *Histoplasma capsulatum* DNA in human tissue. J Clin Microbiol 40:1644–1647
214. Koepsell SA, Hinrichs SH, Iwen PC (2012) Applying a real-time PCR assay for *Histoplasma capsulatum* to clinically relevant formalin-fixed paraffin-embedded human tissue. J Clin Microbiol 50:3395–3397
215. Gomes GM, Cisalpino PS, Taborda CP et al (2000) PCR for diagnosis of paracoccidioidomycosis. J Clin Microbiol 38:3478–3480
216. Buitrago MJ, Merino P, Puente S et al (2009) Utility of real-time PCR for the detection of *Paracoccidioides brasiliensis* DNA in the diagnosis of imported paracoccidioidomycosis. Med Mycol 47:879–882
217. Girois SB, Chapuis F, Decullier E et al (2006) Adverse effects of antifungal therapies in invasive fungal infections: review and meta-analysis. Eur J Clin Microbiol Infect Dis 25:138–149
218. Denning WD (2002) Echinocandins: a new class of antifungal. J Antimicrob Chemoth 49:889–891

219. Bellmann R, Smuszkiewicz P (2017) Pharmacokinetics of antifungal drugs: practical implications for optimized treatment of patients. Infection 45:737–779
220. Herrin KV, Miranda A, Loebenberg D (2005) Posaconazole therapy for systemic coccidioidomycosis in a chimpanzee (*Pan troglodytes*): a case report. Mycoses 48:447–452
221. Ramani R, Chaturvedi V (2007) Antifungal susceptibility profiles of *Coccidioides immitis* and *Coccidioides posadasii* from endemic and non-endemic areas. Mycopathologia 163:315–319
222. Arbona N, Butkiewicz CD, Keyes M et al (2019) Clinical features of cats diagnosed with coccidioidomycosis in Arizona, 2004–2018. J Fel Med Surg:1–9
223. Davidson AP, Shubitz LF, Alcott CJ et al (2019) Selected clinical features of coccidioidomycosis in dogs. Med Mycol 57:S67–S75
224. Foley JP, Legendre AM (1992) Treatment of coccidioidomycosis osteomyelitis with itraconazole in a horse: A brief report. J Vet Intern Med 6:333–334
225. Hector RF, Zimmer BL, Pappagianis D (1990) Evaluation of nikkomycins X and Z in murine models of coccidioidomycosis, histoplasmosis, and blastomycosis. Antimicrob Agents Chemother 34:587–593
226. Bartsch R, Greene R (1997) New treatment of coccidioidomycosis. Vet Forum 4:50–52
227. Johnson SM, Zimmermann CR, Kerekes KM et al (1999) Evaluation of the susceptibility of *Coccidioides immitis* to lufenuron, a chitin synthase inhibitor. Med Mycol 37:441–444
228. Mazepa ASW, Trepanier LA, Foy DS (2011) Retrospective comparison of the efficacy of fluconazole or itraconazole for the treatment of systemic blastomycosis in dogs. J Vet Intern Med 25:440–445
229. Gabal MA (1984) The Effect of Amphotericin-B, 5-fluorocytosine and nystatin on *Histoplasma farciminosum in vitro*. Zbl Vet Med B 31:46–50
230. Travassos LR, Taborda CP (2017) Linear epitopes of *Paracoccidioides brasiliensis* and other fungal agents of human systemic mycoses as vaccine candidates. Front Immunol 8:224
231. Barros MB, de Almeida PR, Schubach AO (2011) *Sporothrix schenckii* and sporotrichosis. Clin Microbiol Rev 24:633–654
232. Berbee ML, Taylor JW (1992) Convergence in ascospore discharge mechanism among pyrenomycete fungi based on 18S ribosomal RNA gene sequence. Mol Phylogenet Evol 1:59–71
233. Guarro J, Gene J, Stchigel AM (1999) Developments in fungal taxonomy. Clin Microbiol Rev 12:454–500
234. Kwon-Chung KJ (1979) Comparison of isolates of *Sporothrix schenckii* obtained from fixed cutaneous lesions with isolates from other types of lesions. J Infect Dis 139:424–431
235. Marimon R, Gene J, Cano J et al (2006) Molecular phylogeny of *Sporothrix schenckii*. J Clin Microbiol 44:3251–3256
236. Marimon R, Gene J, Cano J et al (2008) *Sporothrix luriei*: a rare fungus from clinical origin. Med Mycol 46:621–625
237. Chakrabarti A, Bonifaz A, Gutierrez-Galhardo MC et al (2015) Global epidemiology of sporotrichosis. Med Mycol 53:3–14
238. Rodrigues AM, Cruz Choappa R, Fernandes GF et al (2016) *Sporothrix chilensis* sp. nov. (Ascomycota: Ophiostomatales), a soil-borne agent of human sporotrichosis with mild-pathogenic potential to mammals. Fungal Biol 120:246–264
239. Fukushiro R (1984) Epidemiology and ecology of sporotrichosis in Japan. Zentralbl Bakteriol Mikrobiol Hyg A 257:228–233
240. Bhutia PY, Gurung S, Yegneswaran PP et al (2011) A case series and review of sporotrichosis in Sikkim. J Infect Dev Ctries 5:603–608
241. Vismer HF, Hull PR (1997) Prevalence, epidemiology and geographical distribution of *Sporothrix schenckii* infections in Gauteng, South Africa. Mycopathologia 137:137–143
242. Conti-Díaz IA (1989) Epidemiology of sporotrichosis in Latin America. Mycopathologia 108:113–116
243. Helm MAF, Berman C (1947) The clinical, therapeutic and epidemiological features of the sporotrichosis infection on the mines. In: Sporotrichosis infection on mines of the Witwatersrand. Proceedings of the Transvaal Mine Medical Officers' Association (Dec 1944) Johannesburg. South Africa, The Transvaal Chamber of Mines, pp 59–67

244. Hajjeh R, McDonnell S, Reef S et al (1997) Outbreak of sporotrichosis among tree nursery workers. J Infect Dis 176:499–504
245. Grotte M, Younger B (1981) Sporotrichosis associated with sphagnum moss exposure. Arch Pathol Lab Med 105:50–51
246. Dooley DP, Bostic PS, Beckius ML (1997) Spook house sporotrichosis. A point-source outbreak of sporotrichosis associated with hay bale props in a Halloween haunted-house. Arch Intern Med 157:1885–1887
247. Feeney KT, Arthur IH, Whittle AJ et al (2007) Outbreak of Sporotrichosis, Western Australia. Emerg Infect Dis 13:1228–1231
248. Schubach A, Schubach TM, Barros MB et al (2005) Cat transmitted sporotrichosis, Rio de Janeiro, Brazil. Emerg Infect Dis 11:1952–1954
249. Gremião IDF, Miranda LHM, Reis EG et al (2017) Zoonotic Epidemic of Sporotrichosis: Cat to Human Transmission. PLoS Pathog 13:e1006077
250. Dunstan RW, Reimann KA, Langham RF (1986) Feline sporotrichosis. J Am Vet Med Assoc 189:880–883
251. Rodrigues AM, de Melo TM, de Hoog GS et al (2013) Phylogenetic analysis reveals a high prevalence of *Sporothrix brasiliensis* in feline sporotrichosis outbreaks. PLoS Negl Trop Dis 7:e2281
252. Singer JI, Muncie JE (1952) Sporotrichosis; etiologic considerations and report of additional cases from NewYork. J Med 52:2147–2153
253. Rees RK, Swartzberg JE (2011) Feline-transmitted sporotrichosis: A case study from California. Dermatol Online J 17:2
254. Bove-Sevilla PM, Mayorga-Rodríguez J, Hernández-Hernández O (2008) Sporotrichosis transmitted by a domestic cat. Case report. Med Cutan Iber Lat Am 36:33–35
255. Zamri-Saad M, Salmiyah TS, Jasni S et al (1990) Feline sporotrichosis: an increasingly important zoonotic disease in Malaysia. Vet Rec 127:480
256. Tang MM, Tang JJ, Gill P et al (2012) Cutaneous sporotrichosis: a six-year review of 19 cases in a tertiary referral center in Malaysia. Int J Dermatol 51:702–708
257. Yegneswaran PP, Sripathi H, Bairy I et al (2009) Zoonotic sporotrichosis of lymphocutaneous type in a man acquired from a domesticated feline source: report of a first case in southern Karnataka, India. Int J Dermatol 48:1198–1200
258. Fernández N, Iachini R, Farias L et al (2015) Esporotricosis, una zoonosis en alerta. In: Proceeding soft heInfocus; Nov 5–7; Córdoba, Argentina, Círculo médico de Córdoba, 2015, p 11
259. Córdoba S, Isla G, Szusz W et al (2018) Molecular identification and susceptibility profile of *Sporothrix schenckii sensu lato* isolated in Argentina. Mycoses 61:441–448
260. Mayorga R, Caceres A, Toriello C et al (1978) An endemic area of sporotrichosis in Guatemala. Sabouraudia 16:185–198
261. Haddad VJ, Miot HA, Bartoli LD et al (2002) Localized lymphatic sporotrichosis after fish-induced injury (*Tilapia* sp.). Med Mycol 40:425–427
262. Schubach TM, Schubach AO, Reis RS et al (2002) *Sporothrix schenckii* isolated from domestic cats with and without sporotrichosis in Rio de Janeiro, Brazil. Mycopathologia 153:83–86
263. Duangkaew L, Yurayart C, Limsivilai O et al (2019) Cutaneous sporotrichosis in a stray cat from Thailand. Med Mycol Case Rep 23:46–49
264. Schubach TM, Schubach A, Okamoto T et al (2004) Evaluation of an epidemic of sporotrichosis in cats: 347 cases (1998-2001). J Am Vet Med Assoc 224:1623–1629
265. Pereira SA, Gremião ID, Kitada AA et al (2014) The epidemiological scenario of feline sporotrichosis in Rio de Janeiro, State of Rio de Janeiro, Brazil. Rev Soc Bras Med Trop 47:392–393
266. Rodrigues AM, de Hoog GS, Camargo ZP (2018) Feline Sporotrichosis. In: Seyedmousavi S, de Hoog GS, Guillot J, Verweij PE (eds) Emerging and epizootic fungal infections in animals. Springer Nature, pp 199–232
267. Pereira SA, Passos SR, Silva JN et al (2010) Response to azolic antifungal agents for treating feline sporotrichosis. Vet Rec 166:290–294

268. Civila ES, Bonasse J, Conti-Diaz IA et al (2004) Importance of the direct fresh examination in the diagnosis of cutaneous sporotrichosis. Int J Dermatol 43:808–810
269. Kwon-Chung KJ, Bennet JE (1992) Medical Mycology. Lea & Febiger, Philadelphia, PA
270. Morris-Jones R (2002) Sporotrichosis. Clin Exp Dermatol 27:427–431
271. Mendoza M, Díaz AM, Hung MB et al (2002) Production of culture filtrates of *Sporothrix schenckii* in diverse culture media. Med Mycol 40:447–454
272. Fernandes GF, Lopes-Bezerra LM, Bernardes-Engemann AR et al (2011) Serodiagnosis of sporotrichosis infection in cats by enzyme-linked immunosorbent assay using a specific antigen, SsCBF, and crude exoantigens. Vet Microbiol 147:445–449
273. Morris-Jones R (2002) Sporotrichosis. Clin Exp Dermatol 27:427–431
274. de Miranda LH, Quintella LP, dos Santos IB et al (2009) Histopathology of canine sporotrichosis: a morphological study of 86 cases from Rio de Janeiro (2001-2007). Mycopathologia 168:79–87
275. Zhou X, Rodrigues AM, Feng P et al (2014) Global ITS diversity in the *Sporothrix schenckii* complex. Fungal Divers 66:153–165
276. Rodrigues AM, de Hoog GS, de Camargo ZP (2015) Molecular diagnosis of pathogenic *Sporothrix* species. PloS Negl Trop Dis 9:e0004190
277. Rippon J (1988) Sporotrichosis. In: Rippon J (ed) Medical Mycology- the pathogenic fungi and the pathogenic actinomycetes, 3rd edn. W. B. Saunders Company, Philadelphia, PA, pp 325–352
278. Kauffman CA, Bustamante B, Chapman SW et al (2007) Clinical practice guidelines for the management of sporotrichosis: 2007 update by the Infectious Diseases Society of America. Clin Infect Dis 45:1255–1265
279. Rodrigues AM, de Hoog GS, de Cassia Pires D et al (2014) Genetic diversity and antifungal susceptibility profiles in causative agents of sporotrichosis. BMC Infect Dis 14:219
280. Reis ÉG, Schubach TM, Pereira SA et al (2016) Association of itraconazole and potassium iodide in the treatment of feline sporotrichosis: a prospective study. Med Mycol 54:684–690
281. Viana PG, Figueiredo ABF, Gremião IDF et al (2018) Successful treatment of canine sporotrichosis with terbinafine: case reports and literature review. Mycopathologia 183:471–478
282. Seyedmousavi S, Bosco SMG, de Hoog S et al (2018) Fungal infections in animals: a patchwork of different situations. Med Mycol 56:S165–S187
283. Nascimento RC, Espíndola NM, Castro RA et al (2008) Passive immunization with monoclonal antibody against a 70-kDa putative adhesin of *Sporothrix schenckii* induces protection in murine sporotrichosis. Eur J Immunol 38:3080–3089
284. de Almeida JR, Kaihami GH, Jannuzzi GP et al (2015) Therapeutic vaccine using a monoclonal antibody against a 70-kDa glycoprotein in mice infected with highly virulent *Sporothrix schenckii* and *Sporothrix brasiliensis*. Med Mycol 53:42–50
285. Hartmann K, Day MJ, Thiry E et al (2015) Feline injection-site sarcoma: ABCD guidelines on prevention and management. J Feline Med Surg 17:606–613

Part II

Mycotoxins in Relation to Human and Animal Health

Mycotoxins and Their Inhalatory Intake Risk

7

Elena Piecková

Abstract

Mycotoxins are defined as the secondary metabolites of certain moulds and are toxic to vertebrates (warm-blooded). Their chemical nature is rather variable, non-proteinal and nonvolatile. Primarily, they contaminate plants, and through them, they reach animals and their products. The mycotoxins represent serious health risks for consumers, including genotoxic effect leading to mycotoxicosis, an ill health status. The hygienic limits for mycotoxins in foods and feeds are set legislatively all over the world. The adverse biological effects can be caused by the inhalation of a mycotoxin dose at a minimum level that is one tenth of the elementary one. Mostly, aflatoxins, ochratoxins, *Fusarium* mycotoxins and stachybotryotoxins have been identified in the air of occupational and dwelling environments so far. The precise measurements remain limited due to lack of sensitive and accurate methods of detection. The immunosuppressive, haematotoxic, cytotoxic and inflammatory effects of the complex mixtures of toxicants produced by *Aspergillus versicolor* and *Stachybotrys chartarum* of indoor origin have been discussed in this chapter. Complex toxic fungal metabolites break down the self-cleaning mechanism of the airways, induce systemic damages and are enhanced by simultaneous action of other indoor contaminants (cigarette smoke). They can finally result in the ill health of occupants of damp mouldy dwellings, starting with respiratory disorders and probably culminating in general intoxication especially in children with burning metabolism.

Keywords

Food/feed hygiene · Occupational/indoor exposure · Carcinogenicity · Inflammation · Sick building syndrome

---

E. Piecková (✉)
Faculty of Medicine, Slovak Medical University, Bratislava, Slovakia
e-mail: elena.pieckova@szu.sk

K. Singh, N. Srivastava (eds.), *Recent Trends in Human and Animal Mycology*,
https://doi.org/10.1007/978-981-13-9435-5_7

Fungi (moulds) produce a wide range of toxic chemicals. Most related research has dealt with exposure to such toxicants in animal feeds and human foods. However, animals and humans can also be exposed to high levels of fungal toxins through exposure to air and dust from mouldy environments. Mycotoxins are defined as the chemicals of fungal origin that are toxic to warm-blooded vertebrates. They belong to secondary metabolites of moulds, along with alkaloids and antibiotics. Fungi synthesize these compounds to gain an ecological advantage in their natural habitats. These metabolites (usually) do not serve as a source of energy for their producers, excluding bordering conditions such as starving. Microorganisms employ mostly the polyketide cycle for the mycotoxin synthesis. From the food/feed hygienic point of view, the mycotoxins are counted as endogenous contaminants, which are formed directly in the matrix by toxic mycobiota. Chemically, they represent various structures (lactones, coumarins, etc.) that do not have a proteinaceous nature [1]. This is also the reason of their high thermostability. They are not destructed by culinary processes, including bread baking and coffee roasting; instead, more toxic compounds might be generated from them during these procedures [2]. Despite the low molecular weight of mycotoxins (200–300 kDa), they do not possess volatility, meaning they will not sublimate out of the contaminated material. This is the reason why gas chromatography is not the primarily chosen method for their analysis [2]. The mycotoxins are not commonly involved in pathological process of mycoses [3].

Distribution of mycotoxins through the contaminated substrate is uneven as compared to other chemical contaminants present occasionally (pesticides, heavy metals, etc.). Applying a certain hygienic *positivum* in batches differently contaminated with mycotoxins will help in reaching the final mycotoxin concentration fitting the safety limits. Other mycotoxin distribution specialities are:

- One fungus may synthesize more than one toxin (e.g. *Aspergillus flavus* produces aflatoxins and cyclopiazonic acid as well).
- The same toxin might be produced by different fungi (aflatoxin B1 as the product of *A. flavus*, *A. parasiticus*, *A. nomius*, etc.).

Toxic fungi predominantly colonize plenty of plant materials, including food and feedstuffs, the ones that grow in warm and wet climatic (subtropical) zones. Sometimes, the visible mouldy matrix may not necessarily secrete any mycotoxin; at the same time, every strain of mycotoxigenic fungi cannot produce mycotoxins (e.g. *A. flavus* isolates coming from tropics are toxic, but the strains from moderate climate are usually not) [4]. Such an ability is genetically dependent, and only the quantity of potentially formed mycotoxin can be affected by environmental factors like cultivation temperature, composition of the substrate and so on. In contrast, the visibly non-mouldy substrate may contain high concentrations of mycotoxin(s) like mouldy dried fruits or seeds and can actually account for more severe public health problems. The mycotoxin contamination of substrate on its own does not alter food/feed sensoric quality. Common mycotoxins and their producers are listed in Table 7.1.

Table 7.1 Common mycotoxins and their producers

Mycotoxin	Producers (moulds)
Aflatoxins	*Aspergillus flavus*, *A. parasiticus*, *A. nomius*, *A. argenticus*, etc.
Ochratoxin A	*Penicillium verrucosum*, *P. nordicum*, *A. ochraceus*, *A. carbonarius*, *A. niger*, *A. sclerotioniger*, etc.
Deoxynivalenol	*Fusarium graminearum*, *F. culmorum*, *F. sporotrichioides*, *F. poae*, *F. tricinctum*, etc.
T-2 toxin	*F. sporotrichioides*, *F. poae*
Diacetoxyscirpenol	*F. graminearum*, *F. semitectum*, *F. tricinctum*, *F. oxysporum*, etc.
Nivalenol	*Fusarium nivale*, *F. poae*
Zearalenone	*Fusarium graminearum*, *F. culmorum*
Fumonisin B1	*Fusarium proliferatum*, *F. verticillioides* (syn. *F. moniliforme*), *A. niger*, *A. carbonarius*, etc.

Table 7.2 General toxic effects of common mycotoxins

Toxic effects	Mycotoxin
Dermatotoxic	Trichothecenes, verrucarins, sporidesmins
Estrogenic	Zearalenone
Genotoxic	Aflatoxins, sterigmatocystin, ochratoxin A, zearalenone, patulin, trichothecenes
Haematotoxic	Aflatoxins, ochratoxin A, zearalenone, trichothecenes
Hepatotoxic	Aflatoxins, ochratoxins, rubratoxins, sterigmatocystin, etc.
Immunotoxic	Aflatoxins, ochratoxin A, trichothecenes, patulin
Nephrotoxic	Ochratoxin A
Neurotoxic	Fumonisins, penitrem A, fumitremorgens
Gastrotoxic	Trichothecenes

Mycotoxins cause intoxications in (sensitive) animals and humans, resulting in severe mycotoxicoses. General toxic effects of common mycotoxins are summarized in Table 7.2. Diagnosis of possible mycotoxicosis in veterinary or human cases is quite challenging due to most likely multi-systemic damage of the organism exposed. Its susceptibility to the toxin(s) depends strongly on the individual's genetic, physiological and environmental factors. The most frequent type of mycotoxicoses, chronic primary intoxication, occurs after the long-lasting exposition of an animal/human to low dosages of the toxin(s). Despite unclear macroscopic changes in the organism, mycotoxicosis might be diagnosed by reduced animal productivity especially in the absence of any other disease and/or toxin in the feed. Thus, it is crucially connected to feed hygiene. Delayed chronic toxicity of mycotoxins can itself demonstrate as the secondary mycotoxicosis. They still remain the least defined and are characterized by immunosuppressive effects of mycotoxins with their mutagenic, carcinogenic, estrogenic, teratogenic and embryotoxic potentials. So, in a strict sense, can we speak about the safe levels of mycotoxin in feeds/food? Based on the above statements, the levels are more or less undefinable. However, a hygienically limited mycotoxin intake is given as the total daily intake (TDI) per kg of body weight per day [μg/kg of bw/d]. The TDIs have been set for individual mycotoxins only, while the scenario in real life is much more complicated since people consume mycotoxin mixtures, together with other abiotic and biotic contaminants. Moreover, the individual susceptibility of the organism exposed also matters. However, there are no sufficient estimations on whether the toxicant mixtures act synergistically, additionally or antagonistically.

Currently, a general approach to the study of the mechanism of fungal effects on human beings is becoming a matter of concern. Such an approach includes the immunosuppressive influence of beta-glucans from fungal cell wall as well as toxic and irritative effects of their toxic exo- and endometabolites (mycotoxins and/or volatile organic compounds) [5]. Regarding mycotoxins with rather well-characterized toxicity (carcinogenicity, mutagenicity, teratogenicity, cytotoxicity and immunosuppression), it has been found that an adverse biological effect can be caused by the inhalation of a dose at the level of minimum one tenth of the elementary one [6]. Presence of the fungus does not imply the presence of mycotoxins just as the presence of mycotoxins does not imply the presence of the fungus. Environmental conditions that allow fungal growth are not always the same as those allowing mycotoxin production [7]. However, mycotoxins occur in occupational environments whenever fungi are present [8]. Moreover, it is important to consider that mycotoxins can be present in the environment long after elimination of fungi [9].

Low molecular weight organic compounds (volatile organic compounds, VOCs), namely, alcohols, aldehydes, ketones, aromatic compounds, amines, terpenes, chlorinated hydrocarbons and sulphuric compounds, cause typical mouldy odour and inflammation of airways of sensitive people. Such effects are associated with invisible moulds usually growing under wall papers, carpets or mattresses. There is a positive correlation between the microbial volatile organic compound (mVOC) production and the ability of the fungi to synthesize mycotoxins [10, 11]. *Aspergillus* spp., *Cladosporium* spp. and *Penicillium* spp. belong to the group of strongest producers of such compounds.

Although many reports regarding occupational pulmonary mycotoxicosis resulting from inhaled organic dust contaminated by microbial toxins have been published [12], till date there have been no objective proofs of real clinical diseases borne from mycotoxins, especially produced by *Aspergillus* spp., *Penicillium* spp., *Fusarium* spp., *Trichoderma* spp. and *Stachybotrys chartarum* in the indoor environment. Such epidemiological studies have not been conducted very seriously yet [13].

Aflatoxins are often present in significant quantities in agricultural areas such as damp storage sites for grains and peanuts, in animal feed and in confined houses. They are commonly produced by several *Aspergillus* species in poorly stored grains and peanuts. Aflatoxins are the most carcinogenic among the known chemicals of biological origin. Foodborne aflatoxin exposure has been estimated to cause 25,200 to 155,000 annual human cases of liver cancer worldwide. A Portuguese study measured fungal levels in 7 swine barns and aflatoxin levels in the blood of 28 swine house workers. Twelve species of *Aspergillus* were predominantly detected in the seven swine barns. Detectable blood levels (1–8.9 ng/ml) of aflatoxins were found in 21 out of 28 (75%) swine workers, while all 30 unexposed control workers were found positive for aflatoxin (below detection limit 1 ng/ml) [14].

Aspergillus fumigatus is one of the most ubiquitous saprophytic fungi and is considered to be the species with higher clinical relevance [15]. This species is the most common cause of invasive aspergillosis and a major source of infection-related mortality in immunocompromised patients [16]. One of the most abundantly produced metabolites of *A. fumigatus* is the epipolythiodioxopiperazine (ETP) metabolite gliotoxin, which exhibits a diverse array of biological effects on the immune system [17]. *In vitro*, gliotoxin mediates a broad spectrum of suppressive effects such as inhibition of cytokine production, antigen presentation, production of reactive oxygen species by macrophages and reduction of the activity of cytotoxic T-cells on different cell types of the immune system [18]. It has also been shown that a concentration of 0.2 μg of gliotoxin reduces ciliary movements and damages the respiratory epithelium of the mucous membrane, which may explain why this fungus is considered to injure the respiratory tract when produced *in vivo* [19]. Presence of *A. fumigatus* is well established as a higher risk among others that are also associated with exposure to bioaerosols [20]. This fungus also produces several secondary metabolites during invasive hyphal growth, gliotoxin being one of the most abundantly produced metabolites with a production frequency in environmental strains that ranged from 0% to 33% [17, 21, 22].

In case of fatal infant idiopathic pulmonary haemorrhagic outbreaks in the USA, particular fungal isolates were toxic *in vitro*. It was particularly in this case that epidemiologists stressed tobacco smoke as a factor for increasing health risks of fungal toxins [13]. *Stachybotrys chartarum* produced cytotoxic and immunosuppressive macrocyclic trichothecenes (stachybotryotoxins) and spirocyclic drimanes that caused inflammation and haemorrhages in the respiratory tract and intestines of laboratory animals [23]. The *S. chartarum* isolates from damp schools and dwellings in Denmark produced trichodermal trichothecenes when cultivated onto cardboard and vinyl ceiling [24]. Acute intratracheal exposure of rats to exo- and endometabolites of *S. chartarum*, an atranone chemotype of the fungus that originated from mouldy walls in a Bratislava house, was reported to induce haematological changes, along with cytotoxic and inflammatory pulmonary injury in the animals [25]. Mutagenic and foetotoxic mycotoxins, alternariol and its monomethylether (adversely affected mice) were detected in cellulose tiles overgrown with *Alternaria alternata*. This isolate was also able to grow on a cardboard [26]. Jesenska and Bernat (1994) noticed that *Penicillium expansum* growing on a wallpaper glue produced nephrotoxic citrinin and patulin inhibiting phagocytosis [27]. The metabolite synthesis of fungi depends on the quality of constructing materials [28]. In our study on the trachea of 1-day-old chicks, variation in ciliostatic activity was observed when exposed to chloroform extracts of biomass of building materials (mineral wool, plasterboard and cardboard) inoculated with pure isolates of some moulds of indoor origin (*Penicillium chrysogenum*, *P. palitans*, *Trichoderma viride*, *Stachybotrys* spp. and *A. versicolor*) [29]. Generally, extracts from growth on materials composed of finely divided cellulose were more active than those from growth on mineral wool [30].

The only data available on the ciliostatic activity of indoor mould metabolites involves sterigmatocystin produced from *A. versicolor* and *Chaetomium* spp. [31, 32]. Viable spores of *A. versicolor* and *Penicillium* spp. were isolated from different plasters even after 40 days to 3 months, i.e. till the end of the experiment [33, 34]. Seventy-eight percent of all indoor fungal isolates (namely, *Aspergillus clavatus, A. flavus, A. fumigatus, A. nidulans, A. niger, A. ochraceus, A. restrictus, A. ustus, A. versicolor, Alternaria* spp., *Chaetomium* spp., *Cladosporium cladosporioides, C. sphaerospermum, Phoma* spp., and *S. chartarum*) from Slovakia collected during the last 15 years produced complex metabolites which are able to cease tracheal ciliary beating in 24 h [31, 32, 35–37]. All toxic fungal metabolites are found in endometabolites (micromycetal propagules) and exometabolites (aerosol, detritus and house dust) carriers [34]. It is highly probable that hyphal fragments, dust and material particles that are able to reach alveoli have the highest depository and toxic potential. Most of the fungal spores can be entrapped just in the upper respiratory tract or at most in bronchi, because of their size, morphology and the mode of propagation (slime heads, aggregation etc.). Since the macroorganisms or the host is not only exposed to fungal particles but also to the detritus (other than fungal origin) that carry fungal metabolites, the experimental studies on toxic effects of fungal spore suspensions do not completely replicate the real exposition.

Health-damaging effects of fungal haemolysins (indoor, e.g. stachylysin produced by *S. chartarum* or chrysolysin by *P. chrysogenum*) include activation of histamine and cytokine-producing cells (inflammatory, cold-like SBS symptoms) and vascular tissue lysis (headaches, bleeding and vertigo) [38]. Complex toxic fungal metabolites disrupt self-cleaning mechanism of the airways and induce inflammation and cytotoxic as well as haematological damages, which can be enhanced by simultaneous action of other indoor contaminants like cigarette smoke. They can finally result in the ill health of occupants of damp mouldy dwellings, starting with respiratory disorders and probably ending in general intoxication of macroorganism via lung tissue especially in children with high rate of metabolism [39]. To determine a causal relationship between mycotoxins in the indoor environment and particular human health disorders after their inhalation, it is necessary:

- To estimate the minimal effective concentration of mycotoxin which can induce clinical symptoms in *in vivo* models
- To choose optimal animal or other biological models for studying mycotoxin pathogenicity and pathophysiology (pulmonary deposition)
- To characterize short- and long-term health damages (biomarkers) in people under such conditions [40]

The potential of fungal toxins to cause and develop multifactorial human diseases (immune, degenerative and tumorous) with permanently increasing incidency and prevalency in the modern society cannot be overlooked anymore [14].

References

1. WHO (2001) Safety evaluation of certain mycotoxins in food. WHO food additives series 47. FAO food and nutrition paper 74. WHO, Geneva
2. Berthiller F, Crews C, Dall´Asta C et al (2013) Masked mycotoxins: a review. Mol Nutr Food Res 57:165–186
3. Samson RA, Visagie CM, Houbraken J et al (2014) Phylogeny, identification and nomenclature of the genus *Aspergillus*. Stud Mycol 78:141–173
4. Yu J, Chang PK, Ehrlich KC et al (2004) Clustered pathway genes in aflatoxin biosynthesis. Appl Environ Microbiol 70:1253–1262
5. Perry LP, Iwata M, Tazelaar HD et al (1998) Pulmonary mycotoxicosis: a clinicopathological study of three cases. Modern Pathol 11:432–436
6. Kuhn DM, Ghannoum MA (2003) Indoor mold, toxigenic fungi, and *Stachybotrys chartarum:* Infectious disease perspective. Clin Microbiol Rev 16:144–172
7. Council for Agricultural Science and Technology (2003) Mycotoxins. Task Force Report 139. Risks in plant, animal and human systems. Iowa, USA
8. Mayer Z, Bagnara A, Färber P et al (2003) Quantification of the copy number of nor-1, a gene of the aflatoxin biosynthetic 555 pathway by real-time PCR, and its correlation to the CFU of *Aspergillus flavus* in foods. Int J Food Microbiol 82:143–151
9. Alborch L, Bragulat MR, Abarca ML et al (2011) Effect of water activity, temperature and incubation time on growth and ochratoxin A production by *Aspergillus niger* and *Aspergillus carbonarius* on maize kernels. Int J Food Microbiol 147:53–57
10. Andersen B, Nielsen KF, Jarvis B (2002) Characterisation of morphologically, chemically and physiologically different *Stachybotrys* species from water-damaged buildings. Mycologia 94:392–403
11. Nielsen KF, Hansen MO, Larsen TO et al (1998a) Production of trichothecene mycotoxins on water damaged gypsum boards in Danish buildings. Int Biodeterior Biodegrad 42:1–7
12. Piecková E, Pivovarová Z, Hurbánková M et al (2004) Indoor *Stachybotrys chartarum* and its toxicity. In: Petráš D, Feketeová M, Šabíková J (eds) Indoor climate of buildings ´04: Health, comfort and safety by operation of HVAC-R system. Slovak Soc Environ Technol, Bratislava, pp 167–171
13. Ren P, Ahearn DG, Crow SA Jr (1998) Mycotoxins of *Alternaria alternata* produced on ceiling tiles. J Indust Microbiol Biotechnol 20:53–54
14. Viegas S, Veíga L, Figueredo P et al (2013) Occupational exposure to aflatoxin B1 in swine production and possible contamination sources. J Toxicol Environ Health 76:944–951
15. McCormick A, Loeffler L, Ebel F (2010) *Aspergillus fumigatus*: contours of an opportunistic human pathogenic. Cell Microbiol 12:1535–1543
16. Stanzani M, Orciuolo E, Lewis R et al (2005) *Aspergillus fumigatus* suppresses the human cellular immune response via gliotoxin-mediated apoptosis of monocytes. Blood 105:2258–2264
17. Lewis R, Wiederhold N, Lionakis M et al (2005) Frequency and species distribution of gliotoxin-producing *Aspergillus* isolates recovered from patients at a tertiary-care cancer center. J Clin Microbiol 61:6120–6122
18. Kupfahl C, Geginat G, Hof H (2006) Gliotoxin-mediated suppression of innate and adaptive immune functions directed against *Listeria monocytogenes*. Med Mycol 44:591–599
19. Ramos C, Martínez S, Olivares R (2002) Gliotoxin production in 10 strains of *Aspergillus fumigatus* isolated from clinical cases. Téc Pecu Méx 10:139–148
20. Domingo J, Nadal M (2009) Domestic waste composting facilities: a review of human health risks. Environ Int 35:382–389
21. dos Santos VM, Dorner JW, Carreira F (2003) Isolation and toxigenicity of *Aspergillus fumigatus* from moldy silage. Mycopathologia 156:133–138
22. El-Shanawany AA, Mostafa ME, Barakat A (2005) Fungal populations and mycotoxins in silage in Assiut and Sohag governorates in Egypt, with a special reference to characteristic Aspergilli toxins. Mycopathologia 159:281–289

23. Gravesen S, Frisvad JC, Samson RA (1994) Microfungi. Munksgaard, Copenhagen, p 168
24. Nielsen KF, Thrane U, Larsen TO et al (1998b) Production of mycotoxins on artificially inoculated building materials. Int Biodeter Biodeg 42:9–16
25. Gravesen S, Nielsen PA, Iversen R et al (1999) Microfungal contamination of damp buildings–examples of risk constructions and risk materials. Environ Health Perspect 107:505–508
26. Wilkins K, Piecková E (2002) Detection of ciliostatic activity in fungal growth on building materials. Environ Sci Pollut Res 9:105–106
27. Jesenská Z, Bernát D (1994) Efects of mycotoxins on *in vitro* movement of tracheal cilia from one-day-old chicks. Folia Microbiol 39:155–158
28. Piecková E (2003) *In vitro* toxicity of indoor *Chaetomium* Kunze ex Fr. Ann Agric Environ Med 10:9–14
29. Piecková E, Jesenská Z (1998) Molds on house walls and the effect of their chloroform-extractable metabolites on the respiratory cilia movement on one-day-old chicks *in vitro*. Folia Microbiol 43:672–678
30. Kováčiková Z, Piecková E, Tátrai E et al (2007) Use of the *in vitro* model for the evaluation of toxic effects of metabolites produced by fungi. In: Brebbia CA (ed) Environmental health risk IV. WitPress, Southampton, pp 79–84
31. Kováčiková Z, Tátrai E, Piecková E et al (2008) *In vitro* toxicity of indoor fungi from dwellings in Slovakia: testing on the isolated lung cells. In: Kungolos AG, Brebbia CA, Zamorano M (eds) Environmental toxicology II. WitPress, Southampton, pp 211–218
32. Piecková E, Kolláriková Z (2008) *In vitro* toxicity of indoor moulds from Slovak dwellings. In: Kungolos AG, Brebbia CA, Zamorano M (eds) Environmental toxicology II. WitPress, Southampton, pp 227–234
33. Larsen FO, Clementsen P, Hansen M et al (1998) Volatile organic compounds from the indoor mould *Trichoderma viride* cause histamine release from human bronchoalveolar cells. Inflamm Res 47:S5–S6
34. Korpi A, Pasanen AL, Pasanen P et al (1997) Microbial growth and metabolism in house dust. Int Biodeterior Biodegrad 40:19–27
35. Piecková E, Pivovarová Z, Sternová Z et al (2007) Building materials vs fungal colonization – model experiments. In: Brebbia CA (ed) Environmental health risk IV. WitPress, Southampton, pp 71–78
36. Kováčiková Z, Tátrai E, Piecková E et al (2007) An *in vitro* study of the toxic effects of *Stachybotrys chartarum* metabolites on lung cells. ATLA 1:47–52
37. Piecková E (2015) Domestic environment–indoor mycobiota as a public health risk factor. In: Viegas C, Pinheiro C, Sabino R, Viegas S et al (eds) Environmental mycology in public health. Fungi and mycotoxins risk assessment and management. Elsevier–AP, London, pp 129–146
38. Hendry KM, Cole EC (1993) A review of mycotoxins in indoor air. J Toxicol Environ Health 38:161–182
39. Piecková E (2017) Indoor microbial aerosol and its health effects: Microbial exposure in public buildings–viruses, bacteria, and fungi. In: Viegas C, Viegas S, Gomes AQ, Täubel M, Sabino R (eds) Exposure to microbial agents in indoor and occupational environments. Springer, Cham, pp 237–252
40. Scott PM (2005) Biomarkers of human exposure to ochratoxin A. Food Addit Contam 22:S99–S107

Tenuazonic Acid: A Potent Mycotoxin

8

Ankita Kumari and Neha Nidhi Tirkey

Abstract

The genus *Alternaria* includes many allergenic species, saprophytes, and plant and human pathogens and is an inexhaustible manufacturer of secondary metabolites. In pre- and post-harvest conditions, they are frequent contaminants of crops. A variety of agricultural products including grains are commonly infested by *Alternaria* species. Some of them are psychrophilic in nature and thus are able to colonize even refrigerated commodities. They produce a variety of mycotoxins having acute and chronic effects. *Alternaria alternata* is the most important mycotoxin-producing species that infests cereals and fruits and, hence, has the potential to pose a serious threat to human and animal health. Tenuazonic acid (TeA) is the most studied *Alternaria* mycotoxin and is considered to have the highest toxicity amongst them. It causes haemorrhages in several organs, suppression in weight gain and reduction in feed efficiency in animals. It is known to be a powerful inhibitor of eukaryotic protein synthesis, and its association with oesophageal cancer has been reported in human populations at risk of high exposure to TeA. Since TeA is one of the major mycotoxins in humans and other organisms, it is important to minimize TeA contamination in food and feed to avoid health risks. In this regard, the present review discusses the presence of TeA in various food products and their effect on plants and animals. The review has also tried to integrate the information existing on the toxicology and methods of detection and quantification of TeA toxin. The toxicological database on TeA is still limited, and their risk assessment reports remain inconclusive. Thus, new approaches should be considered to investigate the toxicological interactions of TeA in agricultural products, humans and animals.

A. Kumari (✉) · N. N. Tirkey
Animal Mycology Laboratory, Department of Zoology, Mahila Mahavidyalaya, Banaras Hindu University, Varanasi, India

K. Singh, N. Srivastava (eds.), *Recent Trends in Human and Animal Mycology*,
https://doi.org/10.1007/978-981-13-9435-5_8

Keywords
Alternaria alternata · Tenuazonic acid · Health risk · Human · Animal

8.1 Introduction

The genus *Alternaria* comprises of pathogenic, saprophytic and allergenic species that are present almost everywhere. A wide range of health-endangering secondary metabolites called mycotoxins are also produced by them [1]. They commonly infect a variety of agricultural products, fruits and vegetables [2]. Mycotoxin production by *Alternaria* is favoured in a moist environment (water activity, $a_w = 0.98$) [3]. The germination and growth conditions for *Alternaria alternata* have been reported in some studies with respect to temperature and water activity (Table 8.1) [4]. Various species of the *Alternaria* are reported to produce over 300 toxic metabolites, amongst which alternariol (AOH), altenariolmonomethyl ether (AME), altenuene (ALT), tenuazonic acid (TeA), tentoxin (TEN) and altertoxins I, II and III are of high importance [4, 5]. Various fruits like apples, peppers, mandarins, melons, olives, raspberries and tomatoes have the natural occurrences of AOH, AME and TeA [6]. Previous studies have depicted that TeA, AOH, AME and altertoxins I, II and III (ATX-I, ATX-II, ATX-III) are the most frequently present *Alternaria* mycotoxins [7]. They belong to three structural classes of chemicals [8]:

1. Dibenzopyrone derivatives comprising AOH, AME and ALT.
2. Perylene derivatives comprising ATX-I, ATX-II and ATX-III.
3. Tetramic acid derivatives comprising TeA.

Tenuazonic acid is produced by *Alternaria* spp., *Phoma sorghina*, and *Pyricularia oryzae* and is considered to be the most toxic amongst *Alternaria* mycotoxins [2, 9–12]. TeA inhibits protein biosynthesis and is biologically active, exerting antitumor, antiviral and antibiotic activities [2, 9, 13, 14]. It was associated with the mycotoxicosis, Onyalai, a haematologic disorder that occurred in Africa [9, 15]. The clinical feature of this disease is haemorrhagic bullae on the mucosa of the oronasopharynx. Haemorrhage from ruptured bullae, epistaxis or gastrointestinal bleeding becomes severe and may lead to shock and death. Antibacterial, cytotoxic, insecticidal, phytotoxic and zootoxic properties are exhibited by TeA [15].

Table 8.1 Summary of the range of environmental conditions which allow germination, growth and toxin production by *A. alternata* and *A. tenuissima* (psychrotolerant)

Factors	Germination	Growth	Toxin production
Temperature (°C)	1–35	< 1 and > 35	< 10 and > 35
Water activity (a_w)	0.84–0.995	< 0.85	< 0.90
pH	2.5–10	< 2.5 and > 10	< 2.5 and > 9

Adapted from: *Alternaria* in Food: Ecophysiology, Mycotoxin Production and Toxicology, Lee et al. 2015, Mycobiology [4]

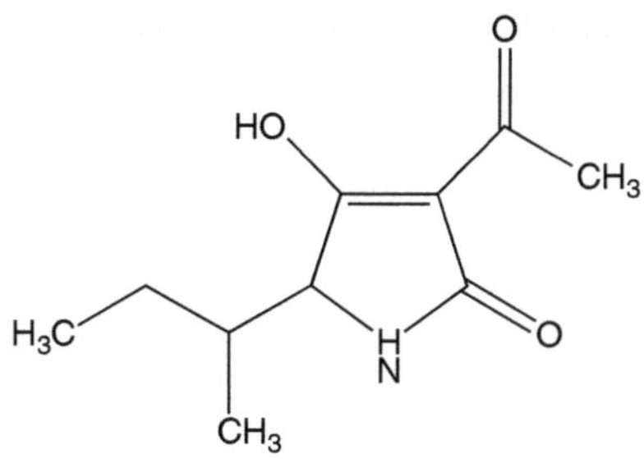

Fig. 8.1 Chemical structure of tenuazonic acid

8.2 Chemical Properties of Tenuazonic Acid

Tenuazonic acid is colourless and viscous oil. The chemical formula/IUPAC name of TeA is 3-acetyl-5-sec-butyl-4-hydroxy-3-pyrrolin-2-one (Fig. 8.1). Its molecular weight is 197 and its molecular formula is $C_{10}H_{15}O_3N$ [15]. TeA is soluble in organic solvents including petroleum ether and is sparingly soluble in water [16]. It is ketonic in nature. Its rotation in methanol is −124°. The rotation of TeA in crude form slowly becomes less negative and crystallizes when kept standing for a long time. Purified TeA is dextrorotatory, $[\alpha]_{5461}$ 22° + 23°. It titrates sharply as a monobasic acid (pKa 3.5), contains no methoxyl group and forms a semicarbazone. Its melting point is 187–189 °C. When boiled with an aqueous alkali, it is converted to an iso-acid form [17]. Tenuazonic acid forms complexes with calcium, magnesium, copper, iron and nickel ions and is usually stored as copper salt [16].

8.3 Tenuazonic Acid as Phytotoxin

A number of phytotoxic metabolites are reported to be produced by phytopathogenic species of *Alternaria*. Many have been characterized chemically and are known to play a significant role in pathogenesis. TeA causes significant harm to crops. Its presence has been observed in a number of plant materials such as groundnuts, olives, sunflower seeds, mandarins, peppers, sorghum, tobacco, rice, melons and linseed [18–20]. Table 8.2 lists the food items and their products that are usually contaminated by TeA [21].

Chen et al. in 2007, for the first time, reported that TeA inhibits photosynthetic activity by working as an inhibitor of photosystem II [18]. Chlorophyll fluorescence studies showed that TeA blocked the flow of electrons from Q_A to Q_B in the acceptor side of photosystem II. TeA was considered as a new photosystem II inhibitor as it had a different binding behaviour within Q_B-niche when compared to other known photosystem II inhibitors. Generation of singlet oxygen mediated the phytotoxic action of TeA [18]. Another report on phytotoxic effect of TeA was published in the year 2017 by Kang et al. [22]. A mutant type of *A. alternata*, HP001, was investigated in the study that produced less TeA [22]. The failure of the mutant species to form the appressorium necessary for causing infection in the host led to the loss of pathogenicity of the organism. Therefore, TeA was reported to be the key virulence factor for the infection of *A. alternata* in the host [22]. TeA not only damages the

Table 8.2 Tenuazonic acid in foodstuffs [25]

Raw food items	Food products
Cereals	Maize, millet, wheat, sorghum, rice, beer, bran (wheat bran), cassava, flakes (corn flakes, oat flakes), flour (rye flour, wheat flour), grit (maize grits)
Fruits	Apple, mandarin, melon, olive, dried fruit
Juices	Apple, apricot, banana, blackcurrant, cherry, grape, orange, pear, pomegranate, strawberry, tomato
Tomatoes	Pulp, puree, sauce, ketchup
Spices	Coriander, cumin, curcuma, curry, ginger, paprika, pepper
Seeds	Sesame, sunflower seed

host plant but is also associated with host recognition, inducing appressoria to infect the host, maintaining ROS content and facilitating the completion of the infection process [22].

Tenuazonic acid, iso-tenuazonic acid (an isomer of TeA) and their salts also show evidences of herbicidal activity with broad spectrum, quick killing and high efficiency. The herbicidal activity of these compounds is enhanced by the addition of adjuvants [19].

8.4 Tenuazonic Acid Toxicity in Animals

Studies regarding the activity of tenuazonic acid on several animals like mice, chickens and dogs have proved its toxicity. It causes haemorrhages in many organs of dogs when given at a dose of 10 mg/kg body weight. Subacute toxicity was observed in chickens when fed at a dose of 10 μg/g of feed. In chickens, when the concentration of TeA was increased gradually from sublethal to lethal levels, it resulted in an increase of internal haemorrhage, suppression in weight gain and reduction in feed efficiency [9, 23, 24]. TeA does not exhibit mutagenic properties in the bacterial system [6, 25]. Precancerous changes were observed in the oesophageal mucosa in mice fed with TeA at a concentration of 25 mg/kg body weight every day for a period of 10 months [9].

In studies carried out by Meronuck and co-workers in 1972, 57 out of 87 isolates of *Alternaria alternata* (Fr.) Keissler cultured on sterilized, moist corn-rice substrate turned out to be fatal when fed to rats. After isolation and characterization, TeA was found to be the main toxin produced. Amongst 23 isolates of toxigenic *Alternaria*, 20 isolates produced tenuazonic acid [26].

Giambrone and co-workers (1978) performed experiments to study the effect of tenuazonic acid on young chickens and reported that TeA administration by daily oesophageal intubation at the levels of 1.25 or 2.50 mg/kg of body weight or when given in the diet at 10 μg/g of feed to 2- to 3-week-old broilers caused a decrease in weight gain and feed efficiency during the second and third weeks of toxin administration [27]. Neither morbidity nor mortality was observed in chickens through administration by either of the routes; however, noticeable gross and histological lesions were evident in different tissues of the birds that received the toxin by either method. Lesions such as enlarged and spotted spleen, minor wearing off of the gizzard, haemorrhage in the intestinal lumen and on the surface of the heart, oedema of the myocardium and haemorrhage with bruising in the musculature of the thigh were observed in maximum number of chickens that were administered with the toxin by oesophageal intubation. Gross haemorrhages were not seen in broilers receiving toxin in the feed; however, gizzard erosion along with pale and spotted spleens was apparent. Kidneys, liver and other affected tissues showed blood vessel congestion and haemorrhage when observed microscopically. Similarly, when 1-week-old white Leghorns were administered with 1.25 or 2.50 mg toxin/kg body weight/day by oesophageal intubation for 21 days, a decrease in their weight gain and feed efficiency was observed. Pathological changes in them were found to be similar to the broilers receiving toxin by oesophageal intubation [27].

Mice being fed orally with AME or TeA for a period of 10 months daily showed moderate and severe dysplasia distinguished by loss of polarity, nuclear pleomorphism and hyperchromasia in microscopic examination of oesophageal mucosa. Electron micrography of in vitro studies of TeA on mucosal epithelial cells revealed condensation of chromatin in some nuclei, granulation and increase in chromatin mass, abnormalities in the nuclear membrane, vacuolization in nucleoplasms and distinct pleomorphism in the nuclei. Thus, dysplastic transformation in the study suggested higher toxicity of TeA [28].

Shigeura and Gordon in 1963 investigated the biological activity of tenuazonic acid, in vivo with rats and in vitro with Ehrlich ascites tumour cells as well as rat liver cells. They demonstrated that TeA inhibited the inclusion of amino acids into proteins, both in vivo and in vitro. In cell-free systems, inhibition of protein synthesis by TeA occurred by the suppression of release of newly formed microsomal proteins into the supernatant fraction [29].

Another study was carried out by Yang and co-workers on the effect of *A. alternata* toxins on insects [30]. The inhibitory effects of the toxins were studied on rose aphid, *Macrosiphum rosivorum*. On application of the purified crude toxin to rose (*Rosa chinensis*) leaves, no effect was observed on the rose plants. However, it affected the reproduction of aphids. When the toxin solution was sprayed over the rose plants, the inhibitory index was found to be approximately 87.99% [30].

Figure 8.2 is a graphical representation of the effects of TeA in plants and in different animal models.

8.5 Tenuazonic Acid Determination: Methods and Techniques

Till date, the *Alternaria* mycotoxins have been quantified by high-performance liquid chromatography (HPLC) coupled with UV (for TeA), fluorescence (for AOH, AME, ALT and ATX-I) [3, 9, 13, 15, 23–25] or electrochemical detection (for AOH, AME and ATX-I) [19–22] and thin-layer chromatography [17, 25, 29] as well as by gas chromatographic methods [18, 21, 31, 32].

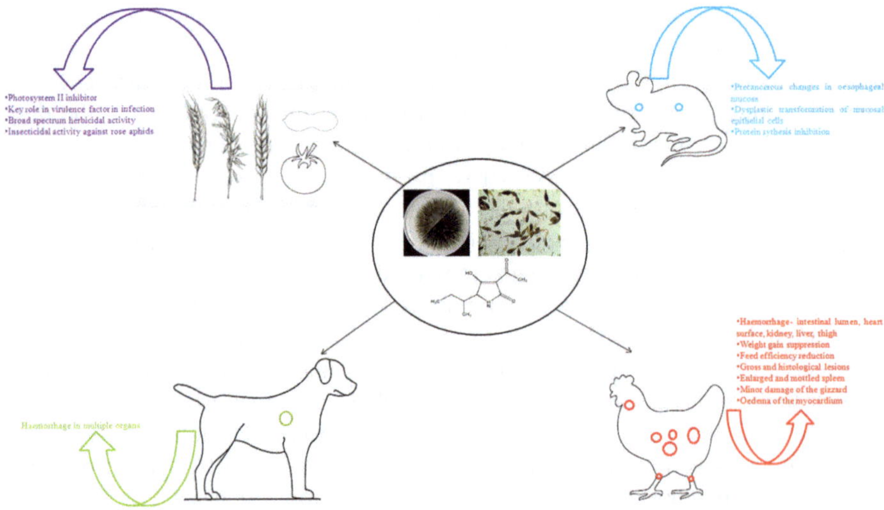

Fig. 8.2 Graphical representation of the effects of tenuazonic acid on plants and animals

For AOH, AME, ALT and ATX-I, HPLC–tandem mass spectrometry (MS/MS) methods have also been reported [19, 20, 25, 26]. Due to the low molecular weight, relatively high acidity (pKa = 3.5) and metal-chelating properties of TeA, data related to its isolation and characterization by LC-MS is scarce [10]. Either ionization in the columns should be lowered, i.e. acidic conditions should be facilitated or ion-pairing techniques should be used to acquire acceptable peak shapes in HPLC. The latter have been applied widely with Zn(II)SO_4 as an eluent modifier [6, 9, 13, 15, 24]; however, Zn(II)SO_4 is not compatible with common MS ion sources, and other ion-pairing systems do not show adequate retention. Adverse ion yields are expected at acidic conditions (ionization suppression chromatography) when MS detectors are used [24].

In Argentina, high levels of *Alternaria* mycotoxin were reported in wheat of the 2004/2005 harvest, with a mean of 2.313 mg/kg of TeA in the positive samples using HPLC. Azcárate et al. (2008) reported that out of 123 *Alternaria* strains isolated from Argentinian wheat, 72% produced TeA [7].

TeA was reported for the first time by Rosett et al. in the year 1957 [33]. Ever since its discovery, the gene responsible for its synthesis remained unidentified for about 60 years, despite several reports on detection of TeA. However, Yun et al. (2015) identified the gene for the biosynthesis of TeA from *Magnaporthe oryzae* [34]. As reports on the NRPS–PKS type fungal enzymes were not very conclusive, the chemical structure of the toxin suggested that TeA was synthesized by PKS and NRPS hybrid enzymes. For the first time, Yun and his co-workers identified the TAS1 gene responsible for the production of NRPS–PKS enzyme that synthesizes TeA [34].

The first ever case of TeA determination in human urine using a stable isotope dilution assay (SIDA) was documented by Asam et al. (2013) [35]. Volunteers were exposed to TeA in their diet. They detected TeA in all the volunteers at levels higher than or equal to the urinary values of other mycotoxins studied. The experiment further revealed that TeA was excreted via urine when ingested through food. It is rapidly absorbed through food and almost completely excreted in urine within 24 hours of intake. Eighty seven to 93% of TeA was eliminated from the body, but the remaining 10% of TeA might contribute to potential health hazards [35].

Chromatographic separation of allo-tenuazonic acid (an isomer of TeA) from TeA is not possible in most analytical methods [32]. Hickert et al. in 2015 developed a method to quantify both TeA and allo-TeA. A QuEChERS (quick, easy, cheap, effective, rugged and safe)-based stable isotope dilution HPLC-MS/MS method included the separation of both the isomers chromatographically and was tested on 20 tomato products from the German market. The presence of both the isomers was found in all the products. TeA was present in a range from 5.3 ± 0.1 to 550 ± 15 μg/kg (average = 120 μg/kg) and allo-TeA in a range from 1.5 ± 0.4 to 270 ± 0.8 μg/kg (average = 58 μg/kg). Both the compounds were separated from a synthetic racemic mixture to assess and compare their cytotoxic effects. TeA exhibited moderate cytotoxic effects on HT-29 cells at concentrations as low as 100 μM, whereas allo- TeA showed no activity [32].

Another method for quantification of TeA in plasma of pig and broiler chicken was devised by Fraeyman et al. (2015) using the liquid chromatography tandem-mass spectrometry (LC-MS/MS) method [36]. The limit of quantification in the plasma of both pig and broiler chicken was 5.0 ng/ml. The developed method was employed to compare and evaluate the toxicokinetic parameters of both pigs and broiler chickens. In both animals, the bioavailability of TeA was found to be 100% when administered orally. However, the rate of absorption and elimination was found to be more time-consuming in broiler chickens than pigs. This might be because the total body clearance in chickens was considerably lower [36].

8.6 Conclusion

Alternaria alternata, a dominant pathogen of numerous plants, fruits and vegetables, secretes TeA with bioactivity against microbes, plants, humans and animals. Study of TeA is imperative because it causes damage to a wide range of economically important food crops. However, information regarding the safety limits of TeA in foodstuffs and feed is still limited. Reliable assessment of the toxic effects of TeA for human and animal consumption is essential to define the guidelines on limits of TeA in a better way.

References

1. Grover S, Lawrence CB (2017) The *Alternaria alternata m*ycotoxin Alternariol suppresses lipopolysaccharide-induced inflammation. Int J Mol Sci 18:1577
2. Weidenbörner M (2001) Encyclopedia of food mycotoxins. Springer, Berlin
3. Magan N, Cayley GR, Lacey J (1984) Effect of water activity and temperature on mycotoxin production by *Alternaria alternata* in culture and on wheat grains. Appl Environ Microbiol 47:1113–1117

4. Lee HB, Patriarca A, Magan N (2015) *Alternaria* in food: ecophysiology, mycotoxin production and toxicology. Mycobiology 43:93–106
5. Robiglio AL, Lopez SE (1995) Mycotoxin production by *Alternaria alternata* strains isolated from red delicious apples in Argentina. Int J Food Microbiol 24:413–417
6. Scott PM, Stoltz DR (1980) Mutagens produced by *Alternaria alternata*. Mutat Res 78:33–40
7. Azcárate MP, Patriarca A, Terminiello L et al (2008) *Alternaria* toxins in wheat during the 2004 to 2005 Argentinean harvest. J Food Prot 71:1262–1265
8. Bottalico A, Logrieco A (1998) Toxigenic *Alternaria* species of economic importance. In: Shinha KK, Bhatnagar D (eds) Mycotoxin in agriculture and food safety. Marcel Dekker Inc, New York, pp 65–108
9. Steyn PS, Rabie CJ (1976) Characterization of magnesium and calcium tenuazonate from *Phoma sorghina*. Phytochemistry 15:1977–1979
10. Umetsu N, Kaji J, Aoyama K et al (1974) Toxins in blast-diseased rice plants. Agric Biol Chem 38:1867–1874
11. Umetsu N, Kaji J, Tamari K (1972) Investigation on the toxin production by several blast fungus strains and isolation of tenuazonic acid as a novel toxin. Agric Biol Chem 36:859–866
12. Iwasaki S, Muro H, Nozoe S et al (1972) Isolation of 3, 4-dihydro-3, 4, 8-trihydroxy-1(2H)-naphthalenone and tenuazonic acid from *Pyricularia oryzae cavara*. Tetrahedron Lett 13:13–16
13. Shephard GS, Thiel PG, Sydenham EW et al (1991) Reversed-phase high-performance liquid chromatography of tenuazonic acid and related tetramic acids. J Chromatogr 566:195–205
14. Schobert R, Schlenk A (2008) Tetramic and tetronic acids: an update on new derivatives and biological aspects. Bioorg Med Chem 16:4203–4221
15. Lurie A, Katz J, Ludwin SK et al (1969) Platelet life-span and sites of platelet sequestration in onyalai. Br Med J 4:146–148
16. The Merck Index online (2018) Tenuazonic acid. Royal Society of Chemistry. https://www.rsc.org/Merck-Index/monograph/m10562?q=authorize. Accessed 20 Feb 2018
17. Shigeura HT (1967) Tenuazonic acid. In: Gottlieb D, Shaw P (eds) Antibiotics, Mechanism of action, vol 1. Springer, Berlin/Heidelberg/New York, pp 360–365
18. Chen S, Xu X, Dai X et al (2007) Identification of tenuazonic acid as a novel type of natural photosystem II inhibitor binding in QB-site of *Chlamydomonas reinhardtii*. Biochimica et Biophys Acta (BBA)-Bioener 1767:306–318
19. Devi PS, Reddy MN, Devamma MN et al (2010) Identification and characterization of tenuazonic acid as the causative agent of *Alternaria alternata* toxicity towards groundnut. Af J Microbiol Res 4:2184–2190
20. Subrahmanyam P, McDonald D, Waliyar F et al (1995) Screening methods and sources of resistance to rust and late leaf spot of groundnut, Information Bulletin no. 47. International Crops Research Institute for the Semi-Arid Tropics, Patancheru
21. Weidenbörner M (2013) Mycotoxins in foodstuffs. Springer, Boston, pp 1–546
22. Kang Y, Feng H, Zhang J et al (2017) TeA is a key virulence factor for *Alternaria alternata* (Fr.) Keissler infection of its host. Plant Physiol Biochem 115:73–82
23. Griffin GF, Chu FS (1983) Toxicity of the *Alternaria* metabolites alternariol, alternariol methyl ether, altenuene, and tenuazonic acid in the chicken embryo assay. Appl Environ Microbiol 46:1420–1422
24. Woody MA, Chu FS (1992) Toxicology of *Alternaria* mycotoxins. In: Chelkowski J, Visconti A (eds) *Alternaria:* biology, plant diseases, and metabolites. Elsevier, Amsterdam, pp 409–434
25. Schrader TJ, Cherry W, Soper K et al (2006) Further examination of the effects of nitrosylation on *Alternaria alternata* mycotoxin mutagenicity *in vitro*. Mutat Res 606:61–71
26. Meronuck RA, Steele JA, Mirocha CJ et al (1972) Tenuazonic acid, a toxin produced by *Alternaria alternata*. Appl Microbiol 23:613–617
27. Giambrone JJ, Davis ND, Diener UL (1978) Effect of tenuazonic acid on young chickens. Poult Sci 57:1554–1558
28. Yekeler H, Bitmiş K, Ozçelik N et al (2001) Analysis of toxic effects of *Alternaria* toxins on esophagus of mice by light and electron microscopy. Toxicol Pathol 29:492–497

29. Shigeura HT, Gordon CN (1963) The biological activity of tenuazonic acid. Biochemist 2:1132–1137
30. Yang FZ, Yang B, Li BB et al (2015) *Alternaria* toxin-induced resistance in rose plants against rose aphid (*Macrosiphum rosivorum*): effect of tenuazonic acid. J Zhejiang Univ-Sci B 16:264–274
31. Logrieco A, Bottalico A, Mulé G et al (2003) Epidemiology of toxigenic fungi and their associated mycotoxins for some Mediterranean crops. Eur J Plant Pathol 109:654–667
32. Hickert S, Krug I, Cramer B et al (2015) Detection and quantitative analysis of the noncytotoxic allo-tenuazonic acid in tomato products by stable isotope dilution HPLC-MS/MS. J Agric Food Chem 63:10879–10884
33. Rosett T, Sankhala RH, Stickings CE et al (1957) Studies in the biochemistry of microorganisms. 103 Metabolites of *Alternaria tenuis* Auct: culture filtrate products. Biochem J 67:390–400
34. Yun CS, Motoyama T, Osada H (2015) Biosynthesis of the mycotoxin tenuazonic acid by a fungal NRPS–PKS hybrid enzyme. Nat Commun 6:8758
35. Asam S, Habler K, Rychlik M (2013) Determination of tenuazonic acid in human urine by means of a stable isotope dilution assay. Anal Bioanal Chem 405:4149–4158
36. Fraeyman S, Devreese M, Broekaert N et al (2015) Quantitative determination of tenuazonic acid in pig and broiler chicken plasma by LC-MS/MS and its comparative toxicokinetics. J Agric Food Chem 63:8560–8567

Part III

Antifungal Therapeutic Candidates

Phytochemicals: New Avenues in Anticandidal Activity

9

Richa Raghuwanshi

Abstract

Human fungal infections have significantly increased in recent years particularly in immunocompromised hosts. Candidiasis is one of them. The emergence of new virulence factors and drug resistance in its aetiological agent *Candida* spp. beckons the need for new drug discoveries, as the present-day drugs are insufficient in combating the disease. Therapeutic alternatives can be the medicinal plants and phytocompounds, which have been in traditional system of medicine for their empirical antifungal properties. They could be reliable alternatives to overcome the disadvantages of antifungal drugs that include undesirable side effects, toxicity, recurrence, drug-drug interactions, and multiple drug resistance. The present article reviews the anticandidal activity of different medicinal plants and phytochemicals under the major classes of secondary metabolites such as phenolics, alkaloids, terpenoids, saponins, flavonoids, proteins, and peptides.

Keywords

Anticandidal activity · Medicinal plants · Phytocompounds

9.1 Introduction

Fungaemia, or fungal blood stream infection, is the most common form of invasive fungal infections [1, 2]. Moulds are hardly reported for causing fungaemia except for the disease caused by *Fusarium*. Candidiasis caused by *Candida* spp. is the most dominant fungal infection affecting human and animals reported globally with a mortality rate of 10–49% in immunocompromised hosts [3]. *Candida albicans*, well

R. Raghuwanshi (✉)
Department of Botany, Mahila Mahavidyalaya, Banaras Hindu University, Varanasi, India
e-mail: richa73@bhu.ac.in

K. Singh, N. Srivastava (eds.), *Recent Trends in Human and Animal Mycology*,
https://doi.org/10.1007/978-981-13-9435-5_9

known for causing a majority of human infections, is a polymorphic fungus whose life cycle may complete as a commensal on the human skin and mucosal surfaces (gut, oral, pharyngeal, and genito-urinary tract) which are quite susceptible to infections or as an aggressive pathogen causing candidiasis [4]. The frequently observed mucosal infections in humans include the chronic mucocutaneous candidiasis, onychomycosis, vaginal candidiasis, and cutaneous candidiasis. The vast majority of candidaemia involves *Candida* spp., while other yeasts such as *Saccharomyces cerevisiae*, *Cryptococcus* spp., *Rhodutorula* spp., and *Trichosporon* spp. although reported in decreasing incidence may also be involved [5–7]. The epithet "the disease of the diseased" is often used, as *Candida* mainly targets the patients under immunosuppressive conditions like AIDS, diabetes, leukaemia, and others such as patients who have undergone organ transplant, on broad-spectrum antibiotics, having indwelling catheters, intravenous drug misuse, and denture wearers. It has also become a problem worldwide due to the huge number of species found in this genus, which easily infect the immunocompromised hosts. Till date, over 40 candidal species have been recovered from patients with life-threatening infections. The occurrence of oral candidiasis at least once in HIV patients is estimated to be 80–95% [8]. Although *C. albicans* remains the predominant species, a variety of other *Candida* species are also involved in human infections. The intrinsic susceptibility pattern for these different species is as diverse as it is for bacteria. Many of the emerging *Candida* species causing infections include the *C. auris*, *C. tropicalis*, *C. glabrata*, *C. dubliniensis*, *C. parapsilosis*, *C. orthopsilosis*, *C. metapsilosis*, *C. krusei*, *C. famata*, *C. guilliermondii*, and *C. lusitaniae* [9–13]. *Candida* sp. is capable of causing life-threatening systemic infections over broad range of body sites due to its high degree of adaptability to different host niches and variable complex environmental factors like oxygen and carbon dioxide levels, pH, osmotic conditions of the cell, availability of nutrients, temperature, etc. [14, 15]. They are a serious risk to human health, due to the resistance developed in them towards the existing antifungal agents [16]. Relapse of *Candida* infections is of great concern. *Candida* spp. also have an ability to form biofilms, which has its own repercussion in the clinical context, as this leads to increased resistance to antimicrobial agents [17].

Phytochemicals have been looked upon as a reliable alternate to cater the needs of developing effective and less toxic novel antifungal agents that would overcome the disadvantages of antimicrobial and antifungal drugs including undesirable side effects, toxicity, recurrence, drug-drug interactions, development of resistance, and ineffectiveness which has made them less successful in therapeutic strategies. Treatment by herbal medicines has gained impetus during the last few decades especially as a remedy to the diseases that are obstinate and incurable in the other systems of medicine as the herbal medicines show advantages like the fewer side effects, patient tolerance, relatively lower expense, and faith by the society due to a long history of use. Natural compounds from plant sources have gained attention as anti-*Candida* therapeutics in the past decade (2004–2015) mainly because they display structural diversity and uniqueness in functional modes of action [18, 19]. Not only in developing countries where botanicals are used as medicinal products, these medicines are also making their place in the integrative healthcare system of the developed nations, known as complementary and alternative system of medicines (CAM).

The efficacy of plants and their extracts is due to the presence of several primary and/or secondary metabolites such as phenolics, polyphenols, tannins, quercetin, flavones, flavonols, alkloids, terpenoids, lectins, polypeptides, and complex mixtures. Although encouraging outputs have been obtained in the area of drug development through molecular modelling and combinatorial and synthetic chemistry, still natural plant-derived compounds continue to be a vast and an invaluable source of medicines for humans.

9.2 Anticandidal Agents and Their Targets

A successful outcome of any treatment is dependent on the fungus (its virulence and susceptibility), the host status (severity of the underlying condition), and the therapy (timing in relation to infection stage, choice of antifungal drug, and dosing). The conventional antifungal agents used to cure candidal infections are azoles (fluconazole and voriconazole), polyenes (amphotericin B and nystatin), allyl amines (naftifine and terbinafine), and echinocandins (anidulafungin and micafungin) as illustrated in Fig. 9.1.

Azole drugs and their derivatives are the most effective antifungal agents used in the treatment of *Candida* and interrelated infections due to their wide-ranging spectrum of activity and high therapeutic values [20]. The azoles inhibit the enzyme P450 demethylase necessary for the ergosterol synthesis. As ergosterol is required for membrane synthesis, this leads to reduced ergosterol formation and growth arrest [21]. The azoles are fungistatic against *Candida*. Itraconazole is a triazole having a broad-spectrum activity and is effective against a wide range of fungal pathogens.

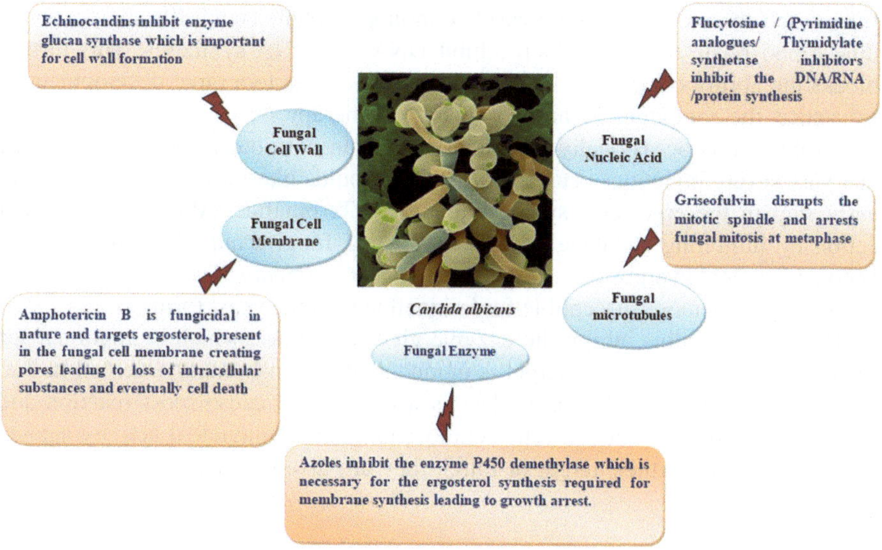

Fig. 9.1 Site of action of different compounds having anticandidal activity

Fluconazole is a triazole, effective in curing mucosal as well as invasive candidiasis in immunocompromised cancer patients [22]. Certain *Candida* species, such as *C. glabrata*, *C. albicans*, *C. tropicalis*, and *C. parapsilosis*, were found to have different degrees of susceptibility against it and were also reported to have developed resistance against it [23]. Although broad-spectrum triazoles are available in the market as conventional medical therapies, the cases reported for invasive candidiasis still remain at its pace mainly due to resistance developed against the antifungal agents and also the emergence of non-albicans strains of *Candida*, such as *C. glabrata*. The pathogen develops resistance against the antifungal agents mainly by modifying the target enzyme, the cytochrome P-450, or even through the failure of azole drugs in getting deposited inside the fungus which is mediated by multidrug resistance (MDR) and candida drug resistance (CDR) genes [24, 25]. Amphotericin B is a hydrophobic polyene antifungal normally used for the treatment of systemic fungal infections. It is fungicidal and targets ergosterol, an essential sterol of fungal cell membrane. Upon binding, it initiates pore formation in the membrane leading to a loss of intracellular substances and eventually resulting in cell death. However, use of amphotericin B, which is also known as the "gold standard", is restricted due to its infusion-related problems and nephrotoxicity [26, 27]. Nystatin is a polyene antifungal effective for the treatment of multiple cutaneous and mucocutaneous fungal infections occurring due to the *Candida* spp. [28]. Nystatin is both fungistatic and fungicidal against *C. albicans* [29].

Flucytosine (5-fluorocytosine) was originally developed as an anticancer agent and thus possesses bone marrow-depressing side effects particularly at higher concentrations. It is almost exclusively used in combination with amphotericin for *Cryptococcus* and other rare yeast infections involving central nervous system (CNS) or other foci where drug penetration is limited. It is taken up by fungal cells via the enzyme cytosine permease and converted first into 5-fluorouracil inside the fungus and then to 5-fluorodeoxyuridine monophosphate (FdUMP) and fluorouridine triphosphate (FUTP), which inhibit DNA and RNA syntheses. This further blocks the ability of fungus to synthesize proteins. Due to a rapid development of resistance, the compound is rarely used as monotherapy.

Terbinafine is occasionally used in combination with other agents to treat rare and very severe infections caused by resistant moulds like *Fusarium*. It inhibits an early step in the ergosterol synthesis pathway. Terbinafine exhibits a fungicidal action by inhibiting squalene epoxidase. Treated fungi accumulate squalene but become deficient in ergosterol, as evident against *C. albicans*. The filamentous form of this fungus is more susceptible for terbinafine than the yeast form.

The echinocandins inhibit the enzyme glucan synthase which is important for the cell wall synthesis. This target is unique as the human eukaryotic cell has no cell wall, and thereby, this drug class is less prone to cause cross-reaction and interfere with the human cell. The echinocandins have fungicidal activity against *Candida*. Owing to their target in fungi, they are also compared to the ß-lactam antibiotics [5, 6].

9.3 Classes of Phytocompounds Showing Anticandidal Activity

Plants in their lifetime produce a vast number of secondary metabolites, which impart them with odours (terpenoids), pigmentation (quinones and tannins), and flavour (terpenoid capsaicin from chilli peppers). These secondary metabolites, which make the plants resistant to a number of pathogens like the viruses, fungi, and bacteria, are also endowed with medicinal properties against a number of human diseases. Traditional medicine listed in pharmacopoeia of many countries is based on formulations made by different plant extracts for treatment of fungal infection, and many of these till date have been tested for in vitro antifungal activity. The World Health Organization (WHO) reports about 80% of the world's population in developing countries to rely on locally available medicinal plants for their primary healthcare. Although the rich plant diversity has been a source of remedy for humans as evidenced since time immemorial, only limited scientific studies exist on the assessment on plant quality, safety, and efficacy. Recent evidence from the pharmaceutical companies shows that, for some complex diseases, natural products still represent an extremely valuable source for the new chemical entities, since they represent privileged structures selected by evolutionary mechanisms over a period of millions of years [30]. The efficacy of plants and their extracts is due to the presence of several primary and/or secondary metabolites such as phenolics, polyphenols, tannins, quercetin, flavones, flavonols, alkloids, terpenoids, lectins, polypeptides, and complex mixtures. Phytocompounds possessing anticandidal activity belonging to the major classes of secondary metabolites, viz. terpenoids, saponins, phenolics, flavonoids, alkaloids, proteins, and peptides, are discussed below.

9.3.1 Terpenoids

Terpenoids are also known as isoprenoids, a subclass of the prenyllipids (terpenes, prenylquinones, and sterols), which represent the small molecular products and are probably the most widespread group of natural products. Antifungal sesquiterpene lactones from the Asteraceae family isolated from *Ajania fruticulosa* (Lebeb.) Poljak and seven xanthanolides from *Xanthium macrocarpum* D.C. were reported to be effective against *C. albicans* and *C. glabrata* [31, 32].

An antimicrobial diterpene 8β–17-epoxylabd-12-ene-15, 16-dial from *Alpinia galanga* synergistically enhanced the antifungal activity of quercetin and chalcone against *C. albicans* [33]. Triterpenoid glycosides obtained from *Solidago virgaurea* and *Bellis perennis* inhibit the growth of human-pathogenic yeasts (*Candida* and *Cryptococcus* species) [34].

9.3.2 Saponins

Saponins are natural plant detergents that act as efficient antimicrobial, cholesterol-lowering anticancerous compounds. Saponins are an important source of constitutive antifungals. Triterpenoid saponins are triterpenes which belong to the group of saponin compounds. These triterpene saponins, together with steroidal saponins, have antifungal properties.

A novel triterpenesaponin, CAY-I, extracted from the *Capsicum frutescens* L. (Solanaceae) plant commonly known as cayenne pepper, showed strong antifungal activity against 16 different fungal strains, including *Candida* spp. The mechanism of fungal growth inhibition was through disruption of the membrane integrity of fungal cells [35].

Two new antifungal triterpene saponins were obtained from ethanolic extracts of the aerial parts of *Clematis tangutica* Skill. (Ranunculaceae), a Tibetan medicinal herb [36]. Similarly, two new dammarane saponins from the methanol extract of the stems of *Anomospermum grandifolium* Eichler (Menispermaceae) showed strong antimycotic activity [37]. Screening of the compounds against *C. albicans* from the rhizomes of *Dioscorea cayenensis* Lam. Holl (Dioscoreaceae) revealed a new steroid saponin with antifungal activity against *C. albicans*, *C. glabrata*, and *C. tropicalis* [38]. Saponins are part of active immune system of plants. Saponins isolated from medicinal plants *Astragalus verrucosus* Moris (Leguminoseae) [39], *A. auriculiformis* (Fabaceae) [40], and *Hedera taurica* Carr. (Araliaceae) exhibited antifungal activity against *C. albicans*, *C. krusei*, and *C. tropicalis* [41], whereas saponins derived from *Solanum chrysotrichum* containing five new spirostan showed antimycotic activity against *C. albicans*. *S. chrysotrichum* also contained 6-α-O-β-D-xylopyranosyl-(1-->3)-β-D-quinovopyranosyl-(25R)-5α-spirostan-3β,23α-ol which was active in a range of 12.5–200 μg/ml against *C. albicans* [42].

Growth of fluconazole-resistant *Candida* strains causing vaginal infections was inhibited by two saponins extracted from *Tribulus terrestris* [43, 44]. The saponinstigogenin-3-O-β-D-xylopyranosyl (1-->2)-[β-D-xylopyranosyl (1-->3)]-β-D-glucopyranosyl (1-->4)-[α-L-rhamnopyranosyl (1-->2)]-β-D-galactopyranoside and tigogenin-3-*O*-β-D-glucopyranosyl (1-->2)-[β-D-xylopyranosyl (1-->3)]-β-D-glucopyranosyl (1-->4)-β-D-galactopyranoside were effective against fluconazole-resistant pathogenic *Candida albicans* (MIC_{80} = 4.4, 9.4 μg/ml), *C. neoformans* (MIC_{80} = 10.7, 18.7 μg/ml), and inherently resistant *C. krusei* (MIC_{80} = 8.8, 18.4 μg/ml) when tested in vitro [45].

9.3.3 Phenolic Compounds

Plants in their entire life span keep synthesizing aromatic substances with different functional groups, most of which are phenols, the hydroxy derivatives of aromatic carbons. Phenolic compounds are mainly helpful in plant defence mechanisms against pathogenic microorganisms, insects, etc. Essential oils extracted from aromatic plants are well documented for their antibacterial and antifungal activities

[46]. These compounds include simple and alkylated phenols, phenolic acid, phenyl propanoids, coumarins, quinines, anthraquinones, xanthones, etc. Four phenolic amides, viz. dihydro-N-caffeoyltyramine, trans-N-feruloyloctopamine, trans-N-caffeoyltyramine, and cis-N-caffeoyltyramine, were extracted in organic solvent (ethyl acetate) from the root bark of *Lyciumchinense* Miller. These compounds were potent antifungals against *C. albicans* with the potency of 5–10 mg/ml showing no toxicity (hemolytic activity) when tested on human erythrocyte cells [47].

Naldoni et al. isolated two phytocompounds benzophenones 7-epiclusianone from pericarp in hexane extract and guttiferone-A from the ethanolic extract of seeds from the fruits of *Garcinia brasiliensis* which showed varying levels of antifungal activity against *C. albicans* [48].

Antifungal activities of three phenolic compounds, 1-galloyl-β-D-glucopyranosyl-(1-- > 4)-β-D-galactopyranoside, 2-methoxy-5-(1′,2′,3′-trihydroxypropyl)-phenyl-1-O-(6″-galloyl)-β-D-glucopyranoside, and 2-methoxy-5-hydroxymethyl-phenyl-1-O-(6″-galloyl)-β-D-glucopyranoside, extracted from the leaves of *Baseonema acuminatum* were reported against *C. albicans* with inhibitory concentration values in the range of 25–100 μg/ml [49].

9.3.4 Flavonoids

Flavonoids or polyphenols are a major class of naturally occurring secondary metabolites in plants. Flavones are phenolic structures with one carbonyl group. Addition of a 3-hydroxyl group yields a flavonol. Flavonoids are also hydroxylated phenolic substances but occur as a C6-C3 unit linked to an aromatic ring. They are classified according to their biosynthetic origin. The classification of polyphenols is a challenging task, as some of them such as chalcones, flavanones, and flavan-3-ols are both intermediates in biosynthesis process and end products accumulating in plant tissues, while other classes such as flavones and flavonols are identified as end products in the biosynthesis [50].

Flavonoids are known to be synthesized by plants in response to microbial infection and thereby found to be effective against a wide array of microorganisms established by in vitro studies [51]. Literature survey shows that most of the testing done on antifungal activity of the flavonoid group, extracted from medicinal plants, mainly belongs to the Fabaceae and Moraceae families. Epidemiological and some clinical studies have reported the majority of polyphenols to exhibit antioxidant and antimicrobial activities including antifungal, antiviral, and antibacterial effects [52].

Arrabidaea brachypoda is well known for its flavonoid content present in the epicuticular wax of the leaves. Alcerito et al. (2002) isolated four flavonoids, 30,40-dihydroxy-5,6,7-trimethoxyflavone, cirsiliol, cirsimaritin, and hispidulin, and reported their antifungal activity against *C. sphaerospermum* [53]. In a study done on different medicinal plants for their antimicrobial activity against *C. albicans* and *S. cerevisiae*, 18 different prenylated flavonoids were purified, of which papyriflavonol A, kuraridin, sophoraflavanone D, and sophoraisoflavanone A exhibited a good antifungal activity. Broussochalcone A was found effective only against

C. albicans [54]. Similarly, the leaves of *Blumea balsamifera* were tested for their antimicrobial activity, of which moderate activity was shown by icthyothereol acetate against the fungi *A. niger*, *T. mentagrophytes*, and *C. albicans*, while cryptomeridiol, lutein, and beta-carotene had low activity against *A. niger*, *T. mentagrophytes*, and *C. albicans* [55]. Flavon 3,4′,5,7-tetraacetyl quercetin purified from heartwood of *Adina cordifolia* revealed moderate antifungal activity against *Cryptococcus neoformans* [56].

9.3.5 Flavonols

Most of the antifungal assays are performed on *C. albicans* strain primarily due to its high prevalence in oral candidiasis as well as in the disseminated candidiasis [57, 58]. The antifungal activity of flavonols, quercetin, myricetin, kaempferol, and quercetin derivative 3-O-beta-glucoside against *C. albicans* has been reported [59, 60]. Quercetin is a plant polyphenol isolated from propolis (bee glue) that inhibits the growth of *Candida albicans*. Quercetin has been reported to exert antifungal activity with MIC values of 197–441 mg/ml in patients suffering from subprosthesis stomatitis, by Herrera [59].

Another quercetin derivative, quercetin 3-O-beta-glucoside, isolated from *Daucaus littoralis* Smith inhibited growth of *C. albicans* (MIC 7.8 mg/ml) [60]. Similarly, prenylated flavanone isolated from the shrub *Eysenhardtia texana* identified as 4′,5,7-trihydroxy-8-methyl-6-(3-methyl-[2-butenyl])-(2S)-flavanone showed inhibitory activity against *C. albicans* [61].

Flavonols such as (R)-roemerine obtained from *Nelumbo nucifera* (and aquatic plant) inhibit the growth of *Candida albicans* (IC_{50}/MIC = 4.5/10 μg/ml) [62]. Galangin, a flavanoid obtained from Argentinean urban propolis, showed strong activity against *C. albicans* and *C. tropicalis* (MIC between 31.2 and 62.5 μg/ml) [63].

The mechanism of action implicated in flavones is through inhibition of efflux pumps, which ultimately results in induction of cell death or apoptosis [58]. Baicalein, one of the flavones, isolated from *Scutellaria biacalensis* has been reported to show same mode of action and induce apoptosis in *C. albicans*. Isoflavones have been known for their antifungal activity, such as glabridin, isolated from *Glycyrrhiza glabra*. It was found to have a broad-spectrum antifungal activity against several *Candida* spp. Its synergistic action with fluconazole affects membrane permeability and ultimately damages the cell envelope [64].

9.3.6 Alkaloids

Alkaloids are naturally occurring chemical compounds containing basic nitrogen atoms. Berberine is well known for its strong antifungal activity against a number of microorganisms including *C. albicans* [65, 66]. A novel alkaloid, 2-(3,4-dimethyl-2,5-dihydro-1H-pyrrol-2-yl)-1-methylethyl pentanoate, isolated by Dabur and his co-workers [67] from the plant *Datura metel* L. (Solanaceae) exhibited strong antifungal activity against *C. albicans* and *C. tropicalis*. Likewise, Slobodníková et al. (2004) illustrated strong in vitro antifungal activity exhibited by the crude extract of *Mahonia aquifolium* (having two main alkaloids, berberine and jatrorrhizine) against 20 strains of *Candida* spp. isolated from chronic vulvovaginal candidiases [68].

Alkaloids isolated from plants of Amaryllidaceae family have a broad-spectrum pharmacological activity. The extracts obtained from bulbs and leaves of *Pancratium illyricum* L. were potent inhibitors of the yeast growth. Alkaloids, lycorine and vittatine, were identified as its components. In vitro inhibition of the collagenase activity exhibited by the bulb extract was much higher than that obtained by the pure alkaloids [69].

A novel alkaloid, 6,8-didec-(1Z)-enyl-5,7-dimethyl-2,3-dihydro-1Hindolizinium, was obtained from the organic extract of *Aniba panurensis* (Meissn) Mez (Lauraceae) by activity-guided fractionation and demonstrated the high toxicity against a resistant strain of *C. albicans* in vitro [70]. The root barks of *Dictamnus dasycarpus* possess two antifungal furoquinoline alkaloids. Similarly, 3-methoxysampangine isolated from *Cleistopholis patens* exhibits significant toxicity against *C. albicans* and *Cryptococcus neoformans* [68].

9.3.7 Peptides and Proteins

Peptides and proteins have come up as new unconventional approaches in antifungal studies. An antifungal protein, AFP-J, purified from *Solanum tuberosum* cv L. Jopung (Solanaceae) was found to strongly inhibit the growth of yeasts, including *C. albicans* [71].

9.3.8 Essential Oils

Essential oils are hydrophobic liquids extracted from plants containing volatile aroma compounds that give fragrance to the plants. The antifungal effects of the essential oils on growth of *C. albicans* from six chemotypes of *Thymus vulgaris* L. and several species of *Satureja montana* L., *Lavandula* hybrid Reverchon, *Origanum vulgare* L., *Lavandula angustifolia* Mill., and *Rosmarinus officinalis* L. belonging to the family Lamiaceae were studied [72]. Essential oil from the *T. vulgaris* thymol chemotype showed the greatest competence. The essential oils of other members of Lamiaceae, viz. *Ocimum*, *Nepeta*, and *Thymbra*, also exhibited antifungal activity. The in vitro antifungal activity of the essential oil of *O. gratissimum* was investigated in order to evaluate its efficacy against *C. albicans*, *C. krusei*, *C. parapsilosis*, and *C. tropicalis*, and the results demonstrated that the essential oil has fungicidal activity against all of the *Candida* species studied. Analysis of the ultra-structure of treated yeast cells revealed changes in the cell wall and in the morphology of some sub-cellular organelles [73].

Leaf oils obtained by hydro-distillation of five endemic species of *Psiadia* of the Asteraceae family (native to Mauritius) were studied for anticandidial activity; out of which *Psiadia lithospermifolia* Lam. was observed to significantly inhibit the growth of *C. pseudotropicalis* [74]. This activity has been attributed to the presence of δ-elemene, farnesene, α-curcumene, selina-4,7(11)-diene, and β-bisabolene, some of which have established antifungal profiles. Antifungal essential oils from the Cupressaceae family including *Juniperus comunis* L. were found to be active against the *Candida* strains [75].

9.3.9 Crude Extracts

Studies reported on antifungal activity by plant extracts have mostly been done with organic solvents having high polarity like petroleum ether, chloroform, ethyl acetate, ethanol, and water. The polarity of organic solvent may affect not only the quality of the extracted compounds but also its quantity in any given plant species.

The crude extracts of plants, viz. *Clausena anisata*, *Sclerocariya birrea*, *Turraea holstii*, *Sterculia africana*, *Acacia robusta* sub spp. *usambarensis*, *Cyphosterma hildebrandti*, *Elaeodendron buchannanii*, *A. nilotica*, *Jatropha multifida*, and *Pteridium aquilinum*, have been reported for their strong antifungal activity against the susceptible yeasts, viz. *Cryptococcus neoformans*, *C. krusei*, *C. tropicalis*, and *C. parapsilosis* [76]. Crude extracts obtained from *Schinus terebintifolius*, *O. gratissimum*, *Cajanus cajan*, and *Piper aduncum* were found to be active against *C. albicans* at MIC of 1.25 mg/ml, whereas that of *Bixaorellana*, *O. gratissimum*, and *Syzygium cumini* extracts when tested against *C. neoformans* had MIC of 0.078 mg/ml [77].

Anticandidal studies done by Krisch et al. (2009) using extracts of fruit skin and seeds of blackcurrant (*Ribes nigrum*), gooseberry (*R. uva-crispa*), and their hybrid plant (jostaberry, *Ribes* x *nidigrolaria*) when tested for inhibitory action against the 12 human-pathogenic *Candida* isolates showed inhibitory activity against all except *C. albicans*, *C. krusei*, *C. lusitaniae*, and *C. pulcherrima* [78]. Root and tuber extracts of *A. racemosus* were checked for anticandidal activity against *C. albicans*, *C. tropicalis*, *C. krusei*, *C. guillermondii*, *C. parapsilosis*, and *C. stellatoida*, isolated from vaginal thrush patients. High degree of inhibitory activity was found against all the candidal strains. Fresh leaves extract of *Aloe vera* can inhibit both the growth and the germ tube formation in *C. albicans* [79]. Roots of *Salvadora persica* have been used as chewing sticks (miswak), in many parts of the world particularly in Saudi Arabia for centuries as oral hygiene [80]. The antimycotic effect of the extract at a concentration of 15% and above has been reported and is probably due to its richness in chlorine, trimethylamine, alkaloid resin, and sulphur compounds [80].

Although the literature has abundant reports on the anticandidal activity of the plant extracts, however systematic reviews on the natural products describing them as an alternative to antifungal drugs are still scanty. The antifungal activity of crude extracts of several important medicinal plants reported against *Candida* has been listed in the Table 9.1 [81–84].

Table 9.1 Anticandidal activity shown by plant crude extract obtained in different solvents

S.No	Plant name	Solvent used for extraction
1.	*Abutilon theophrasti*	ME
2.	*Acalypha indica* L	PE, CH, EA, AC, ME
3.	*Acalypha indica* L.	ET, AQ
4.	*Achillea fragrantissima*	PE, ME
5.	*Allium cepa* Var.	ET, AQ
6.	*Allium cepa* Var. aggregatum L.	ET, AQ
7.	*Allium sativum* L.	ET, AQ
8.	*Allium schoenoprasum* L.	ET, AQ
9.	*Aloe barbadensis*	ET, AQ
10.	*Aloe vera*	ET
11.	*Aloysia triphylla*	Essential oil
12.	*Alstonia scholaris*	ME
13.	*Anethum graveolens*	Essential oil
14.	*Annona cornifolia*	HE, ME, ET
15.	*Anthemis nobilis*	Essential oil
16.	*Argemone mexicana*	ME
17.	*Artemisia dracunculus*	Essential oil
18.	*Artemisia sieberi*	Essential oil
19.	*Azadirachta indica* A. Juss	ET, AQ
20.	*Azadirachta indica*	–
21.	*Camellia sinensis*	AQ
22.	*Camellia sinensis* (L.) O. Ktze	ET, AQ
23.	*Capsicum annuum* L.	ET, AQ
24.	*Cassia alata* L.	ET, AQ
25.	*Cassia alata* Linn	ET, AQ
26.	*Cassia fistula* L.	ET, AQ
27.	*Cassia occidentalis* L.	ET, AQ
28.	*Cassia roxburghii*	AQ
29.	*Cinnamomum* spp.	ET, AQ
30.	*Cinnamomum verum*	HE, ME, ET, AQ
31.	*Citrullus colocynthis*	ET
32.	*Citrus aurantifolia*	ET, AQ
33.	*Coffea arabica* L.	ET, AQ
34.	*Communis hominis*	Essential oil
35.	*Cuminum cyminum*	ET
36.	*Curcuma longa* L.	ET, AQ
37.	*Curcuma longa* L.	ME
38.	*Cymbopogon citratus*	ET, AQ
39.	*Cymbopogon martini*	Essential oil
40.	*Cymbopogon winterianus*	Essential oil
41.	*Cynomorium coccineum*	ME
42.	*Cyperus articulatus*	Essential oil
43.	*Datura alba*	ME
44.	*Dorycnium herbaceum*	EA, AC, ET
45.	*Ecballium elaterium*	ET

(continued)

Table 9.1 (continued)

S.No	Plant name	Solvent used for extraction
46.	*Echinophora platyloba*	ET
47.	*Eucalyptus globulus*	Essential oil
48.	*Eugenia uniflora*	AC, AQ
49.	*Foeniculum vulgare*	Essential oil
50.	*Heracleum persicum*	ME, ET
51.	*Inula viscosa*	ET, AQ
52.	*Larrea nitida*	DCM
53.	*Lavandula angustifolia*	Essential oil
54.	*Lavandula stoechas*	Essential oil
55.	*Lavandula pedunculata*	ET
56.	*Lavandula stoechas*	HE, EA, DCM, ME, AQ
57.	*Lawsonia inermis* L.	ET, AQ
58.	*Lippia alba*	Essential oil
59.	*Luehea paniculata*	ET
60.	*Manilkara zapota* L.	AQ
61.	*Matricaria chamomilla* L.	ET
62.	*Melaleuca alternifolia* Cheel.	Essential oil
63.	*Melaleuca alternifolia*	ET, AQ
64.	*Melilotus albus*	EA, AC, ET
65.	*Mentha arvensis*	Essential oil
66.	*Mentha piperita*	Essential oil
67.	*Mentha* sp.	Essential oil
68.	*Mentha spicata*	Essential oil
69.	*Metasequoia glyptostroboides*	HE, EA,ME
70.	*Mikania glomerata*	Essential oil
71.	*Moringa oleifera*	ET
72.	*Muntingia calabura*	ET, AQ
73.	*Murraya koenigii*	–
74.	*Myrtus communis*	–
75.	*Nigella sativa*	–
76.	*Ocimum sanctum* L.	ET, AQ
77.	*Origanum vulgare* L.	Essential oil
78.	*Pelargonium asperum*	ET, AQ
79.	*Pelargonium roseum*	Essential oil
80.	*Pimenta pseudocaryophyllus*	HE, EA, DCM, ET, AQ
81.	*Piper betle* L.	ET, AQ
82.	*Piper betle* L.	AQ
83.	*Piper longum*	ET
84.	*Psidium guajava* L.	ET, AQ
85.	*Psoralea corylifolia* L.	ET, AQ
86.	*Punica granatum*	ET, AQ
87.	*Punica granatum*	ME
88.	*Pyrostegia venusta*	ME, ET
89.	*Quercus infectoria*	ET, AQ

Solvent: *AC* acetone, *AQ* aqueous, *CH* chloroform, *DCM* dichloromethane, *EA* ethyl acetate, *ET* ethanol, *HE* hexane, *ME* methanol, *PE* petroleum ether

9.4 Ayurvedic Remedies for *Candida*

In Ayurveda, which is the traditional system of medicine in India, *C. albicans* is called "krumi" (parasite) and is treated as a parasitic infection. It is classified into three types based on the body element as *vata* (airy element of body) provoking *C. albicans* infections, *pitta* (Fiery element of body) provoking *C. albicans*, and *kapha* (watery element of body) provoking *C. albicans.* The Ayurvedic treatment of candidiasis works on two basic principles, of which first is to insure the integrity of the digestive system and second is to build immunity back into the body. For digestion, herbs like ginger (*Zingiber officinale*), black pepper (*Piper longum*), and hing (*Ferula asafoetida*) and a blend of herbs like trikatu (Trikatu is a classic Ayurvedic herbal blend of pippali, ginger, and black pepper) are recommended which can stimulate the digestive fire. Herbs that work well in building strength and immunity in body are ashwaganda (*Withania somnifera*), brahmi (*Bacopa monnieri*), and gudduchi (*Tinospora cordifolia*). Once the underlying body imbalance is controlled, then herbs like vidanga (*Embelia ribes*), neem (*Azadirachta indica*), pomegranate (*Punica granatum*), and tulsi (*Ocimum sanctum*) work well in destroying the yeast [85, 86].

Ayurvedic treatments emphasize on the systemic reduction of *C. albicans* with anti-parasitic herbs, coupled with the increasing population of healthy gut bacteria using probiotic foods and supplements and practices that strengthen the immune system. While undergoing treatment of *vata candida*, one should ensure the avoidance of milky products, sugar, fermented foods like breads, and mucous-producing foods. Cold and raw foods should also be avoided. *C. albicans* are most likely to colonize in the human gut where they ferment sugars to use as energy sources.

Gymnema sylvestre is a common herb used in Ayurveda to counter the negative effects of sugar consumption. *Gymnema* increases insulin secretion and muscle cell receptivity to insulin. Muscle cell receptivity aids in the absorption of sugar and, thereby, decreases its availability to *C. albicans*. There are heating anti-parasitical herbs which are more beneficial for *vata* and *kapha* type *C. albicans* such as thyme and oregano, while cooling anti-parasitic herbs such as dandelion root and burdock root are more beneficial for *pitta*. A course of anti-parasitic spices and herbs like cinnamon (*Cinnamomum*), clove (*Syzygium aromaticum*), cardamom (*Elettaria cardamomum*), rosemary (*Rosmarinus officinalis*), and thyme (*Thymus vulgaris*) can be helpful for reducing *vata* type *C. albicans*. Neem oil can be beneficial for *pitta* type candida. For *kapha* type candida, the remedies include heating spices and herbs such as cayenne pepper, black pepper, and ginger. Similarly, seed extracts of *Nigella sativa* reduced the infection of *C. albicans*.

9.5 Synergistic Anticandidal Activity of Plant Extracts

Synergism occurs when compounds interact in a manner which enhance, or potentiate, each other's effect more efficiently than the individual's effect [87]. Herbal remedies mostly contain different plant species, and therefore, the pharmacological effects of such mixtures result from the additive effects of different classes of compounds with diverse mechanisms of action. Synergistic effects of drugs may be due to complex formations that become more effective to inhibit the growth of fungal pathogen by hampering cell wall synthesis, interfering with enzyme actions, or causing cell death [88]. There have been reports that the total contents of herbal product show a significantly better effect than an equivalent dose of a single isolated active ingredient or a single constituent herb [87, 89]. The concept of using mono drugs to treat infections and diseases with multicausal aetiology or complex pathophysiology is changing [90].

In the last few years, research is more focused on searching antifungal phytochemicals, having a wide range of structural classes with synergistic interactions and fewer side effects. Synergistic antifungal activity between natural products and conventional antifungals has been tested in many studies, one of which reported that the methanolic extract of *Terminalia catappa* leaves exhibited good activity against five fungi when tested along with nystatin and amphotericin B [91]. The results showed that maximum antifungal activity was found against *Candida epicola* NCIM3367.

Ethanolic extract of *Echinophora platyloba* and azole drugs when tested against 27 clinical isolates of *C. albicans* isolated from women suffering with chronic recurrent vaginitis showed synergistic anticandidal activity with MIC_{90} and MFC values between 3.1–6.25 mg/ml and 6.2–12.5 mg/ml, respectively [92]. Synergistic anticandidal activity of ethanol extract of *Hyptis martiusii* combined with metronidazole against three *Candida* spp., viz. *C. albicans*, *C. krusei*, and *C. tropicalis*, has been reported by Santos et al. (2013) [93]. Castano et al. (2011) reported synergistic anticandidal activity of essential oils and extracts from aromatic and medicinal plants in combination with antifungal drugs and itraconazole and *Piper bredemeyeri* Jacq (FIC range 0.09–0.13) combination exhibited the best synergistic effect against *C. albicans* [94].

Multidrug therapy is considered advantageous because interactions between the substances accelerate the immune stimulatory, protective and repair mechanisms [95], expands the antimicrobial spectrum, prevents the emergence of resistant mutants, etc. [96].

9.6 Conclusion

Identification of new chemo types for drug development remains an urgent need in antifungal therapeutics. A future research direction may involve translating in vitro analysis to in vivo study and understand the underlying mechanisms of pharmacokinetics as well as signalling pathways involved in the binding of compounds to target sites like plasma membrane and tissue transporters. A number of antifungal compounds reported till date are tested for their in vitro activities and not for in vivo activities. Therefore, such studies should be subjected to animal and human system to determine their effectiveness in whole-organism. In vitro testing and method of extraction should also be standardized so that the search could be more systematic as phytocompounds show enormous potential for antifungal activity against *Candida*.

References

1. Leroy O, Gangneux JP, Montravers P et al (2009) Epidemiology, management, and risk factors for death of invasive *Candida* infections in critical care: a multicenter, prospective, observational study in France (2005–2006). Crit Care Med 37:1612–1618
2. Vincent JL, Rello J, Marshall JE et al (2009) International study of the prevalence and outcomes of infection in intensive care units. JAMA 302:2323–2329
3. Pfaller MA, Diekema DJ (2007) Epidemiology of invasive candidiasis: a persistent public health problem. Clin Microbiol Rev 20:133–163
4. Prieto D, Correia I, Pla J et al (2016) Adaptation of *Candida albicans* to commensalism in the gut. Future Microbiol 11:567–583
5. Arendrup MC, Fuursted K, Gahrn-Hansen B et al (2008) Semi-national surveillance of fungaemia in Denmark 2004-2006: increasing incidence of fungaemia and numbers of isolates with reduced azole susceptibility. Clin Microbiol Infect 14:487–494
6. Arendrup MC, Fuursted K, Gahrn-Hansen B et al (2005) Semi national surveillance of fungemia in Denmark: notably high rates of fungemia and numbers of isolates with reduced azole susceptibility. J Clin Microbiol 43:4434–4440
7. Arendrup MC (2014) Update on antifungal resistance in *Aspergillus and Candida*. Clin Microbiol Infect 20(6):42–48
8. Borg-von Zepelin M, Meyer I, Thomssen R et al (1999) HIV-Protease inhibitors reduce cell adherence of *Candida albicans* strains by inhibition of yeast secreted aspartic proteases. J Invest Dermatol 113:747–751
9. Kim J, Sudbery P (2011) *Candida albicans*, a major human fungal pathogen. J Microbiol 49:171–177
10. Brunke S, Hube B (2013) Two unlike cousins: *Candida albicans* and *Candida glabrata* infection strategies. Cell Microbiol 15:701–708
11. Ferreira AV, Prado CG, Carvalho RR et al (2013) *Candida albicans* and non-*C. albicans Candida* species: comparison of biofilm production and metabolic activity in biofilms, and putative virulence properties of isolates from hospital environments and infections. Mycopathologia 175:265–272
12. Mayer FL, Wilson D, Hube B (2013) *Candida albicans* pathogenicity mechanisms. Virulence 4:119–128
13. Sardi JCO, Scorzoni L, Bernardi T et al (2013) *Candida* species: current epidemiology, pathogenicity, biofilm formation natural antifungal products and new therapeutic options. J Med Microbiol 62:10–24
14. Calderone RA, Fonzi WA (2001) Virulence factors of *Candida albicans*. Trends Microbiol 9:327–355
15. Hube B (2004) From commensal to pathogen: stage- and tissue-specific gene expression of *Candida albicans*. Curr Opin Microbiol 7:336–341
16. Gullo FP, Sardi JCO, Santos VAFFM et al (2012) Antifungal activity of maytenin and pristimirin. J Evid Based Complement Altern Med 2012:1–6
17. Henriques M, Azeredo J, Oliveira R (2006) *Candida albicans* and *Candida dubliniensis*: comparison of biofilm formation in terms of biomass and activity. Br J Biomed Sci 63:5–11
18. Kamba AS, Hassan LG (2010) Phytochemical screening and antimicrobial activities of *Euphorbia balsamifera* leaves, stems and roots against some pathogenic microorganisms. Af J Pharm Pharmacol 4:645–652
19. Toure A, Bahi C, Ouattara K et al (2011) Phytochemical screening and *in vitro* antifungal activities of extracts of leaves of *Morinda morindoides* (Morinda, Rubiaceae). J Med Plants Res 5:6780–6786
20. Chojnacka EBML, Staniszewska M (2015) Tetrazole activity against *Candida albicans*. The role of KEX2 mutations in the sensitivity to (±)-1-[5-(2-chlorophenyl)-2H-tetrazol- 2-yl] propan-2-yl acetate. Bioorg Med Chem Lett 25:2657–2663

21. Pfaller MA, Diekema DJ, Sheehan DJ (2006) Interpretive breakpoints for fluconazole and *Candida* revisited: a blueprint for the future of antifungal susceptibility testing. Clin Microbiol Rev 19:435–447
22. Grant SM, Clissold SP (1990) Fluconazole: a review of its pharmacodynamic and pharmacokinetic properties, and therapeutic potential in superficial and systemic mycoses. Drugs 39:877–916
23. Sanguinetti M, Posteraro B, Lass-Florl C (2015) Antifungal drug resistance among *Candida* species: Mechanisms and clinical impact. Mycoses 5:2–13
24. White TC, Holleman S, Dy F et al (2002) Resistance mechanisms in clinical isolates of *Candida albicans*. Antimicrob Agents Chemother 46:1704–1713
25. Shokri H (2014) Genotypic variation and antifungal susceptibly of *Candida zeylanoides* clinical isolates. J Mycol Med 24:179–184
26. Grasela TH, Goodwin SD, Walawander MK (1990) Prospective surveillance of intravenous amphotericin B use patterns. Pharmacotherapy 10:341–348
27. Fanos V, Cataldi L (2000) Amphotericin B-induced nephrotoxicity: a review. J Chemother 12:463–470
28. Agarwal S, Thakur K, Kanga A et al (2008) Catheter-related candidemia caused by *Candida lipolytica* in a child with tubercular meningitis. Ind J Pathol Microbiol 51:298–300
29. Lupan L, Bandula R, Vasilescu M et al (1996) Spectroscopic study on nystatin conformation modification generated by its interaction with the solvent. J Anal Chem 355:409–411
30. Koehn FE, Carter GT (2005) The evolving role of natural products in drug discovery. Nat Rev Drug Discov 4:206–220
31. Meng JC, Hu YF, Chen JH et al (2001) Antifungal highly oxygenated guaianolides and other constituents from *Ajania fruticulosa*. Phytochemistry 58:1141–1145
32. Lavault M, Landreau A, Larcher G et al (2005) Antileishmanial and antifungal activities of xanthanolides isolated from *Xanthium macrocarpum*. Fitoterapia 76:363–366
33. Sabanero M, Quijano L, Rios T et al (1995) Encelin: a fungal growth inhibitor. Planta Med 61:185–186
34. Starks CM, Williams RB, Goering MG et al (2010) Antibacterial clerodane diterpenes from Goldenrod (*Solidago virgaurea*). Phytochemistry 71:104–109
35. Renault S, De Lucca AJ, Boue S et al (2003) CAY-I, a novel antifungal compound from cayenne pepper. Med Mycol 41:75–82
36. Du Z, Zhu N, Ze-Ren-Wang-Mu N et al (2003) Two new antifungal saponins from the Tibetan herbal medicine *Clematis tangutica*. Planta Med 69:547–551
37. Plaza A, Cinco M, Tubaro A et al (2003) New triterpene glycosides from the stems of *Anomospermum grandifolium*. J Nat Prod 66:1606–1610
38. Sauton M, Mitaine AC, Miyamoto T et al (2004) A new steroidal saponin from *Dioscorea cayenensis*. Chem Pharm Bull 52:1353–1355
39. Pistelli L, Bertoli A, Lepori E et al (2002) Antimicrobial and antifungal activity of crude extracts and isolated saponins from *Astragalus verrucosus*. Fitoterapia 73:336–339
40. Mandal P, Sinha SP, Mandal NC (2005) Antimicrobial activity of saponins from *Acacia auriculiformis*. Fitoterapia 76:462–465
41. Mel'nichenko EG, Kirsanova MA, Grishkovets VI et al (2003) Antimicrobial activity of saponins from *Hedera taurica* Carr. Mikrobiol Z 65:8–12
42. Zamilpa A, Tortoriello J, Navarro V et al (2002) Five new steroidal saponins from *Solanum chrysotrichum* leaves and their antimycotic activity. J Nat Prod 65:1815–1819
43. Bedir E, Khan IA, Walker LA (2002) Biologically active steroidal glycosides from *Tribulus terrestris*. Pharmazie 57:491–493
44. Zhang JD, Xu Z, Cao YB et al (2006) Antifungal activities and action mechanisms of compounds from *Tribulus terrestris* L. J Ethnopharmacol 103:76–84
45. Zhang JD, Cao YB, Xu Z et al (2005) *In vitro* and *in vivo* antifungal activities of the eight steroid saponins from *Tribulus terrestris* L. with potent activity against fluconazole-resistant fungal pathogens. Biol Pharm Bull 28:2211–2215

46. Kumar M, Sarma P, Dkhar MS et al (2017) Assessment of chemically characterised *Gaultheria fragrantissima* Wall. essential oil and its major component as safe plant based preservative for millets against fungal, aflatoxin contamination and lipid peroxidation during storage. J Food Sci Technol 55:111–119
47. Lee DG, Park Y, Kim MR et al (2004) Anti-fungal effects of phenolic amides isolated from the root bark of *Lycium chinense*. Biotechnol Lett 26:1125–1130
48. Naldoni FJ, Claudino AL, Cruz JW Jr et al (2009) Antimicrobial activity of benzophenones and extracts from the fruits of *Garcinia brasiliensis*. J Med Food 12:403–407
49. De Leo M, Braca A, De Tomasi N et al (2004) Phenolic compounds from *Baseonema acuminatum* leaves: isolation and antimicrobial activity. Planta Med 70:841–846
50. Andrae-Marobela K, Ghislain FW, Okatch H et al (2013) Polyphenols: a diverse class of multi-target anti-HIV-1 agents. Curr Drug Metab 14:392–413
51. Cowan MM (1999) Plant products as antimicrobial agents. Clin Microbiol Rev 12:564–582
52. Yigit D, Yigit N, Mavi A (2009) Antioxidant and antimicrobial activities of bitter and sweet apricot (*Prunus armeniaca* L.) kernels. Braz J Med Biol Res 42:346–352
53. Alcerito T, Barbo FE, Negri G et al (2002) Foliar epicuticular wax of *Arrabidaea brachypoda*: flavonoids and antifungal activity. Biochem Syst Ecol 30:677–683
54. Sohn HY, Son KH, Kwon CS et al (2004) Antimicrobial and cytotoxic activity of 18 prenylated flavonoids isolated from medicinal plants: *Morus alba* L., *Morus mongolica* Schneider, *Broussnetia papyrifera* (L.) Vent, *Sophora flavescens* Ait and *Echinosophora koreensis* Nakai. Phytomedicine 11:666–672
55. Ragasa CY, Co AL, Rideout JA (2005) Antifungal metabolites from *Blumea balsamifera*. Nat Prod Res 19:231–237
56. Rao MS, Duddeck H, Dembinski R (2002) Isolation and structural elucidation of 3,4′,5,7-tetraacetyl quercetin from *Adina cordifolia* (Karam ki Gaach). Fitoterapia 73:353–355
57. Lalla RV, Patton LL, Dongari-Bagtzoglou A (2013) Oral candidiasis: pathogenesis, clinical presentation, diagnosis and treatment strategies. J Calif Dental Assoc 41:263–268
58. Serpa R, Franca EJ, Furlaneto-Maia L et al (2012) *In vitro* antifungal activity of the flavonoid baicalein against *Candida* species. J Med Microbiol 61:1704–1708
59. Herrera CL, Alvear M, Barrientos L et al (2010) The antifungal effect of six commercial extracts of *Chilean propolis* on *Candida* spp. Ciencia E Invest Agraria 37:75–84
60. Yousefbeyk F, Gohari AR, Hashemighahderijani Z et al (2014) Bioactive terpenoids and flavonoids from *Daucus littoralis* Smith subsp. hyrcanicus Rech. f, an endemic species of Iran. DARU J Pharmaceu Sci 22:12
61. Wachter GA, Hoffmann JJ, Furbacher T et al (1999) Antibacterial and antifungal flavanones from *Eysenhardtia texana*. Phytochemistry 52:1469–1471
62. Agnihotri VJK, ElSohly HN, Khan SI et al (2008) Constituents of *Nelumbo nucifera* leaves and their antimalarial and antifungal activity. Phytochem Lett 1:89–93
63. Aguero MB, Svetaz L, Baroni V et al (2014) Urban propolis from San Juan province (Argentina): Ethnopharmacological uses and antifungal activity against *Candida* and dermatophytes. Ind Crops Prod 57:166–173
64. Liu W, Li LP, Zhang JD et al (2014) Synergistic antifungal effect of glabridin and fluconazole. PLoS one 9:pe103442
65. Greathouse GA, Walkins GH (1938) Berberine as a factor in the resistance of *Mahonia trifoliate* and *M. swaseya* to *Phymatotrichum* root-rot. Am J Bot 25:743–748
66. Freile M, Giannini M, Sortino M et al (2006) Antifungal activity of aqueous extracts and of Berberine isolated from *Berberis heterophylla*. Acta Farm Bona 25:83–88
67. Dabur R, Chhillar AK, Yadav V et al (2005) *In vitro* antifungal activity of 2-(3,4-dimethyl-2,5-dihydro-1H-pyrrol-2-yl)-1-methylethyl pentanoate, a dihydro–pyrrole derivative. J Med Microbiol 54:549–552
68. Slobodníková L, Kost'álová D, Labudová D et al (2004) Antimicrobial activity of *Mahonia aquifolium* crude extract and its major isolated alkaloids. Phytother Res 18:674–676

69. Bonvicini F, Antognoni F, Iannello C et al (2014) Relevant and selective activity of *Pancratium illyricum* L. against *Candida albicans* clinical isolates: a combined effect on yeast growth and virulence. BMC Complement Altern Med 14:409
70. Klausmeyer P, Chmurny GN, McCloud TG et al (2004) A novel antimicrobial indolizinium alkaloid from *Aniba panurensis*. J Nat Prod 67:1732–1735
71. Park Y, Choi BH, Kwak JS et al (2005) Kunitz-type serine protease inhibitor from potato (*Solanum tuberosum* L. cv. Jopung). J Agric Food Chem 53:6491–6496
72. Giordani R, Regli P, Kaloustian J et al (2004) Antifungal effect of various essential oils against *Candida albicans*. Potentiation of antifungal action of amphotericin B by essential oil from *Thymus vulgaris*. Phytother Res 18:990–995
73. Nakamura CV, Ishida K, Faccin LC et al (2004) *In vitro* activity of essential oil from *Ocimum gratissimum* L. against four *Candida* species. Res Microbiol 155:579–586
74. Salgueiro LR, Piato E, Goncalves MJ et al (2004) Active antifungal substances from natural resources. Planta Med 70:572–575
75. Cavaleiro C, Pinto E, Goncalves MJ et al (2006) Antifungal activity of *Juniperus* essential oils against dermatophyte, *Aspergillus* and *Candida* strains. J Appl Microbiol 100:1333–1338
76. Hamza OJM, Beukel CJPB, Matee MIN et al (2006) Antifungal activity of some Tanzanian plants used traditionally for the treatment of fungal infections. J Ethnopharmacol 108:124–132
77. Braga FG, Bouzada MLM, Fabri RL et al (2007) Antileishmanial and antifungal activity of plants used in traditional medicine in Brazil. J Ethnopharmacol 111:396–402
78. Krisch J, Ordogh L, Galgoczy L et al (2009) Anticandidal effect of berry juices and extracts from *Ribes* species. Cent Eur J Biol 4:86–89
79. Bernardes I, Felipe Rodrigues MP, Bacelli GK et al (2012) *Aloe vera* extract reduces both growth and germ tube formation by *Candida albicans*. Mycoses 55:257–261
80. Al Bagieh NH, Idowu A, Salako NO (1994) Effect of aqueous extract of miswak on the *in vitro* growth of *Candida albicans*. Microbios 80:107–113
81. Aqil F, Zahin M, Ahmad I (2010) Antifungal activity of medicinal plant extracts and phytocompounds: a review. In: Ahmad I et al (eds) Combating fungal infections. Springer, Berlin/Heidelberg, pp 449–484
82. Sharanappa R, Vidyasagar GM (2013) Anti-candida activity of medicinal plants: a review. Inter J Pharm Pharm Sci 5:9–16
83. Rathod T, Padalia H, Chanda S (2015) The potential of plant extracts against multidrug resistant *Candida* species – a review. In: Mendez-Vilas A (ed) The battle against microbial pathogens: basic science, technological advances and educational programs, pp 246–256
84. Seleem D, Pardi V, Murata RM (2017) Review of flavonoids: a diverse group of natural compounds with anti- *Candida albicans* activity *in vitro*. Arch Oral Biol 76:76–83
85. Katoch M, Salgotra A, Singh G (2014) Endophytic fungi found in association with *Bacopa monnieri* as potential producers of industrial enzymes and antimicrobial bioactive compounds. Braz Arch Biol Technol 57:714–722
86. Sharma D, Roy V, Saraf A (2017) An update on phytochemicals analysis and medicinal prospects of Indian herb *Withania somnifera*. Ambient Sci 4:1–6
87. Williamson EM (2001) Synergy and other interactions in phytomedicines. Phytomedicine 8:401–409
88. Rakholiya K, Chanda S (2012) *In vitro* interaction of certain antimicrobial agent in combination with plant extracts against some pathogenic bacterial strains. Asian Pac J Trop Biomed 2:S1466–S1470
89. Nahrstedt A, Butterweck V (2010) Lessons learned from herbal medicinal products: the example of St. John's wort. J Nat Prod 73:1015–1021
90. Wagner H, Merzenich GU (2009) Synergy research: approaching a new generation of phytopharmaceuticals. Phytomedicine 16:97–110
91. Chanda S, Rakholiya K, Dholakia K et al (2013) Antimicrobial, antioxidant and synergistic properties of two nutraceutical plants: *Terminalia catappa* L. and *Colocasia esculenta* L. Turk J Biol 37:81–91

92. Avijgan M, Mahboubi M, Nasab MM et al (2014) Synergistic activity between *Echinophora platyloba* DC ethanolic extract and azole drugs against clinical isolates of *Candida albicans* from women suffering chronic recurrent vaginitis. J de Mycol Med 24:112–116
93. Santos KKA, Matias EFF, Sobral-Souza CE et al (2013) Trypanocide, cytotoxic, and anticandida activities of natural products: *Hyptis martiusii* Benth. Eur J Integr Med 5:427–431
94. Castano VT, Royero JC, Londono BZ et al (2011) Anti-*Candida albicans* activity, cytotoxicity and interaction with antifungal drugs of essential oils and extracts from aromatic and medicinal plants. Asocia Colom De Infectol 15:160–167
95. Wagner H (2006) Multitarget therapy–the future of treatment for more than just functional dyspepsia. Phytomedicine 13:122–129
96. Pankey G, Ashcraft D, Patel N (2005) *In vitro* synergy of daptomycin plus rifampin against *Enterococcus faecium* resistant to both linezolid and vancomycin. Antimicrob Agents Chemother 49:5166–5168

Recent Advances in the Development of Coumarin Derivatives as Antifungal Agents

10

Rajesh Kumar Sharma and Diksha Katiyar

Abstract

Coumarin is a privileged scaffold found in numerous pharmaceutically important natural products and synthetic molecules. The compounds bearing coumarin moiety exhibit broad spectrum of biological properties such as antibacterial, antiviral, anticancerous, anti-inflammatory, antihyperglycemic, and antipyretic activities. Coumarins are also well-known for their antifungal properties. In recent past, several literature reports have been published which highlight the importance of coumarin motif in the area of antifungal drug development. The present contribution provides an overview of synthetic and natural coumarins which have demonstrated potent antifungal activity, reported during 1992–2017. Structure Activity Relationship (SAR) may help medicinal chemists in the rational design and synthesis of new compounds based on coumarin scaffold for the treatment of fungal infections.

Keywords

Coumarin · Antifungal · Treatment · Fungal infection · Structure activity relationship

10.1 Introduction

Fungi are ubiquitous. It is estimated that approximately 611,000 species of fungi exist on earth. Though most of them are harmless, about 600 species are pathogenic to humans and cause variety of diseases ranging from mild skin rashes to life-threatening infections such as cryptococcal meningitis [1–3]. Recent estimates

R. K. Sharma · D. Katiyar (✉)
Department of Chemistry, MMV, Banaras Hindu University, Varanasi, India
email: diksha@bhu.ac.in

K. Singh, N. Srivastava (eds.), *Recent Trends in Human and Animal Mycology*,
https://doi.org/10.1007/978-981-13-9435-5_10

suggest that about 1.2 billion of world population is infected with fungal infections [4]. Medically important fungal infections can be broadly classified as i) superficial mycoses, ii) subcutaneous mycoses, and iii) systemic (invasive) mycoses caused by pathogenic and opportunistic fungi. The incidence of invasive fungal infections has dramatically increased over the past few decades and has caused substantial morbidity and mortality, especially in immunocompromised patients, such as those undergoing organ transplants or anticancer chemotherapy and patients with AIDS [5–8]. *Candida* spp., *Cryptococcus neoformans*, and *Aspergillus* spp. are the three main opportunistic pathogens responsible for most of these infections. Of these, *Candida* sp. is the most common accounting for about 75% of major systemic infections. The remaining 13% and 6% infections are due to *Aspergillus* spp. and *Cryptococcus* spp., respectively [9]. Cryptococcal meningitis caused by *C. neoformans* is a serious infection of brain and spinal column. It is a leading cause of mycological morbidity and mortality in AIDS patients. This mycosis is regarded as one of the "AIDS-defining illnesses," meaning that patients with cryptococcosis and serological proof of HIV infection are considered to have AIDS [10]. An estimated 950,000 cases of cryptococcal meningitis occur globally per annum. Of these, sub-Saharan Africa alone accounts for approximately 720,000 cases killing more HIV/AIDS patients than tuberculosis [11–13]. In recent years, the emergence of new pathogens, many of which are resistant to antifungal agents such as non-fumigatus *Aspergillus* spp., non-albicans *Candida* spp., zygomycetes, and hyaline molds (*Fusarium* and *Scedosporium*), is also an increasing concern of clinicians [14]. These infections cause greater mortality as their diagnosis is more difficult and most of these fungi are more resistant to standard antifungal drugs than the organisms in the past. Therefore, the management of these infections requires new diagnostic methods and antifungal agents [15–17]. The current situation on antifungal market is dubious due to the limited therapeutic options. Owing to the eukaryotic nature of fungal cell, the development of antifungal drug is more difficult than antibacterial drug. Moreover, despite the increasing medical need, the big pharmaceutical companies are also not focusing on the antifungal drug research due to low profit involved [18, 19]. The only novel antifungal drug class to reach the clinic in more than 30 years is the echinocandins [20]. The current arsenal of antifungal drugs can be divided into five different classes based on their mode of action, including the polyenes (e.g., amphotericin B and nystatin), echinocandins (e.g., caspofungin and micafungin), azoles (e.g., fluconazole, voriconazole, ketoconazole, itraconazole, and posaconazole), nucleoside analog (e.g., 5-flucytosine), and cell mitosis inhibitors (e.g., griseofulvin) [21–23] (Fig. 10.1). In spite of successfully treating the number of disease cases and reducing the loads of fungal infections, these drugs are not ideal mainly because of their association with number of side effects and development of drug-resistant strains [3, 24]. Therefore, development of new, more effective, safe, and affordable antifungal agents that are not only active against drug-resistant fungal strains but also possess a new mode of action is the top priority in health-care field [25].

The oxygenated heterocycles play an important role in designing new class of structural entities for medicinal purposes. Majority of pharmaceutical products that

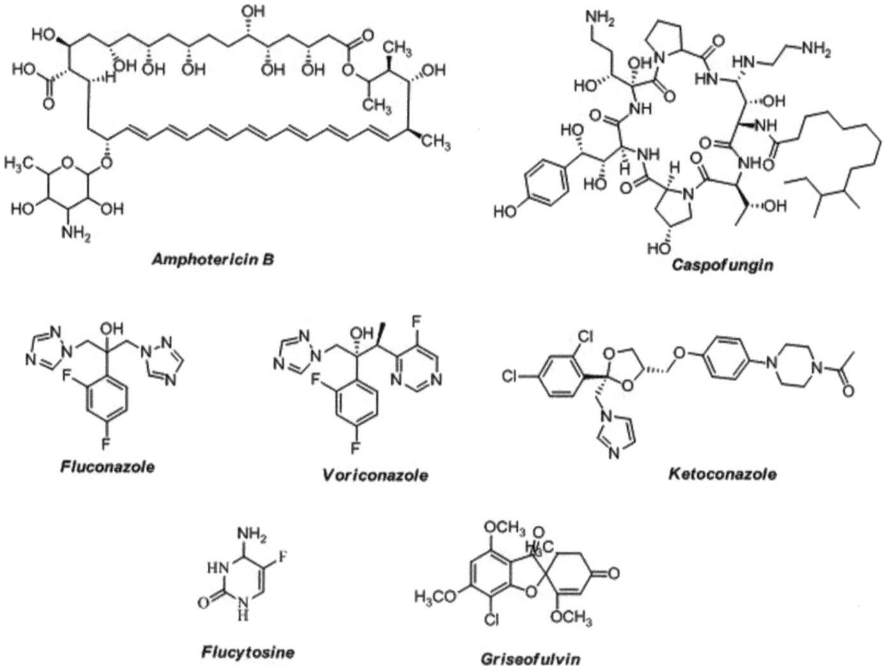

Fig. 10.1 Chemical structures of some antifungal drugs

mimic natural products with biological activity are heterocycles. Over recent years, there is an increasing demand of new heterocyclic scaffold for numerous biological, medicinal, and pharmacological investigations [26–28]. Among them, coumarins (also known as 1,2-benzopyrone) constitute an important class of natural products. It was first isolated from tonka bean (*Dipteryx odorata* Wild) in 1820. Coumarins are present in several plant families in notable amounts such as Rutaceae, Umbelliferae, Clusiaceae, Guttiferae, Apiaceae, and Gramineae. It is also found in essential oils and has been used as fragrance in food and cosmetic products. To date, more than 1300 coumarin compounds have been identified from natural resources, mainly green plants [29]. They are structurally diverse and are broadly classified as, simple coumarins, furanocoumarins, pyranocoumarins, coumarinolignans, biscoumarins, and triscoumarins. The pharmacological, biochemical properties and therapeutic potential of coumarins generally depend upon pattern of substitutions on the basic chemical structure [29]. Coumarin derivatives have been reported to possess wide range of biological activities. A number of excellent reviews have been published on the occurrence and biological activities of coumarins over the past decades [30–36]. It is well established now that many coumarin compounds have significant medicinal value. For instance, coumarin derivative imperatorin shows anti-inflammatory activity [37]; warfarin, a synthetic derivative of dicoumarol, exhibits anticoagulant property and acts as vitamin K antagonist [38]; osthol exhibits a wide range of antifungal activity [39]; umbelliferone and scopoletin display

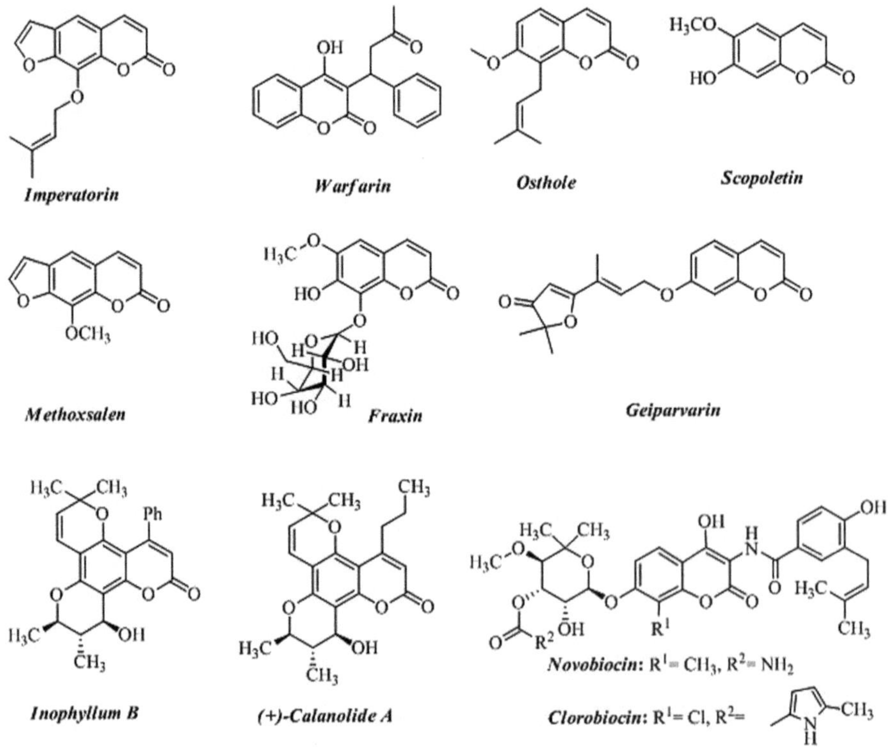

Fig. 10.2 Some coumarin-based drugs from natural sources

significant activity against *Mycobacterium tuberculosis* $H_{37}Rv$ [40]; methoxsalen displays potent mechanism-based microsomal P_{450} inhibitory activity in vitro [41]; and fraxin shows antioxidative effect through inhibition of cyclo AMP phosphodiesterase enzyme [42]. Some natural coumarins, for instance, geiparvarin, inophyllums, and calanolides are also known for their anticancer and antiviral potential [43, 44], while aminocoumarin antibiotics novobiocin and clorobiocin are potent inhibitors of bacterial gyrase [45] (Fig. 10.2).

Many review articles highlighting the anticancerous [46–48], antiviral [49], antitubercular [50, 51], anti-inflammatory [52, 53], antimicrobial [54], anti-Alzheimer's [55], and cholinesterase inhibitory [56, 57] activities of coumarins are available in literature. But authors could not find a comprehensive discussion on the antifungal properties of coumarin derivatives in the literature. Therefore, in this contribution, an overview of recent literature in this research field is presented as per the following scheme:

1. Naturally occurring coumarins.
2. Synthetically derived coumarins.
   2.1. Triazole derivatives.

2.2. Pyrazole and thiazole derivatives.
2.3. Pyridine derivatives.
2.4. Pyrimidine, thiadiazine, and piperazine derivatives.
2.5. Metal complexes.
2.6. Miscellaneous derivatives.

10.1.1 Naturally Occurring Coumarins

A wide range of novel coumarin-based natural products have been reported during past decades showing antifungal properties (Fig. 10.3). Kofinas et al. (1998) isolated eight coumarin-based compounds from the aerial parts of *Tordylium apulum*, an annual herb widely used as a spice in Greece [58]. Out of these, seven compounds exhibited antifungal properties. Among them, umbelliferone **1**, xanthotoxin

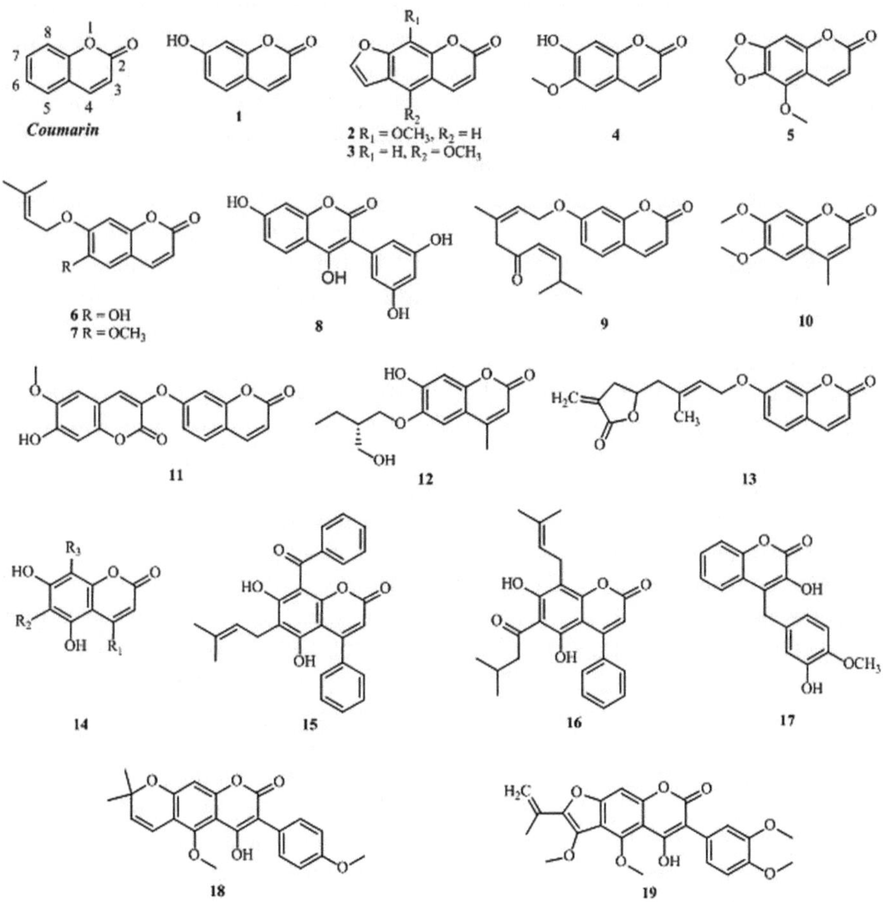

Fig. 10.3 The coumarin core and chemical structures of naturally occurring coumarins

2, and iso-bergapten **3** were most active with inhibition zones of 9, 12.4, and 13 mm, respectively, against *Cladosporium cucumerinum*, but all derivatives were inactive against both *C. albicans* and *Bacillus subtilis*. Duke and coworkers also reported furanocoumarins **2** and **3** from the ethyl acetate extract of leaves of *Ruta graveolens* along with one quinolone and four quinolone alkaloids. All the isolated compounds were evaluated against seven fungal species. Among them, compounds **2** and **3** showed moderate activity against *Fusarium oxysporum* [59]. Scopoletin **4** is found in many edible plants and fruits. This compound was isolated along with other phenolic compounds from the seed kernels of *Melia azedarach* L by Carpinella and coworkers. The study showed a synergistic effect of **4** when used in combination with other synthetic and natural substances in evoking the antifungal response against *Fusarium verticillioides*. The complete inhibition in the growth of the pathogen was observed when **4** was added at a concentration of up to 5% of its MIC value [60]. Stein et al. extracted five coumarin derivatives from the hexane extracts of aerial parts of three *Pterocaulon* spp. (*P. alopecuroides*, *P. balansae*, and *P. polystachyum*) belonging to family Asteraceae from South Brazil [61]. Among different components, ayapin **5**, prenyletin **6**, and prenyletin-methyl-ether **7** showed good antifungal activity against the four tested fungal pathogens. The mixture of compounds **6** and **7** in the ratio of 7:3 exhibited good activity against *Microsporum gypseum* and *Trichophyton mentagrophytes* (MIC = 21.87 μg/ml) and *Trichophyton rubrum* with MIC value of 43.75 μg/ml. Asphodelin A **8** was isolated from the bulbs and roots of *Asphodelus microcarpus*, collected in Egypt [62]. This compound displayed antimicrobial activity against *Staphylococcus aureus*, *Escherichia coli*, and *Pseudomonas aeruginosa* with MIC values of 4–16 mg/ml. The compound was also tested for its antifungal activity against *C. albicans* and *Botrytis cinerea* (MIC = 64 and 128 mg/ml, respectively). Recently, bioactivity-guided fractionation of *Baccharis darwinii*, a traditional medicinal plant from Argentina, resulted in the isolation of diversinin **9**, which displayed fungicidal activity against *Microsporum gypseum*, *Trichophyton rubrum*, and *Trichophyton mentagrophytes* with MIC value of 15.60 μg/ml [63]. Antimicrobial activity of seven coumarin derivatives from *Mexican tarragon* (*Tagetes lucida*) was investigated by Cespedes et al. [64]. Among these compounds, **8** and **10**, which were identified as scoparone and its derivative, respectively, displayed superior activity against *T. mentagrophytes* and *Rhizoctonia solani* as compared to standard drug ketoconazole.

Bis-coumarin (daphnoretin **11**), umbelliferone **1**, and scopoletin **4**, isolated from the dichloromethane extract of the aerial parts of *Loeselia mexicana*, an annual herb, commonly known as *espinosilla* in Mexico, were investigated for antifungal properties against three mycelial fungi and one yeast. All the three compounds showed significant activity against all the tested fungal strains with MIC values ranging from 12.5 to 100 μg/ml [65]. A new prenylated coumarin, pavietin **12**, was isolated from the leaves of *Aesculus pavia* by Curir et al. [66] and analyzed for various fungal pathogens. It exhibited appreciable inhibitory activity against *Aesculus*-specific fungal parasite *Guignardia aesculi* at the three concentrations tested (50, 100, and 200 mM), suggesting that this compound displayed a defensive role against this fungal parasite.

γ-Lactone coumarin (excavarin-A) **13**, isolated by Kumar et al. [67] from the leaves of *Clausena excavata*, was found to be active against 15 fungal strains pathogenic against plants and human. It was most active against the human pathogen, *Candida tropicalis*, and the plant pathogens *Rhizoctonia solani* and *Sclerotinia sclerotiorum*. The activity of excavarin-A for plant pathogens, *Colletotrichum gloeosporioides*, *Lasiodiplodia theobromae*, *F. oxysporum*, and *Rhizopus stolonifer*, and human pathogens, *Aspergillus fumigatus* and *Mucor circinelloides*, was stronger (MIC = 0.625 mg/ml) than that of standard antimicrobial agent nystatin (MIC = 1.25 mg/ml). Montagner et al. reported the antifungal activity of 35 natural coumarins against the fungal strains *C. albicans* (ATCC 14053), *A. fumigatus* (ATCC 16913), and *Fusarium solani* (ATCC 36031), using the broth microdilution method [68]. The result showed that all of these compounds had weak antifungal activity. Among them, osthenol **14** displayed the most effective activity with MIC value of 125 μg/ml for *F. solani* and 250 μg/ml for *C. albicans* and *A. fumigatus*. Sandjo et al. described antifungal activity of ten coumarin derivatives from *Pedilanthus tithymaloides* (Euphorbiaceae) [69]. Out of these, nine compounds were found to be capable of inhibiting conidial germination in the phytopathogenic fungus at low concentrations. The zone of inhibition of the most active compound **15** was 20 mm. Marcondes et al. isolated coumarin derivative mammeisin **16** from *Kielmeyera elata*. The compound was tested against four species of *Candida* and displayed antifungal activity against all of them at very high concentration of 512 μg/ml [70]. It also showed activity against fluconazole-resistant *C. tropicalis*. Sribuhom et al. reported isolation of 22 compounds from the ethanolic extract of stem of *Alyxia schlechteri* [71]. Two of these compounds were new and identified as alyterinin **17**, a benzyl coumarin derivative, and alyterinone, a germacrane sesquiterpene, while others were known compounds. Compound **17** was evaluated for antifungal activity against *Pythium insidiosum*. Ayine-Tora et al. isolated two coumarin derivatives, robustic acid **18** and thonningine-C **19**, from *Millettia thonningii*. These compounds exhibited strong activity against *C. albicans* at low concentrations of 1.0 and 0.5 mg/ml, respectively. The molecular modeling studies showed that these compounds could inhibit the lanosterol 14α-demethylase and disrupt the synthesis of some sterols important for survival of fungus [72].

10.1.2 Synthetically Derived Coumarins

Owing to the variety and relevance of pharmacological properties of coumarins, considerable efforts have been paid to the synthesis of biologically active compounds based on coumarin moiety. In recent years, some works have manifested that coumarin backbone in combination with nitrogen-containing heterocyclic scaffold could significantly increase the antimicrobial efficiency and broaden their antimicrobial spectrum. Majority of the research papers published in this field reported the antifungal activity of the synthesized compounds along with antibacterial activity; for the brevity of the topic, only the antifungal activity of the compounds is discussed herewith.

10.1.2.1 Triazole Derivatives

Keeping in mind the pharmaceutical importance of triazole moiety, many derivatives bearing triazolyl substituent mainly at C-3, C-4, C-5, and C-7 positions of coumarin were synthesized and evaluated for their antifungal activity (Fig. 10.4). Shi and Zhou synthesized and evaluated coumarin triazole derivatives for their antifungal properties in vitro against three fungi by twofold serial dilution technique [73]. The bioactivity assay showed that some synthesized coumarin triazoles (**20–24**) and bis-triazoles (**25–29**) displayed comparable or even better antifungal efficacy as compared to reference drug fluconazole. Interestingly, coumarin bis-triazole compounds exhibited stronger antifungal efficacy than their corresponding mono-triazole derivatives. Compounds **20** and **23**, bearing two carbon chain linker, showed antifungal potency at concentrations lower than 4 μg/ml against *C. albicans* and *S. cerevisiae* which was comparable to fluconazole (MIC = 1 and 2 μg/ml, respectively). Moreover, these compounds inhibited the growth of *A. fumigatus* at lower concentrations of 16 and 4 μg/ml, respectively, as compared to fluconazole (MIC = 128 μg/ml). It was also observed that coumarin triazole hydrochlorides **24** and **29** displayed stronger antifungal activity in comparison with their poor water-soluble aralkyl triazole precursors. This study also showed the importance of

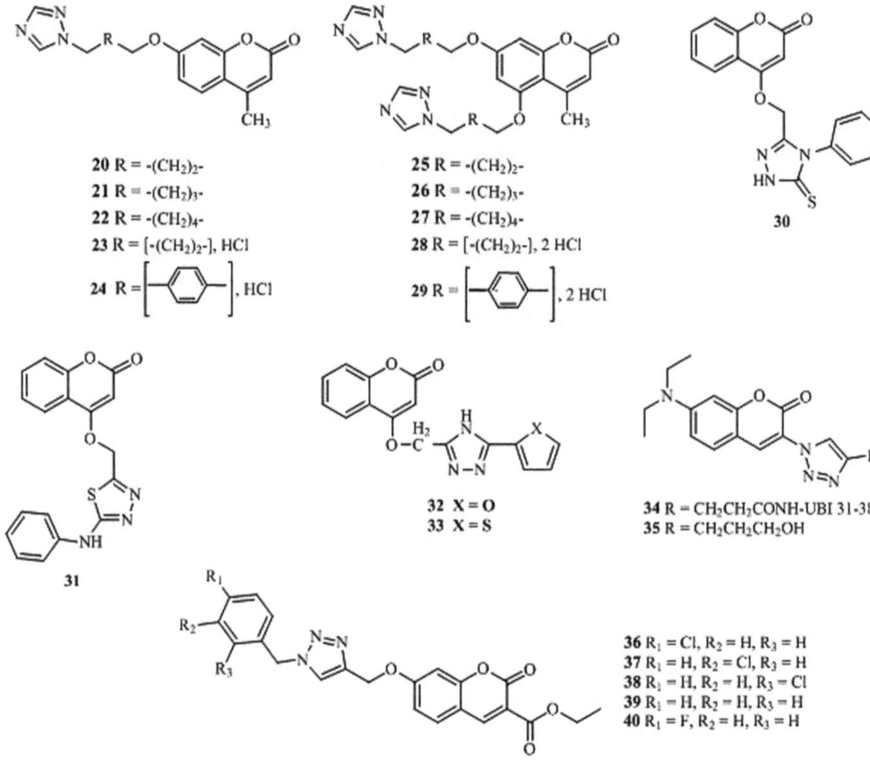

Fig. 10.4 Chemical structures of triazole derivatives of coumarin

triazole ring in exerting antimicrobial potential as its replacement by coumarin ring resulted in reduction and/or complete loss of antifungal properties. Al-Amiery et al. synthesized 4-oxytriazolyl coumarin derivatives and evaluated compounds **30** and **31** against *Aspergillus niger* and *C. albicans*. These compounds exhibited antifungal activity in the range of 0.125–1 μg/ml [74]. Panda et al. synthesized a series of 4-triazolylmethoxy coumarin derivatives and evaluated their antimicrobial properties [75]. Compounds **32** and **33**, bearing furan and thiophene moiety, respectively, showed good antifungal activity with MFC values in the range of 12.5–25 μg/ml.

Antimicrobial peptides (AMPs) are currently being investigated as potential source of novel therapeutics against an increasing number of microorganisms resistant to conventional antibiotics [76]. The conjugation of an AMP to other bioactive compounds is an interesting approach for the development of new derivatives with increased antimicrobial efficiency and broader spectra of action. Ubiquicidin (UBI) is an antimicrobial cationic peptide consisting of 59 amino acid residues, UBI 1-59. Its fragment UBI 31–38 has been shown to be microbicidal against a broad spectrum of pathogens. Ferreira et al. synthesized a peptide-coumarin conjugate **34** by coupling the fully protected alkyne-decorated peptidyl resin, prepared by coupling 4-pentynoic acid to the peptide UBI 31–38 during the solid phase synthesis, to the 3-azido-7-diethylaminocoumarin [77]. The peptide-coumarin conjugate **34** exhibited moderate to excellent antifungal activities against the tested strains *Cryptococcus gattii* and *C. neoformans* (0.04–0.18 μmol/ml) with MIC values comparable to the standard drug fluconazole (MIC =0.003–0.15 μmol/ml). In addition, the conjugate **34** efficiently inhibited the growth of a fluconazole-resistant strain of *C. gattii* (L27/01F) at a concentration of 0.09 μmol/ml. In contrast, peptide fragment UBI 31–38 and the non-peptide derivative **35** showed low activities against the strains of *C. gattii* and *C. neoformans* with MIC values of >0.23 and > 0.75 μmol/ml, respectively.

Recently, a novel series of eight coumarin derivatives was synthesized by Shaikh and coworkers via click chemistry approach. Compounds **36** (chloro group at *para*), **37** (chloro group at *meta*), **38** (chloro group at *ortho*), and **39** with MIC values 25 μg/ml were found to be equipotent against *C. albicans* when compared with miconazole. Compound **40** with fluoro group at *para* position (MIC = 12.5 μg/ml) was found to be twofold more active compared with miconazole (MIC = 25 μg/ml) and equipotent to fluconazole against *C. albicans* [78]. The molecular docking study showed that these derivatives have a high affinity toward the active site of *C. albicans* enzyme P450 cytochrome lanosterol 14α-demethylase.

10.1.2.2 Pyrazole and Thiazole Derivatives

Pyrazole is a key structural motif present in a number of pharmacologically active molecules exhibiting a wide range of biological activities such as antimicrobial [79], anticancer [80], anti-inflammatory [81], anticonvulsant, and antipyretic activities [82]. Enhancement in pharmacological profile of pyrazoles has also been reported when pyrazole moiety is coupled with coumarin nucleus [83]. Prompted by this fact, many coumarin-pyrazole hybrids have been investigated for their antifungal efficacy in recent past (Fig. 10.5). Renuka and Kumar reported the antifungal

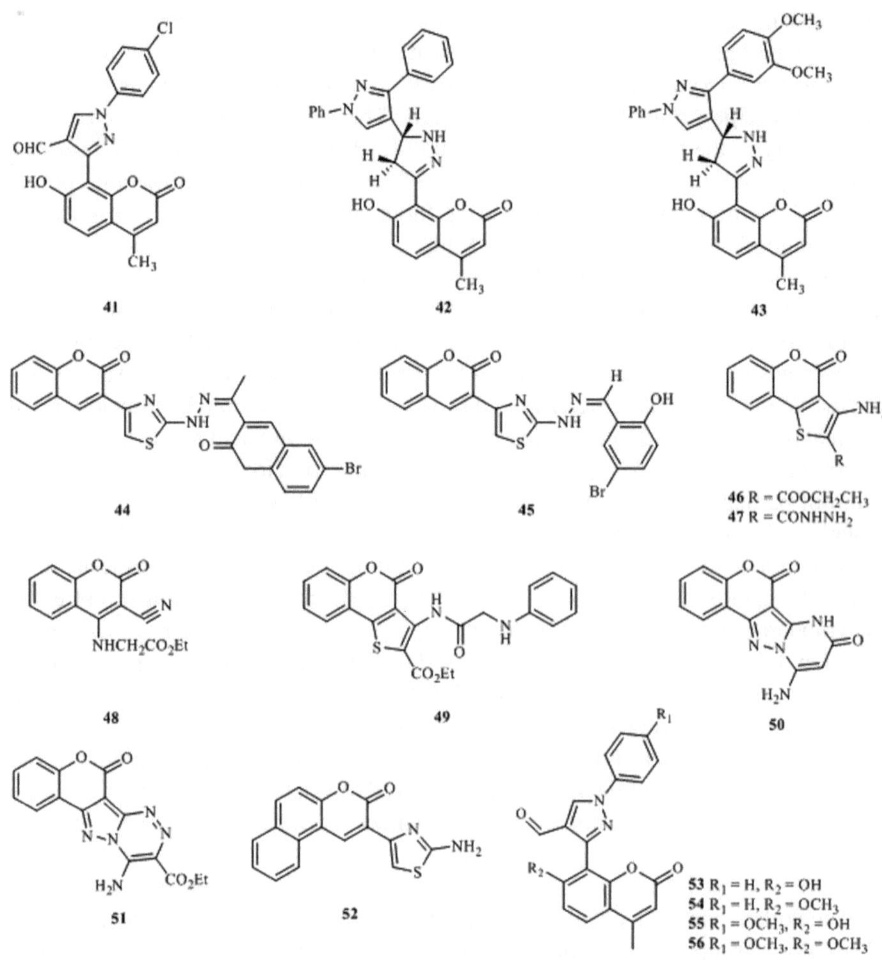

Fig. 10.5 Chemical structures of pyrazole and thiazole derivatives of coumarin

activity of a series of coumarin-pyrazole hybrids against *C. neoformans*, *A. niger*, *A. flavus*, and *C. albicans* [84]. In this series, compound **41** bearing chloro group exhibited superior activity (15 μg/ml) against *C. neoformans* compared to reference drug amphotericin B (MIC = 25 μg/ml), while for other strains its activity (MIC = 25–60 μg/ml) was comparable to amphotericin B (MIC = 25–50 μg/ml).

Dongamanti et al. synthesized a series of coumarin derivatives bearing pyrazoline moiety at C-8 position and evaluated their antifungal activity against *A. niger*, *Penicillium italicum*, and *F. oxysporum* using griseofulvin as a standard drug [85]. SAR has shown that unsubstituted pyrazole derivative **42** showed the highest activity against the fungi, while the electron-rich pyrazole bearing methoxy groups **43** showed lower activity than **42**. Furthermore, compounds bearing bromo, chloro, and fluoro groups also displayed inferior activity as compared to standard drug, indicating that substituents on the pyrazole are detrimental to the observed activity.

Compounds containing thiazole ring are also known for their remarkable medicinal value due to their potential chemotherapeutic properties [86]. Arshad et al. (2011) prepared compounds containing both pharmacophore coumarin and thiazole moieties and reported their antifungal activity against *C. albicans* (Fig. 10.5). Unfortunately, the activity of these compounds was not good; only two compounds of this series **44** and **45** inhibited growth of the fungus at concentrations 31 μM and 35 μM, respectively, which was higher than the fluconazole (10 μM). The study revealed that introduction of bromide and hydroxyl groups enhanced the activity profile of the compounds. The antifungal activity of the remaining compounds of this series was in the range of 86–331 μM [87]. El-Dean et al. synthesized a number of compounds containing thieno-coumarin and pyrazolo-coumarin frameworks [88]. Compounds **46–51** demonstrated superior activity against the tested fungal strains *C. albicans*, *Trichophyton rubrum*, *A. flavus*, *F. oxysporum*, *Scopulariopsis brevicaulis*, and *Geotrichum candidum* with zone of inhibition 9–13 mm compared to reference drug clotrimazole (20–44 mm). Compound **46** bearing aminothienocarboxylate group displayed activity only against *C. albicans* and *Scopulariopsis*, while replacement of the ester group with carbohydrazide group in compound **47** increases activity. The glycinate derivative **48** elicited activity against four fungal strains, whereas compound **49** inhibited the growth of five fungal strains with zone of inhibition in the range of 10–13 mm. Interestingly, replacement of the thieno ring in compound **46** by pyrazole ring led to most active derivative **51**, which showed activity against all the tested strains with zone of inhibition 9–14 mm. Biocide additives have been used to prevent or slow down the growth of organisms on the surface coatings [89]. In a recent study, El-Wahab et al. reported synthesis and antimicrobial activity of 2-aminothiazole derivative of coumarin **52**, which could be applied to a polyurethane varnish as biocide additive [90]. Enhancement in the physical and mechanical properties of polyurethane varnish was observed on incorporation of compound **52**. The molecular modeling study revealed that the compound is biologically safe, active, and fulfills Lipinski's rule of five [91]. In the continuing efforts to develop potent antifungal molecules, Reddy et al. synthesized a series of coumarin derivatives with pyrazole-4-carbaldehyde substituent at C-8 position. Compounds **53–56** exhibited good antifungal activity with zone of inhibition 36–46 mm against *A. niger* [92].

10.1.2.3 Pyridine Derivatives

The antimicrobial activity of pyridine derivatives against broad range of microbes makes it an important skeleton in medicinal chemistry and drug development against microbes [93]. Recently, coumarins substituted with pyridine ring at C-3 and C-7 positions were reported as antifungal agents (Fig. 10.6). Lad et al. prepared a series of coumarin derivatives bearing bipyridine moiety at the 3-position and reported their antimicrobial properties [94]. Compounds **60** (MIC = 200 mg/ml) and **61** (MIC = 250 mg/ml), bearing (4``, 4`) linkage in bipyridine moiety, showed better antifungal activity against *C. albicans* compared to griseofulvin (MIC = 500 mg/ml), whereas compounds **57**, **58**, **59**, and **62** were found to be equipotent to griseofulvin. However, none of these compounds showed significant activity against *A. niger*.

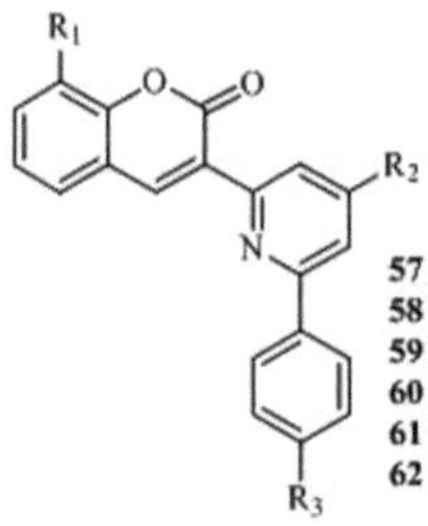

57 R_1 = H, R_2 = 3-pyridinyl, R_3 = CH_3
58 R_1 = OCH_3, R_2 = 3-pyridinyl, R_3 = Cl
59 R_1 = H, R_2 = 4-pyridinyl, R_3 = CH_3
60 R_1 = OCH_3, R_2 = 4-pyridinyl, R_3 = CH_3
61 R_1 = OCH_3, R_2 = 4-pyridinyl, R_3 = OCH_3
62 R_1 = OCH_3, R_2 = 4-pyridinyl, R_3 = Cl

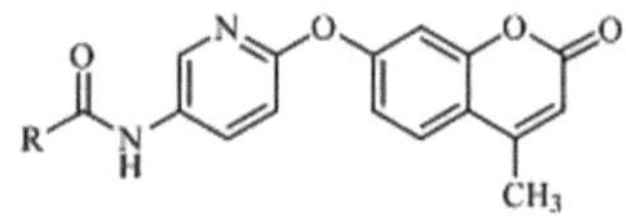

63 R = CH_3
64 R = CF_3
65 R = CH_2CN

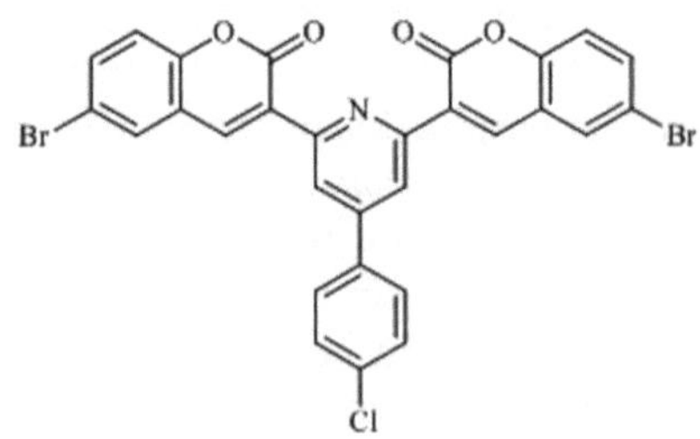

66

Fig. 10.6 Chemical structures of pyridine derivatives of coumarin

Chai and coworkers prepared a series of 7-*O*-substituted pyridyl-4-methyl coumarin derivatives and examined their antifungal activity against various fungal strains (Fig. 10.6). Among these, compounds **63**, **64**, and **65** bearing CH_3, CF_3, and CH_2CN substituents were found to be equipotent to the positive control drug fluconazole (MIC_{80} = 0.25 μg/ml). The study revealed that the antifungal activity was improved as the side chain became smaller and shorter. The amide group was also reported to be essential for activity [95].

Kenchappa et al. reported 3-(6-(2-Oxo-2H-chromen-3-yl)-4-phenylpyridin-2-yl)-2H-chromen-2-one derivatives as antifungal agents [96]. In this series, compound **66** bearing halogen substituents on the coumarin and phenyl ring exhibited potent antifungal activity against *A. flavus*, *C. albicans*, *Microspora griseous*, and *Aspergillus terreus* with MIC values in the range of 12.60–13.95 μg/ml (Fig. 10.6).

10.1.2.4 Pyrimidine, Thiadiazine, and Piperazine Derivatives

Pyrimidine scaffold is an important heterocyclic core that provides large number of compounds having wide spectrum of pharmacological activities [97, 98]. A recent review has highlighted the significance and biological importance of pyrimidine derivatives including their clinical applications in the antimicrobial drug development [99]. Encouraged by these observations, Ghashang et al. prepared coumarin derivatives containing pyrimidine moiety at C-3 position and screened the compounds for their antifungal activity against *A. flavus*, *Rhizopus schipperae*, *A. niger*, and *C. albicans* by serial plate dilution method [100]. All the synthesized compounds showed moderate activity against all the tested fungal strains. Compounds **67–71** were found to be equipotent with the reference drug amphotericin B

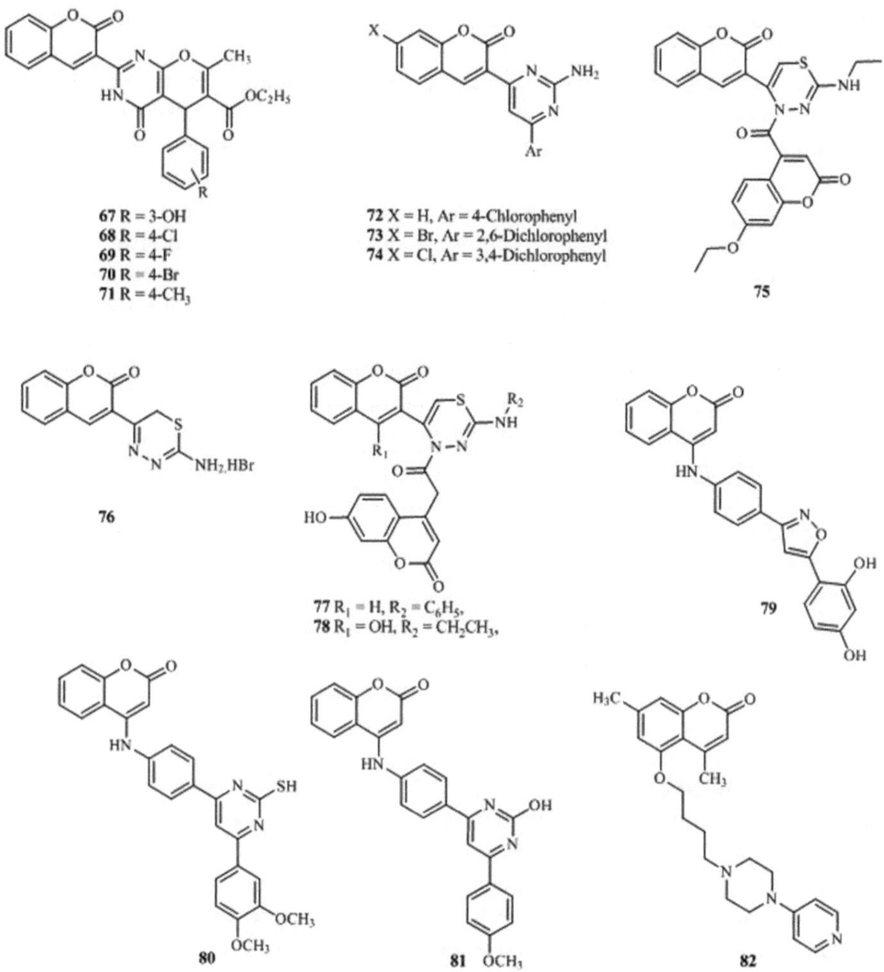

Fig. 10.7 Chemical structures of pyrimidine, thiadiazine, and piperazine derivatives of coumarin

(MIC = 6.25 μg/ml) (Fig. 10.7). Imran and Khan (2015) synthesized a series of coumarin-based pyrimidine derivatives and tested their antifungal potential against *C. albicans*, *A. niger*, *A. flavus*, *Monascus purpureus*, and *Penicillium citrinum* [101]. Though compounds **72–74** inhibited the growth of fungus to some extent, none of these compounds exhibited comparable antifungal activity to the standard antifungal ketoconazole (MIC = 12.5–25 μg/ml) even at higher concentrations.

In recent years, interest in thiadiazines and thiazolidinones has increased due to their wide range of biological activity [102]. Cacic et al. prepared a series of novel coumarin-based disubstituted and trisubstituted 1,3,4-thiadiazines (Fig. 10.7). The antifungal activity of these compounds was found to be dependent upon type of fungal species. Compound **75** showed best activity against *A. flavus*, while compounds **76** and **77** proved efficacious against *A. ochraceus* (MIC_{50} = 0.01 μg/ml), and compound **78** showed the best antifungal activity against *F. verticillioides* with MIC_{50} value of 0.01 μg/ml [103].

Patel et al. had synthesized a series of coumarin containing isoxazole, pyrimidinthione, and pyrimidin-2-one moieties with an electron donating/withdrawing functionality at phenyl ring [104]. These compounds were tested on four fungal strains (*A. niger*, *C. albicans*, *A. fumigatus*, and *Aspergillus clavatus*). Among the isoxazoles, compound **79** showed highest activity against *A. niger*, *A. fumigatus*, and *A. clavatus*. In case of pyrimidinthione and pyrimidin-2-one derivatives, compounds **80** and **81** bearing methoxy group on phenyl ring emerged as the most active compounds, respectively. Compound **80** and **81** exhibited antifungal activity against all the tested fungal strains with MIC in the range of 3.12–6.25 μg/ml and 6.25–12.5 μg/ml, respectively. In this study, pyrimidinthiones were found to be more active than isoxazole and pyrimidin-2-one derivatives (Fig. 10.7).

Number of studies has revealed that the incorporation of a piperazine moiety could enhance the bioactivity of various biologically active compounds [105]. The piperazine analogs have been shown to possess potent antiproliferative, antibacterial, and antifungal activities [106]. Microwave-assisted hybridization of coumarin with piperazine moiety has furnished coumarin-piperazine hybrids. The antifungal activities of these synthesized compounds were evaluated against three strains of *Candida* spp. (*C. albicans* ATCC 10231, *C. albicans* ATCC 2091, and *C. parapsilosis* ATCC 22019) [107]. Among these, compound **82** bearing 4-pyridyl substituted piperazinyl ring exhibited moderate activity against the tested strains with MIC value of 62.5 μg/ml (Fig. 10.7).

10.1.2.5 Metal Complexes

Medical applications of transition metal ions such as copper (II), iron (II), iron (III), and platinum (II) have been known for many years [108–110]. Coordination of metal ions to therapeutic agents to enhance their efficacy and bioactivity is an interesting area of research in medicinal chemistry. According to Tweedy's theory [111], chelation reduces the polarity of the metal atom and such a chelation could enhance the lipophilic character of the central metal atom, which subsequently favors its permeation through the lipid layers of the cell membrane and blocking the metal binding sites on enzymes of microorganism. Numerous studies have been carried out to investigate the pharmacological properties of metal-bioactive ligand complexes. Recent reviews have highlighted the progress in this field [112–116]. Recently, several metal complexes of coumarin-based ligands have been synthesized and the effect of complexation on the antifungal profile of ligands was studied (Fig. 10.8). Creaven et al. prepared a series of Ag (I) complexes of coumarin-3-carboxylic acid and assessed their antimicrobial activity [117]. The anti-candidal activity of each of the complexes and their respective ligands was determined using a clinical isolate of *C. albicans*. Though the free ligands, with the exception of coumarin 3-carboxylic acid, were ineffective in preventing the growth of the fungus, the Ag (I) complexes of the hydroxylated coumarin acids displayed moderate activity. Of these complexes, **83** (MIC_{80} = 34.1 μM) showed comparable activity to the commercially available fungicide ketoconazole (MIC_{80} = 25 μM). In other works, the

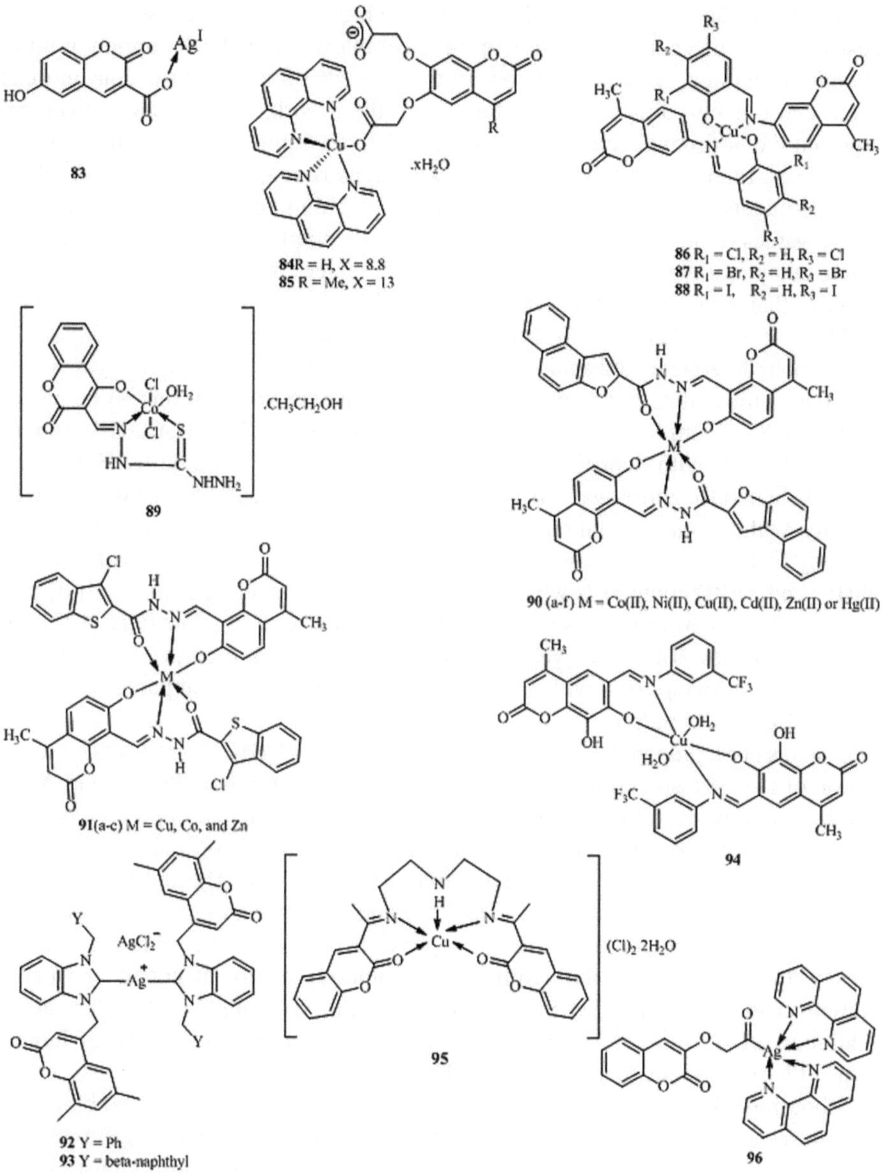

Fig. 10.8 Chemical structures of metal complexes of coumarin

same group demonstrated the activity of a number of coumarin-based Cu (II) and Mn (II) complexes against *C. albicans*. Complexes **84** and **85** were found to be more active than their metal-free ligands. Complex **84** displayed anti-candidal activity (MIC_{80} = 22 μM) comparable to that of ketoconazole (MIC_{80} = 25 μM) [118]. When administered to *C. albicans*, **84** and **85** inhibited respiration, reduced the levels of ergosterol in the membrane, and altered cytochrome content. These results suggested that the mechanism of action of these complexes could be disruption of mitochondrial function, which is different from the mode of action of the conventional azole and polyene drugs. As an extension to their work, Creaven and coworkers reported the antifungal activity of coumarin-derived Schiff bases and their Cu (II) complexes. Interestingly, the complexes with ligands bearing dichloro- and dibromosubstituents (**86** and **87**, respectively) exhibited high anti-candidal activity with MIC_{50} values of 3.6 and 4.4 μM. From this series, complex **88** emerged as the most active compound with activity equipotent to amphotericin B (MIC_{50} = 0.7 μM) and superior than ketonazole (MIC_{50} = 4.7 μM) [119].

Mosa et al. (2011) evaluated Co (III) complex **89** with 4-hydroxycoumarin-3-thiocarbohydrazone ligand against the filamentous fungi *A. niger*, *A. fumigatus*, and *A. flavus*. This compound displayed activity against the tested strains with zone of inhibition 8–17 mm [120]. Metal complexes of the Co (II), Ni (II), Cu (II), Zn (II), Cd (II), Hg (II) with Schiff's base of 4-methyl coumarin prepared by Halli et al. were evaluated for antifungal efficacy [121]. In this series, complexes **90a-f** showed activity with a zone of inhibition of 12.5–50 mm. The study revealed that the potency of ligand was enhanced on coordination with metal salts. Cu(II), Co(II), Ni(II), and Zn(II) complexes with Schiff base ligand containing both benzo[*b*]thiophene and coumarin moieties were also screened for antifungal activity [122]. Of these, complexes **91a-c** displayed inferior activity against *A. niger*, *C. albicans*, and *Cladosporium oxysporum* with MIC values 25–75 μg/ml as compared to fluconazole (MIC = 12.5 μg/ml).

Karatas et al. prepared Ag(I) N-heterocyclic carbene (NHC) complexes by interaction of the corresponding imidazolium or benzimidazolium chlorides and Ag_2O and reported their antifungal activity against *C. albicans* and *C. tropicalis*. Among these, two complexes **92** and **93** showed activity at 25 μg/ml concentration [123].

Patil et al. reported the antifungal activity of Co (II), Ni (II), and Cu (II) complexes of coumarin Schiff bases against *Candida* spp., *A. niger*, and *Rhizopus* spp. [124]. The Cu (II) complexes showed superior activity than the other metal complexes. In this study, all the synthesized complexes exhibited potent antifungal activity than their respective Schiff bases. The Cu (II) complex **94** with zone of inhibition 11–15 mm was found to be equipotent to fluconazole (zone of inhibition = 13–16 mm) at concentration of 200 μg/ml.

Abou-hussein and Linert assessed the antifungal activity of mono- and binuclear complexes of Co (II), Ni (II), Zn (II), and VO (IV) with Schiff base ligand derived from the condensation of 3-acetylcoumarine and diethylenetriamine against *F. oxysporum* [125]. The results revealed that complexes of Cu (II) displayed the maximum inhibition against the growth of the selected fungi in contrast to Co (II), Ni (II), Zn (II), and VO (IV) complexes. The Cu complex **95** inhibited the growth of the fungus with 21 mm zone of inhibition.

Mujahid et al. prepared Ag (I) complexes of coumarin-derived oxyacetate ligands. In addition, authors also prepared their 1,10-phenanthroline (phen) adducts in order to improve the solubility and biological activities of the complexes [126]. The ligands, their metal complexes, and 1,10-phenanthroline adducts were tested for their anti-candidal activity against *C. albicans* strain ATCC 10231 using amphotericin B as reference. Though the metal-free ligands and most of the Ag (I) complexes were found to be inactive, low anti-candidal activity (79 to 97 μM) was exhibited by some of the complexes. On the other hand, the phen adduct **96** showed better anti-candidal activity than the Ag (I) salt and some of them inhibited the growth of *C. albicans* by 50% at a concentration range of 7–10 μM. This was comparable to MIC_{50} values of amphotericin B 4.3 (μM) and ketoconazole (4.7 μM).

10.1.2.6 Miscellaneous Derivatives

Antifungal activity of a series of imino and amino derivatives of 4-hydroxycoumarins against *Aspergillus glaucus*, *A. niger*, *C. albicans*, *F. oxysporum*, *Penicillium verrucosum*, *Trichoderma longibrachiatum*, *Trichoderma harzianum*, and *Trichoderma viride* was studied by Vukovic et al. [127]. While prepared imines showed antifungal activity in the range of 31.25–125 μg/ml, amino derivatives displayed lower potency against tested fungal strains, with MIC values in the range of 62.5–250 μg/ml. The structure of some of the active compounds **97–101** are shown in Fig. 10.9. Coumarin derivatives of various aromatic and heterocyclic amines were synthesized by Sandhya et al. (2011) and antifungal activity of the synthesized compounds was tested against *C. albicans* and *A. niger* using griseofulvin as reference drug [128].

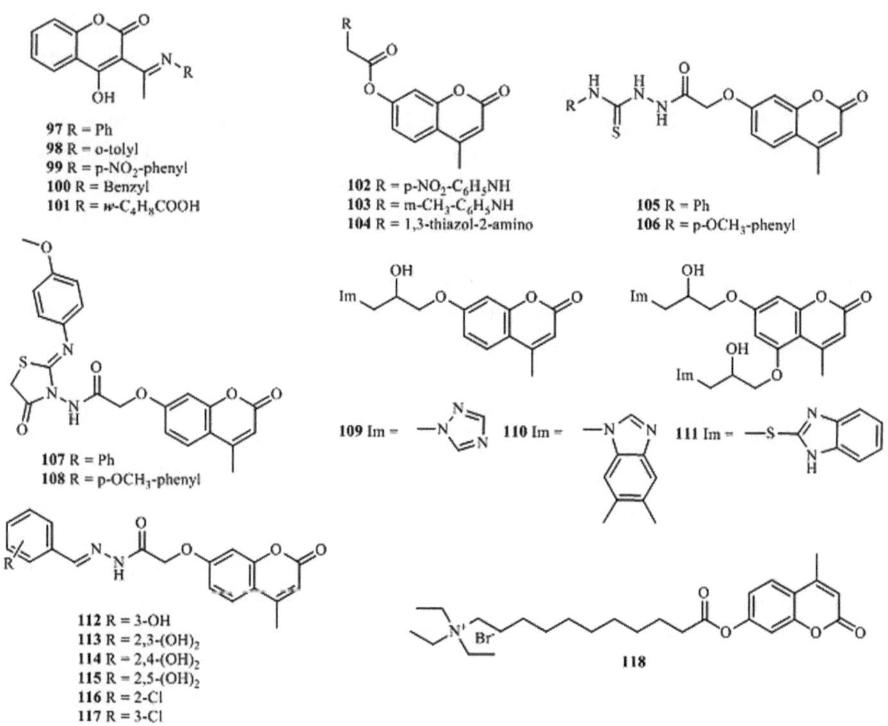

Fig. 10.9 Chemical structures of miscellaneous coumarin derivatives

None of the compounds of this series showed significant activity against the tested strains. Compounds **102–104** showed inhibition zone in the range 10.1–13.4 mm at the concentration of 10 μg/ml. This study demonstrated that coumarin derivatives containing a substituted hydroxy group at the C-7 position could enhance the antifungal activities. Šarkanj et al. synthesized derivatives of 4-methyl-7-hydroxycoumarin, substituted in position seven with thiosemicarbazide and 4-thiazolidinone moieties, and evaluated their antioxidant and antifungal activities [102]. Antifungal activity was determined against four common foodborne mycotoxigenic fungi: *A. flavus*, *Aspergillus ochraceus*, *Fusarium graminearum*, and *F. verticillioides*. Compounds **105**, **106**, **107**, and **108** containing substituted phenyl ring in their structure proved to be the best antifungals for this series. Further, 4-thiazolidinones showed better antifungal activity on all four examined fungal species than thiosemicarbazides. The mold *F. graminearum* was found to be the most susceptible toward the tested compounds, whereas *F. verticillioides* was the least susceptible. The results revealed that substitution of a starting compound in position seven by thiosemicarbazide and 4-thiazolidinone moieties increased the antifungal efficacy of the compounds.

In a quest to develop better antifungal agents, Damu et al. incorporated triazolyl ethanol, an important fragment in fluconazole, into coumarin skeleton (Fig. 10.9) [129]. To gain insight into SAR, various modifications were carried out using different kinds of azole rings such as triazole, benzotriazole, benzimidazole, and thiolbenzimidazole. All these coumarin azole alcohols effectively inhibited the growth of the tested fungal strains to some extent, except for benzimidazole derivatives. Both mono-azole and bis-azole coumarin derivatives **109–111** showed superior activity (MIC = 2–4 μg/ml) against *Candida utilis* compared to reference fluconazole (MIC = 1 μg/ml). Moreover, coumarin triazole alcohol **109** also elicited comparable anti-*C. albicans* and anti-*C. mycoderma* activity to fluconazole (MIC = 1 and 4 μg/ml). Molnar et al. examined a series of coumarinyl Schiff bases **112–117** for their antifungal and metal chelating activity (Fig. 10.9). Antifungal activity was performed against four common mycotoxin-producing foodborne fungi, *A. flavus*, *A. ochraceus*, *F. graminearum*, and *F. verticillioides*. Compounds bearing dihydroxyphenyl moiety **113–115** exhibited good antifungal activity with $MIC_{100} > 0.1$ μg/ml [130].

Despite advances in therapeutic modalities, aspergillosis remains a leading cause of mortality. Therefore, in an effort to explore potent anti-aspergillus molecules, Singh and coworkers investigated the pharmacological properties of *N*, *N*, *N*-triethyl-11-(4-methyl-2-oxo-2H-benzopyran-7-yloxy)-11-oxoundecan-1-aminium bromide **118**, a synthetic coumarin derivative (Fig. 10.9). Compound **118** exhibited potent activity against pathogenic aspergilli (MIC_{90} = 15.62 mg/ml) and resulted in complete inhibition of proteins belonging to key metabolic pathways of cell replication and also inhibited the riboflavin biosynthesis of *A. fumigatus* [131, 132]. Singh et al. (2014) also examined the safety and antifungal efficacy of **118** using mouse model where it showed LD_{50} cut-off 2000 mg/kg body weight and resulted in significant reduction in colony counts in vital organs of the animals. Its application also resulted in reduction in the levels of serum biochemical parameters with respect to infected-untreated animals [133].

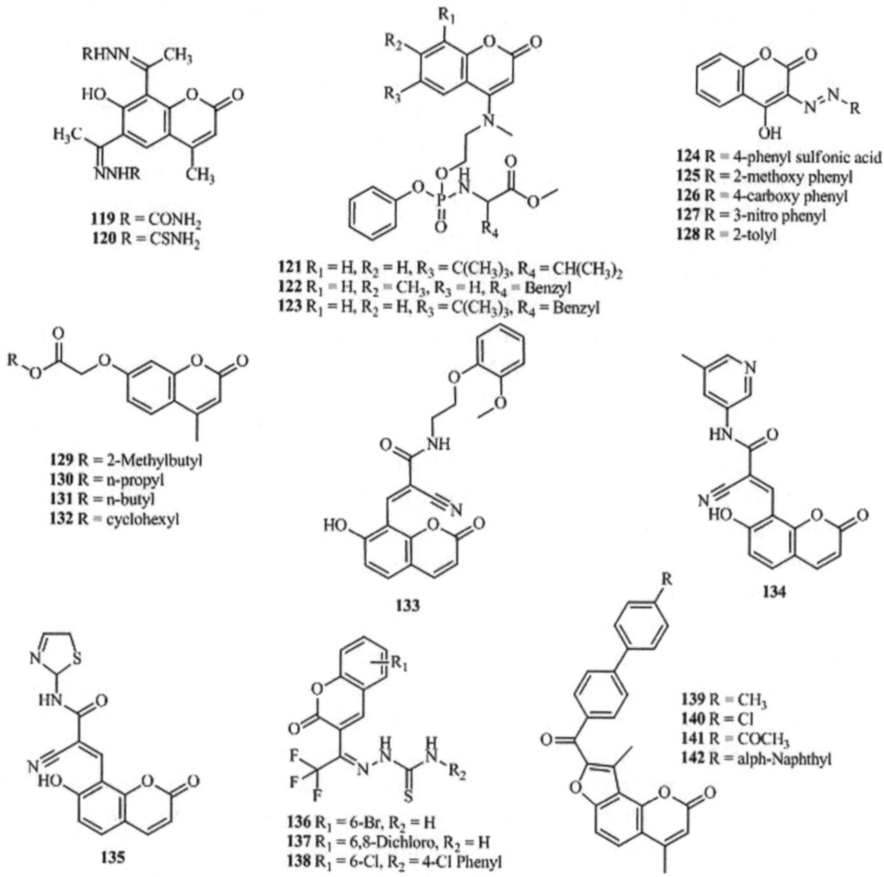

Fig. 10.10 Chemical structures of miscellaneous coumarin derivatives

Encouraged by the biological importance of coumarins, Nagamallu et al. (2016) prepared a series of novel coumarin-based pyrazole, hydrazone, hydrazinecarboxamide, and hydrazinecarbothioamide analogs via Vilsmeier-Haack formylation reaction (Fig. 10.10) [134]. All the compounds were investigated for their in vitro antifungal activity against *A. niger*, *A. flavus*, and *C. albicans* using fluconazole as standard drug. All these compounds exerted modest antifungal activity with MIC value 12.5–100 μg/ml. Compounds **119** and **120** bearing CONH$_2$ and CSNH$_2$ groups, respectively, emerged as the most active compounds of the series and inhibited the growth of the tested fungi at the concentration of 12.5–25 μg/ml.

The present day need in antifungal drug research involves development of antifungal agents effective against those targets which are absent in human host. Chitin, a linear β-(1–4)-linked polymer of N-acetylglucosamine (GlcNAc), is an important structural component of the fungal cell wall that is responsible for imparting shape, strength, and rigidity to the cell wall. Chitin synthase plays an important role in the biosynthesis of chitin that is absent in plant and human. Thus, chitin synthase is a valuable and attractive target to design new fungicide [135, 136]. Naturally occurring polyoxins and nikkomycins are the most potent chitin synthase inhibitors [137, 138]. However, despite excellent in vitro results, clinical utility of these inhibitors is compromised by their metabolic instability and poor cellular uptake. Recently, phosphoramidate derivatives have been used as important frameworks in drug and prodrug design in a number of fields including antifungal drug research [139, 140]. Keeping the biological importance of phosphonamidates and coumarins in mind, Ji et al. (2016) synthesized coumarin-phosphoramidate hybrids and reported their antifungal and chitin synthase inhibitory activity [141]. Most of the target compounds exhibited good chitin synthase inhibitory activity (Fig. 10.10). Among them, compound **123** with IC_{50} of 0.08 mM was the most active and exhibited stronger chitin synthase inhibitory activity than the reference polyoxin B (IC_{50} = 0.16 mM). Compounds **121**, **122**, and **123** also inhibited the growth of *A. flavus* at the concentration of 1–2 mg/ml. SAR study into these compounds has shown that tert-butyl group attached to coumarin ring and methyl, propyl, and benzyl substituents on phosphoramide moiety increase potency of the compounds against chitin synthase and fungi. Some bioactive 3-arylazo-substituted 4-hydroxycoumarins [142] and 4-methylcoumarin esters [143] were also reported for their antifungal activity. Compounds **124–128** belonging to the former series and compounds **129–132** belonging to the latter series demonstrated moderate antifungal activity. A series of 8-substituted-7-hydroxycoumarin derivatives has also been reported as antifungal agents (Fig. 10.10). Of these, compounds **133** (MIC = 4 μg/ml), **134** (MIC = 5 μg/ml), and **135** (MIC = 4–5 μg/ml) showed superior activity compared to fluconazole (MIC = 12–14 μg/ml). Recently, Yang et al. synthesized some novel trifluoromethyl coumarin thiosemicarbazones [144]. In this series, compounds **136–138** exhibited good antifungal activity with 98.98% inhibitory index as compared to the standard drug azoxystrobin having 50% inhibitory index. New furocoumarin-biaryl derivative using Suzuki coupling was synthesized by Dongamanti et al. using green chemistry approach [145]. The inhibition zone against tested fungi for the derivatives **139–142** was comparable to standard drug clotrimazole (Fig. 10.10).

Recently, Guerra et al. studied the antifungal effects of 4-acetatecoumarin **143** both alone and in combination with antifungal drugs (Fig. 10.11) [146]. The compound inhibited the *Aspergillus* spp. virulence factors (mycelia growth and germination of conidia) and also affected the structure of the fungal cell wall. When applied in combination with azoles, both synergistic and additive effects were observed. Encouraged by the reported bioactive significance of isoxazole derivatives, Pang et al. designed and synthesized a novel series of coumarin-oxazole hybrids [147]. Among these, compounds **144–147** bearing halogen substituents demonstrated moderate activity against *C. albicans* with inhibition zone of 10 mm. Khajuria et al. (2017) reported one-pot synthesis of coumarin-pyridone conjugates. Out of all the synthesized compounds, compound **148** was found to be moderately active against all the tested fungal organisms with MIC value 60 μg/ml [148]. Tiwari et al. (2017) reported a series of widely substituted 3-[(dicyclohexylamino) (substituted phenyl/heteryl)-methyl]-4-hydroxy-2H-chromen-2-ones. Two compounds

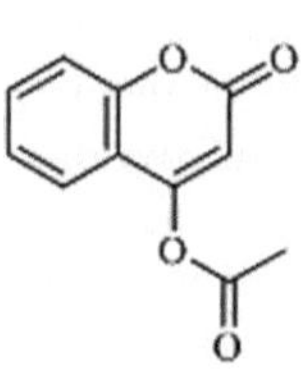

143

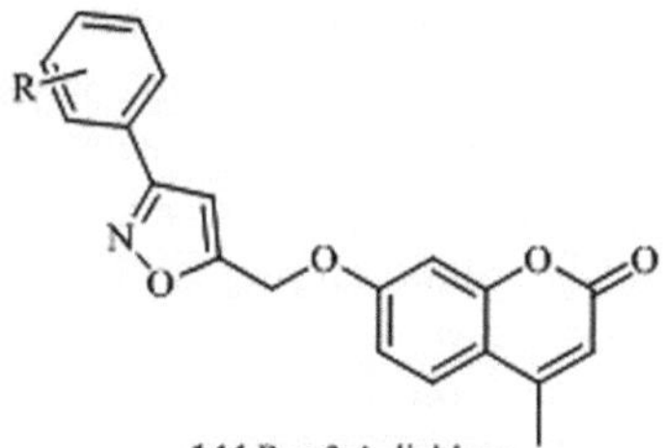

144 R = 3,4-dichloro
145 R = 3,5-dichloro
146 R = 3,4-difluoro
147 R = 3,5-difluoro

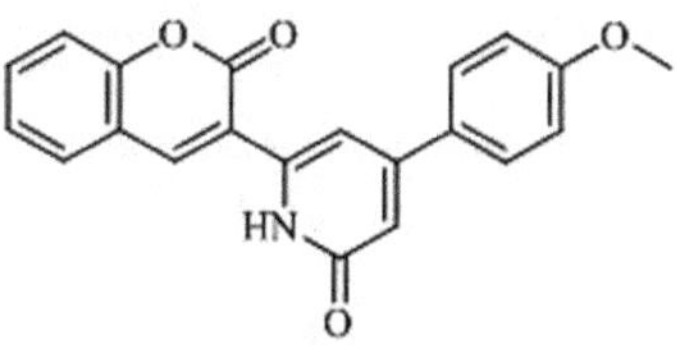

148

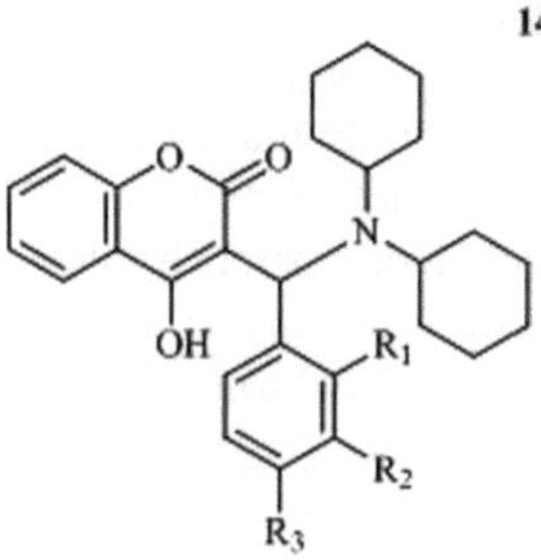

149 R_1 = F, R_2 = H, R_3 = F
150 R_1 = H, R_2 = OCH_3, R_3 = OH

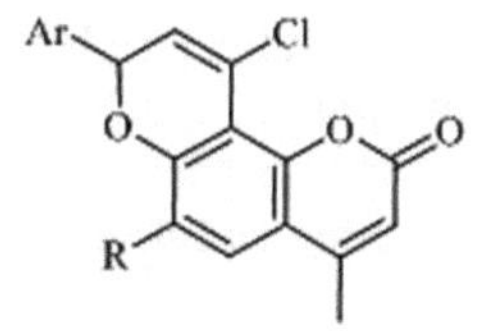

151 R = H, Ar = 3,4-Dimethoxyphenyl
152 R = Cl, Ar = 3,4-Dimethoxyphenyl
153 R = H, Ar = 4-Fluorophenyl

Fig. 10.11 Chemical structures of miscellaneous coumarin derivatives

149 and **150** were found to be the most active antifungal agents with MIC values 12–30 μg/ml against *Candida*, *Fusarium*, *Aspergillus*, and *Cryptococcus* spp. [149].

Pyrano-fused coumarins are an important group of compounds that occur widely in natural products. They exhibit broad spectrum of biological properties and have attracted considerable interest over recent years for the medicinal chemistry applications [150]. Recently, Dongamanti et al. (2015) synthesized pyranocoumarins by the condensation of acetyl and hydroxyl coumarin derivative with aromatic aldehydes and reported their antifungal activity against *F. oxysporum*, *A. niger*, and *A. flavus* [151]. Compounds **151–153** showed moderate to high antifungal action against the tested fungi compared to the standard drug, clotrimazole (Fig. 10.11).

10.2 Conclusion

The wide natural occurrence and diverse bioactivities of coumarins have drawn the attention of organic and medicinal chemists for decades. A great deal of efforts has been invested in the past decades for the development of novel antifungal agents. In this chapter, we have classified the coumarin derivatives on the basis of structure and functionalization, which could assist researches in future design of this class of compounds as antifungal agents. In recent years, several natural coumarin derivatives have been isolated, purified, and tested for antifungal activity. Owing to the recent trend to create hybrid molecules with improved biological activity, much attention has been paid to the synthesis of antifungal hybrid compounds by incorporating another heterocyclic motif such as triazole, pyrazole, thiazole, pyridine, pyrimidine, and pyran moieties either as a substituent or a fused component into coumarin nucleus. Furthermore, many metal complexes of coumarins have been synthesized and variety of modifications has been carried out on coumarin nucleus with a focus on their antifungal potential. However, despite all these efforts, none of these compounds have progressed further into drug development mode. Majority of the coumarin-based antifungal agents reported in the literature were screened in vitro against target organisms, but their in vivo activity and mode of action are not investigated. Therefore, more intense research pertaining to detail mechanism of action and animal studies of coumarin compounds are required to make them suitable candidates for clinical trials. It is hoped that the antifungal research reports on development of different coumarin derivatives summarized in this article provide an insight into the SAR of the compounds which would certainly help the medicinal chemists in designing and development of clinically relevant coumarin-based antifungal agents.

References

1. Mora C, Tittensor DP, Adl S et al (2011) How many species are there on Earth and in the ocean? PLoS Biol 9:e1001127
2. Brown GD, Denning DW, Levitz SM (2012) Tackling human fungal infections. Science 336:647
3. Brown GD, Denning DW, Gow N et al (2012) Hidden killers: human fungal infections. Sci Transl Med 4:165rv13
4. Denning DW, Bromley MJ (2015) How to bloster the antifungal pipeline. Science 347:1414–1416
5. Cuenca-Estrella M, Bernal-Martinez L, Buitrago MJ et al (2008) Update on the epidemiology and diagnosis of invasive fungal infection. Int J Antimicrob Agents 32:S143–S147
6. Patterson TF (2005) Advances and challenges in management of invasive mycoses. Lancet 366:1013–1025
7. Rodloff C, Koch D, Schaumann R (2011) Epidemiology and antifungal resistance in invasive candidiasis. Eur J Med Res 16:187–195
8. Brandt ME, Park BJ (2013) Think fungus-prevention and control of fungal infections. Emerg Infect Dis 19:1688–1689
9. Azie N, Neofytos D, Pfaller M et al (2012) The PATH (Prospective Antifungal Therapy) Alliance® registry and invasive fungal infections: update 2012. Diagn Microbiol Infect Dis 73:293–300
10. Armstrong-James D, Meintjes G, Brown GD (2014) A neglected epidemic: fungal infections in HIV/AIDS. Trends Microbiol 22:120–127
11. Park BJ, Wannemuehler KA, Marston BJ et al (2009) Estimation of the current global burden of cryptococcal meningitis among persons living with HIV/AIDS. AIDS 23:525–530
12. Jarvis JN, Casazza JP, Stone HH et al (2013) The phenotype of the *Cryptococcus*-specific CD4+ memory T-cell response is associated with disease severity and outcome in HIV-associated cryptococcal meningitis. J Infect Dis 207:1817–1828
13. Rajasingham R, Smith RM, Park BJ et al (2017) Global burden of disease of HIV-associated cryptococcal meningitis: an updated analysis. Lancet Infect Dis 17:873–881

14. de Pauw BE, Picazo JJ (2008) Present situation in the treatment of invasive fungal infection. Int J Antimicrob Agents 32:S167–S171
15. Ellis M (2001) Invasive fungal infections: evolving challenges for diagnosis and therapeutics. Mol Immunol 38:947–957
16. Lai CC, Tan CK, Huang YT et al (2008) Current challenges in the management of invasive fungal infections. J Infect Chemother 14:77–85
17. Chen SCA, Playford EG, Sorrell TC (2010) Antifungal therapy in invasive fungal infections. Curr Opin Pharmacol 10:522–530
18. Groll AH, De Lucca AJ, Walsh TW (1998) Emerging targets for the development of novel antifungal therapeutics. Trends Microbiol 6:117–124
19. Orme M, Sjöqvist F (2010) Clinical pharmacology in research, teaching and health care. Basic Clin Pharmacol Toxicol 107:531–559
20. Shapiro RS, Robbins N, Cowen LE (2011) Regulatory circuitry governing fungal development, drug resistance and disease. Microbiol Mol Biol Rev 75:213–267
21. Richardson M, Lass-Florl C (2008) Changing epidemiology of systemic fungal infections. Clin Microbiol Infect 14:5–24
22. Denning DW (2003) Echinocandin antifungal drugs. Lancet 362:1142–1151
23. Allen D, Wilson D, Drew R et al (2015) Azole antifungals: 35 years of invasive fungal infection management. Expert Rev Anti-Infective Ther 13:787–798
24. Sanglard D (2002) Resistance of human fungal pathogens to antifungal drugs. Curr Opin Microbiol 5:379–385
25. Roemer T, Krysan DJ (2014) Antifungal drug development: challenges, unmet clinical needs, and new approaches. Cold Spring Harb Perspect Med 4:a019703
26. Pour M, Spulak M, Buchta V et al (2001) 3-Phenyl-5-acyloxymethyl-2H,5H-furan-2-ones: synthesis and biological activity of a novel group of potential antifungal drugs. J Med Chem 44:2701–2706
27. Dua R, Shrivastava S, Sonwane SK et al (2011) Pharmacological significance of synthetic heterocycles scaffold: a review. Adv Biol Res 5:120–144
28. Kathiravan MK, Salake AB, Chothe AS et al (2012) The biology and chemistry of antifungal agents: a review. Bioorg Med Chem 20:5678–5698
29. Hault JRS, Paya M (1996) Pharmacological and biochemical actions of simple coumarins: natural products and therapeutic potential. Gen Pharmacol 27:713–722
30. Murray RDH (1995) Coumarins. Nat Prod Rep 12:477–505
31. Garazd MM, Garadz YL, Khilya VP (2003) Neoflavones. 1. Natural distribution and spectral and biological properties. Chem Nat Compd 39:54–121
32. Borges F, Roleira F, Milhazes N et al (2005) Simple coumarins and analogues in medicinal chemistry: occurrence, synthesis and biological activity. Curr Med Chem 12:887–916
33. Riveiro ME, Kimpe ND, Moglioni A et al (2010) Coumarins: old compounds with novel promising therapeutic perspectives. Curr Med Chem 17:1325–1338
34. Kontogiorgis C, Detsi A, Hadjipavlou-Litina D (2012) Coumarin-based drugs: a patent review (2008 – present). Expert Opin Ther Pat 22:437–454
35. Barot KP, Jain SV, Kremer L et al (2015) Recent advances on therapeutic journey of coumarins: current status and perspectives. Med Chem Res 24:2771–2798
36. Gaudino EC, Tagliapietra S, Martina K et al (2016) Recent advances and perspectives in the synthesis of bioactive coumarins. RSC Adv 6:46394–46405
37. Huang GJ, Deng JS, Liao JC et al (2012) Inducible nitric oxide synthase and cyclooxygenase-2 participate in anti-inflammatory activity of imperatorin from *Glehnla littoralis*. J Agric Food Chem 60:1673–1681
38. Gomez-Outes A, Suárez-Gea ML, Calvo-Rojas G et al (2012) Discovery of anticoagulant drugs: a historical perspective. Curr Drug Discov Technol 9:83–104
39. Wang CM, Zhou W, Li CX et al (2009) Efficacy of osthol, a potent coumarin compound, in controlling powdery mildew caused by *Sphaerotheca fuliginea*. J Asian Nat Prod Res 11:783–791

40. Shin E, Choi KM, Yoo HS et al (2010) Inhibitory effects of coumarins from the stem barks of *Fraxinus rhynchophylla* on adipocyte differentiation in 3T3-L1 cells. Biol Pharm Bull 33:1610–1614
41. Tinel M, Belghiti J, Descatoire V et al (1987) Inactivation of human liver cytochrome P-450 by the drug methoxsalen and other psoralen derivatives. Biochem Pharmacol 36:951–955
42. Whang WK, Park HS, Ham I et al (2005) Natural compounds, fraxin and chemicals structurally related to fraxin protect cells from oxidative stress. Exp Mol Med 37:436–446
43. Carotti A, Carrieri A, Chimichi S et al (2002) Natural and synthetic geiparvarins are strong and selective. Bioorg Med Chem Lett 12:3551–3555
44. Newman RA, Chen W, Madden TL (1998) Pharmaceutical properties of related calanolide compounds with activity against human immunodeficiency virus. J Pharm Sci 87:1077–1080
45. Maxwell A (1997) DNA gyrase as a drug target. Trends Microbiol 5:102–109
46. Musa MA, Cooperwood JS, Khan MOF (2008) A review of coumarin derivatives in pharmacotherapy of breast cancer. Curr Med Chem 15:2664–2679
47. Thakur A, Singla R, Jaitak V (2015) Coumarins as anticancer agents: a review on synthetic strategies, mechanism of action and SAR studies. Eur J Med Chem 101:476–495
48. Dandriyal J, Singla R, Kumar M et al (2016) Recent developments of C-4 substituted coumarin derivatives as anticancer agents. Eur J Med Chem 119:141–168
49. Hassan MZ, Osman H, Ali MA et al (2016) Therapeutic potential of coumarins as antiviral agents. Eur J Med Chem 123:236–255
50. Keri RS, Sasidhar BS, Nagaraja BM et al (2015) Recent progress in the drug development of coumarin derivatives as potent anti tuberculosis agents. Eur J Med Chem 100:257–269
51. Hu YQ, Xu Z, Zhang S et al (2017) Recent developments of coumarin-containing derivatives and their anti-tubercular activity. Eur J Med Chem 136:122–130
52. Grover J, Jachak SM (2015) Coumarins as privileged scaffold for anti-inflammatory drug development. RSC Adv 5:38892–38905
53. Fylaktakidou KC, Hadjipavlou-Litina DJ, Litinas KE et al (2004) Natural and synthetic coumarin derivatives with anti-inflammatory/antioxidant activities. Curr Pharm Des 10:3813–3833
54. Al-Majedy YK, Kadhum AAH, Al-Amiery AA et al (2017) Coumarins: the antimicrobial agents. Sys Rev Pharm 8:62–70
55. Patil PO, Bari SB, Firke SD et al (2013) A comprehensive review on synthesis and designing aspects of coumarin derivatives as monoamine oxidase inhibitors for depression and Alzheimer's disease. Bioorg Med Chem 21:2434–2450
56. Anand P, Singh B, Singh N (2012) A review on coumarins as acetylcholinesterase inhibitors for Alzheimer's disease. Bioorg Med Chem 20:1175–1180
57. de Souza LG, Renno MN, Figueroa-Villar JD (2016) Coumarins as cholinesterase inhibitors: a review. Chem Biol Interact 254:11–23
58. Kofinas C, Chinou I, Loukis A et al (1998) Flavonoids and bioactive coumarins of *Tordylium apulum*. Phytochemistry 48:637–641
59. Oliva A, Meepagala KM, Wedge DE et al (2003) Natural fungicides from *Ruta graveolens* L. leaves, including a new quinolone alkaloid. J Agric Food Chem 51:890–896
60. Carpinella MC, Ferrayoli CG, Palacios SM (2005) Antifungal synergistic effect of scopoletin, a hydroxyl coumarin isolated from *Melia azedarach* L. fruits. J Agric Food Chem 53:2922–2927
61. Stein AC, Alvarez S, Avancini C et al (2006) Antifungal activity of some coumarins obtained from species of *Pterocaulon* (Asteraceae). J Ethnopharmacol 107:95–98
62. El-Seedi HR (2007) Antimicrobial arylcoumarins from *Asphodelus microcarpus*. J Nat Prod 70:118–120
63. Kurdelas RR, Lima B, Tapia A et al (2010) Antifungal activity of extracts and prenylated coumarins isolated from *Baccharis darwinii* Hook and Arn. (Asteraceae). Molecules 15:4898–4907
64. Céspedes CL, Avila JG, Martínez A et al (2006) Antifungal and antibacterial activities of Mexican tarragon (*Tagetes lucida*). J Agric Food Chem 54:3521–3527

65. Navarro-García VM, Rojas G, Avilés M et al (2011) *In vitro* antifungal activity of coumarin extracted from *Loeselia mexicana* Brand. Mycoses 54:e569–e571
66. Curir P, Galeotti F, Dolci M et al (2007) Pavietin, a coumarin from *Aesculus pavia* with antifungal activity. J Nat Prod 70:1668–1671
67. Kumar R, Saha A, Saha D (2012) A new antifungal coumarin from *Clausena excavata*. Fitoterapia 83:230–233
68. Montagner C, de Souza SM, Groposoa C et al (2008) Antifungal activity of coumarins. Z Naturforsch C 63:21–28
69. Sandjo LP, Foster AJ, Rheinheimer J et al (2012) Coumarin derivatives from *Pedilanthus tithymaloides* as inhibitors of conidial germination in *Magnaportheoryzae*. Tetrahedron Lett 53:2153–2156
70. Marcondes HC, de Oliveira TT, Taylor JG et al (2015) Antifungal activity of coumarin mammeisin isolated from species of the *Kielmeyera* Genre (Clusiaceae or Guttiferae). J Chem Article ID 241243:1–4
71. Sribuhom T, Sriphana U, Thongsri Y et al (2015) Chemical constituents from the stems of *Alyxia schlechteri*. Phytochem Lett 11:80–84
72. Ayine-Tora DM, Kingsford-Adaboh R, Asomaning WA et al (2016) Coumarin antifungal lead compounds from *Millettia thonningii* and their predicted mechanism of action. Molecules 21:1369–1382
73. Shi Y, Zhou CH (2011) Synthesis and evaluation of a class of new coumarin triazole derivatives as potential antimicrobial agents. Bioorg Med Chem Lett 21:956–960
74. Al-Amiery AA, Kadhum AA, Mohamad AB (2012) Antifungal activities of new coumarins. Molecules 17:5713–5723
75. Panda SS, Malik R, Chand M et al (2012) Synthesis and antimicrobial activity of some new 4-triazolylmethoxy-2H-chromen-2-one derivatives. Med Chem Res 21:3750–3756
76. Bahar AA, Ren D (2013) Antimicrobial peptides. Pharmaceuticals 6:1543–1575
77. Ferreira SZ, Carneiro HC, Lara HA et al (2015) Synthesis of a new peptide–coumarin conjugate: a potential agent against cryptococcosis. ACS Med Chem Lett 6:271–275
78. Shaikh MH, Subhedar DD, Khan FA et al (2016) 1,2,3-Triazole incorporated coumarin derivatives as potential antifungal and antioxidant agents. Chin Chem Lett 27:295–301
79. Gilbert AM, Failli A, Shumsky J et al (2006) Pyrazolidine-3,5-diones and 5-hydroxy-1*H*-pyrazol-3(2*H*)-ones, inhibitors of UDP-*N*-acetylenolpyruvyl glucosamine reductase. J Med Chem 49:6027–6036
80. Magedov IV, Manpadi M, Slambrouck SV et al (2007) Discovery and investigation of antiproliferative and apoptosis-inducing properties of new heterocyclic podophyllotoxin analogues accessible by a one-step multicomponent synthesis. J Med Chem 50:5183–5192
81. Szabo G, Fischer J, Kis-Varga A et al (2008) New celecoxib derivatives as anti-inflammatory agents. J Med Chem 51:142–147
82. Sener A, Sener MK, Bildmci I et al (2002) Studies on the reactions of cyclic oxalyl compounds with hydrazines or hydrazones: synthesis and reactions of 4-benzoyl-1- (3-nitrophenyl)-5-phenyl-1*H*-pyrazole-3-carboxylic acid. J Heterocyclic Chem 39:869–875
83. Abdelhafez OM, Amin KM, Batran RZ et al (2010) Synthesis, anticoagulant and PIVKA-II induced by new 4-hydroxycoumarin derivatives. Bioorg Med Chem 18:3371–3378
84. Renuka N, Kumar K (2013) Synthesis and biological evaluation of novel formyl-pyrazoles bearing coumarin moiety as potent antimicrobial and antioxidant agents. Bioorg Med Chem Lett 23:6406–6409
85. Dongamanti A, Bommidi VL, Sidda R et al (2015) Microwave-assisted synthesis of some new coumarin–pyrazoline hybrids and their antimicrobial activity. J Serb Chem Soc 80:305–313
86. Kashyap SJ, Garg VK, Sharma PK et al (2012) Thiazoles: having diverse biological activities. Med Chem Res 21:2123–2132
87. Arshad A, Osman H, Bagley MC et al (2011) Synthesis and antimicrobial properties of some new thiazolyl coumarin derivatives. Eur J Med Chem 46:3788–3794

88. El-Dean AM, Zaki RM, Geies AA et al (2013) Synthesis and antimicrobial activity of new heterocyclic compounds containing thieno [3, 2-c] coumarin and pyrazolo [4, 3-c] coumarin frameworks. Russ J Bioorganic Chem 39:553–564
89. Chiou BS, Shoen PE (2002) Effect of crosslinking on thermal and mechanical properties of polyurethanes. J Appl Polym Sci 83:212–223
90. El-Wahab HA, El-Fattah MA, El-Khalik NA et al (2014) Synthesis and characterization of coumarin thiazole derivative 2-(2-amino-1, 3-thiazol-4-yl)-3H-benzo[*f*] chromen-3-one with anti-microbial activity and its potential application in antimicrobial polyurethane coating. Prog Org Coat 77:1506–1511
91. Lipinski CA, Lombardo F, Dominy BW et al (1997) Experimental and computational approaches to estimate solubility and permeability in drug discovery and development settings. Adv Drug Deliv Rev 23:4–25
92. Reddy VN, Yamini L, Rao YJ et al (2017) Synthesis of pyrazole-4-carbaldehyde derivatives for their antifungal activity. Med Chem Res 26:1664–1674
93. Al-Tel TH, Al-Qawasmeh RA, Zaarour R (2011) Design, synthesis and *in vitro* antimicrobial evaluation of novel imidazo[1,2-a] pyridine and imidazo[2,1-b][1,3]benzothiazole motifs. Eur J Med Chem 46:1874–1881
94. Lad HB, Giri RR, Brahmbhatt DI (2013) An efficient synthesis of some new 3-bipyridinyl substituted coumarins as potent antimicrobial agents. Chin Chem Lett 24:227–229
95. Chai X, Yu S, Wang X et al (2013) Synthesis and antifungal activity of novel 7-*O*-substituted pyridyl-4-methyl coumarin derivatives. Med Chem Res 22:4654–4662
96. Kenchappa R, Bodke YD, Chandrashekar A et al (2017) Synthesis of some 2,6-bis(1-coumarin-2-yl)-4-(4-substituted phenyl) pyridine derivatives as potent biological agents. Arab J Chem 10:S1336–S1344
97. Selvam TP, James CR, Dniandev PV et al (2012) A mini review of pyrimidine and fused pyrimidine marketed drugs. Res Pharm 2:1–9
98. Jain KS, Chitre TS, Miniyar PB et al (2006) Biological and medicinal significance of pyrimidines. Curr Sci 90:793–803
99. Sharma V, Chitranshi N, Agarwal AK (2014) Significance and biological importance of pyrimidine in the microbial world. Int J Med Chem 2014:1–31
100. Ghashang M, Mansoor SS, Aswin K (2014) Pentafluorophenylammonium triflate (PFPAT) catalyzed facile construction of substituted chromeno [2, 3-d] pyrimidinone derivatives and their antimicrobial activity. J Adv Res 5:209–218
101. Imran M, Khan SA (2015) Synthesis and antimicrobial activity of some 2-amino-4-(7-substituted/unsubstituted coumarin-3-yl)-6-(chlorosubstitutedphenyl) pyrimidines. Trop J Pharm Res 14:1265–1272
102. Sarkanj B, Molnar M, Cacic M et al (2013) 4-Methyl-7-hydroxycoumarin antifungal and antioxidant activity enhancement by substitution with thiosemicarbazide and thiazolidinone moieties. Food Chem 139:488–495
103. Čačić M, Pavić V, Molnar M et al (2014) Design and synthesis of some new 1, 3, 4-thiadiazines with coumarin moieties and their antioxidative and antifungal activity. Molecules 19:1163–1177
104. Patel D, Kumari P, Patel NB (2017) Synthesis and biological evaluation of coumarin based isoxazoles, pyrimidinthiones and pyrimidin-2-ones. Arab J Chem 10:S3990–S4001
105. Rokosz LL, Huang CY, Reader JC et al (2005) Surfing the piperazine core of tricyclic farnesyltransferase inhibitors. Bioorg Med Chem Lett 15:5537–5543
106. Mandala D, Valeru A, Pochampalli J et al (2013) Synthesis, antimicrobial activity and molecular modeling of novel 4-(3-(4-benzylpiperazin-1-yl)propoxy)-7-methoxy-3-substituted phenyl-*2H*-chromen-2-one. Med Chem Res 22:5481–5489
107. Ostrowska K, Grzeszczuk D, Maciejewska D et al (2016) Synthesis and biological screening of a new series of 5-[4-(4-aryl-1-piperazinyl) butoxy] coumarins. Monatsh Chem 147:1615–1627
108. Berthon G (1995) Handbook of metal–ligand interactions in biological fluids, vol 1 and 2. Marcel-Dekker Inc, New York

109. Nagy L, Csintalan G, Kálmán E et al (2005) Applications of metal ions and their complexes in medicine. Acta Pharm Hung 73:221–236
110. Savić ND, Milivojevic DR, Glišić BĐ et al (2016) A comparative antimicrobial and toxicological study of gold(III) and silver(I) complexes with aromatic nitrogen-containing heterocycles: synergistic activity and improved selectivity index of Au(III)/Ag(I) complexes mixture. RSC Adv 6:13193–13206
111. Tweedy BG (1964) Plant extracts with metal ions as potential antimicrobial agents. Phytopathology 55:910–914
112. Ahmad S, Isab AA, Ali S et al (2006) Perspectives in bioinorganic chemistry of some metal based therapeutic agents. Polyhedron 25:1633–1645
113. Grazul M, Budzisz E (2009) Biological activity of metal ions complexes of chromones, coumarins and flavones. Coord Chem Rev 253:2588–2598
114. Kioseoglou E, Petanidis S, Gabriel C et al (2015) The chemistry and biology of vanadium compounds in cancer therapeutics. Coord Chem Rev 301–302:87–105
115. Yang Y, Ouyang R, Xu L et al (2015) Review: bismuth complexes: synthesis and applications in biomedicine. J Coord Chem 68:379–397
116. Ndagi U, Mhlongo N, Soliman ME (2017) Metal complexes in cancer therapy–an update from drug design perspective. Drug Des Devel Ther 11:599–616
117. Creaven BS, Egan DA, Kavanagh K et al (2006) Synthesis, characterization and antimicrobial activity of a series of substituted coumarin-3-carboxylatosilver (I) complexes. Inorg Chim Acta 359:3976–3984
118. Creaven BS, Egan DA, Karcz D et al (2007) Synthesis, characterization and antimicrobial activity of copper (II) and manganese (II) complexes of coumarin-6, 7-dioxyacetic acid (cdoaH_2) and 4-methylcoumarin-6, 7-dioxyacetic acid (4-MecdoaH_2): X-ray crystal structures of [Cu (cdoa)(phen)$_2$] · 8.8H_2O and [Cu (4-Mecdoa)(phen)$_2$] · 13H_2O (phen=1, 10-phenanthroline). J Inorg Biochem 101:1108–1119
119. Creaven BS, Devereux M, Karcz D et al (2009) Copper (II) complexes of coumarin-derived Schiff bases and their anti-*Candida* activity. J Inorg Biochem 103:1196–1203
120. Mosa AI, Emara AAA, Yousef JM et al (2011) Novel transition metal complexes of 4-hydroxy-coumarin-3-thiocarbohydrazone: pharmacodynamic of Co(III) on rats and antimicrobial activity. Spectrochim Acta A 81:35–43
121. Halli MB, Sumathi RB, Kinni M (2012) Synthesis, spectroscopic characterization and biological evaluation studies of Schiff's base derived from naphthofuran-2-carbohydrazide with 8-formyl-7-hydroxy-4-methyl coumarin and its metal complexes. Spectrochim Acta A 99:46–56
122. Raj KM, Mruthyunjayaswamy BHM (2014) Synthesis, spectroscopic characterization, electrochemistry and biological evaluation of some metal (II) complexes with ONO donor ligand containing benzo[*b*]thiophene and coumarin moieties. J Mol Struct 1074:572–582
123. Karataş MO, Olgundeniz B, Günal S et al (2016) Synthesis, characterization and antimicrobial activities of novel silver(I) complexes with coumarin substituted *N*-heterocyclic carbene ligands. Bioorg Med Chem 24:643–650
124. Patil SA, Prabhakara CT, Halasangi BM et al (2015) DNA cleavage, antibacterial, antifungal and anthelmintic studies of Co(II), Ni(II) and Cu(II) complexes of coumarin Schiff bases: synthesis and spectral approach. Spectrochim Acta A 137:641–651
125. Abou-hussein AA, Linert W (2015) Synthesis, spectroscopic studies and inhibitory activity against bacteria and fungi of acyclic and macrocyclic transition metal complexes containing a triamine coumarine Schiff base ligand. Spectrochim Acta A 141:223–232
126. Mujahid M, Trendafilova N, Arfa-Kia AF et al (2016) Novel silver (I) complexes of coumarin oxyacetate ligands and their phenanthroline adducts: biological activity, structural and spectroscopic characterisation. J Inorg Biochem 163:53–67
127. Vukovic N, Sukdolak S, Solujic S et al (2010) Substituted imino and amino derivatives of 4-hydroxycoumarins as novel antioxidant, antibacterial and antifungal agents: synthesis and *in vitro* assessments. Food Chem 120:1011–1018

128. Sandhya B, Giles D, Mathew V et al (2011) Synthesis, pharmacological evaluation and docking studies of coumarin derivatives. Eur J Med Chem 46:4696–4701
129. Damu GL, Cui SF, Peng XM et al (2014) Synthesis and bioactive evaluation of a novel series of coumarin azoles. Bioorg Med Chem Lett 24:3605–3608
130. Molnar M, Šarkanj B, Cacic M et al (2014) Antioxidant properties and growth-inhibitory activity of coumarin Schiff bases against common foodborne fungi. Der Pharma Chemia 6:313–320
131. Gupta S, Singh S, Kathuria A et al (2012) Ammonium derivatives of chromenones and quinolinones as lead antimicrobial agents. J Chem Sci 124:437–449
132. Singh S, Gupta S, Singh B et al (2012) Proteomic characterization of *Aspergillus fumigatus* treated with an antifungal coumarins for identification of novel target molecules of key pathways. J Proteome Res 11:3259–3268
133. Singh S, Dabur R, Gatne MM et al (2014) *In vivo* efficacy of a synthetic coumarin derivative in a murine model of aspergillosis. PLoS One 9:e103039
134. Nagamallu R, Srinivasan B, Ningappa MB et al (2016) Synthesis of novel coumarin appended bis (formylpyrazole) derivatives: studies on their antimicrobial and antioxidant activities. Bioorg Med Chem Lett 26:690–694
135. Yeagera AR, Finney NS (2004) Second-generation dimeric inhibitors of chitin synthase. Bioorg Med Chem 12:6451–6460
136. Magellan H, Boccara M, Drujon T et al (2013) Discovery of two new inhibitors of *Botrytis cinerea* chitin synthase by a chemical library screening. Bioorg Med Chem 21:4997–5003
137. Lenardon MD, Munr CA, Gow NAR (2010) Chitin synthesis and fungal pathogenesis. Curr Opin Microbiol 13:416–423
138. Jackson KE, Pogula PK, Patterson SE (2013) Polyoxin and nikkomycin analogs: recent design and synthesis of novel peptidyl nucleosides. Heterocycl Commun 19:375–386
139. Disney MD, Matray T, Gryaznov SM et al (2001) Binding enhancement by tertiary interactions and suicide inhibition of a *Candida albicans* group I intron by phosphoramidate and 2'-*O*-methyl hexanucleotides. Biochemistry 40:6520–6526
140. Subramanyam C, Ramana KV, Rasheed S et al (2013) Synthesis and biological activity of novel diphenyl N-substituted carbamimidoyl phosphoramidate derivatives. Phosphorus Sulfur Silicon 188:1228–1235
141. Ji Q, Ge Z, Ge Z et al (2016) Synthesis and biological evaluation of novel phosphoramidate derivatives of coumarin as chitin synthase inhibitors and antifungal agents. Eur J Med Chem 108:166–176
142. Sahoo J, Kumar PS (2015) Biological evaluation and spectral characterization of 4-hydroxy coumarin analogues. J Taibah Univ Med Sci 10:306–319
143. Medimagh-Saidana S, Romdhane A, Daami-Remadi M et al (2015) Synthesis and antimicrobial activity of novel coumarin derivatives from 4-methylumbelliferone. Med Chem Res 24:3247–3257
144. Yang G, Xu C, Zhao M et al (2016) Microwave assisted one-pot synthesis of novel trifluoromethyl coumarin thiosemicarbazones and their antifungal activities. Curr Microwave Chem 3:60–67
145. Dongamanti A, Bommidi VL, Madderla S (2016) An efficient microwave-assisted Suzuki Cross-Coupling on coumarin derivatives in water and evaluation of antimicrobial activity. Lett Org Chem 13:76–84
146. Guerraa FQS, Araújob RSA, Sousaa JP et al (2017) A new coumarin derivative, 4-acetatecoumarin, with antifungal activity and association study against *Aspergillus* spp. Braz J Microbiol 311:1–7
147. Pang GX, Niu C, Mamat N et al (2017) Synthesis and *in vitro* biological evaluation of novel coumarin derivatives containing isoxazole moieties on melanin synthesis in B16 cells and inhibition on bacteria. Bioorg Med Chem Lett 27:2674–2677
148. Khajuria R, Mahajan S, Ambica (2017) Expeditious synthesis of coumarin-pyridone conjugates molecules and their anti-microbial evaluation. J Chem Sci 129:1549–1557

149. Tiwari SV, Seijas JA, Vazquez-Tato MP et al (2017) Facile synthesis of novel coumarin derivatives, antimicrobial analysis, enzyme assay, docking study, ADMET prediction and toxicity study. Molecules 22:1172–1190
150. Pratap R, Ram VJ (2014) Natural and synthetic chromenes, fused chromenes, and versatility of dihydrobenzo[*H*]chromenes in organic synthesis. Chem Rev 114:10476–10526
151. Dongamanti A, Bommidi VL, Sidda R et al (2015) Microwave-assisted synthesis of substituted 4-chloro-8-methyl-2- phenyl-1,5-dioxa-2H-phenanthren-6-ones and their antimicrobial activity. Med Chem Res 24:1487–1495

Index

K. Singh, N. Srivastava (eds.), *Recent Trends in Human and Animal Mycology*,
https://doi.org/10.1007/978-981-13-9435-5

B

C

G

H

N

FSC
www.fsc.org
MIX
Papier aus verantwortungsvollen Quellen
Paper from responsible sources
FSC® C105338

If you have any concerns about our products,
you can contact us on
ProductSafety@springernature.com

In case Publisher is established outside the EU,
the EU authorized representative is:
Springer Nature Customer Service Center GmbH
Europaplatz 3, 69115 Heidelberg, Germany

Printed by Libri Plureos GmbH
in Hamburg, Germany